Atmospheric Science: An Introduction

Atmospheric Science: An Introduction

Ela Dean

Callisto Reference

www.callistoreference.com

Callisto Reference,
118-35 Queens Blvd., Suite 400,
Forest Hills, NY 11375, USA

Visit us on the World Wide Web at:
www.callistoreference.com

ISBN: 978-1-64116-564-8 (Hardback)

Cataloging-in-Publication Data

Atmospheric science : an introduction / Ela Dean.
p. cm.
Includes bibliographical references and index.
ISBN 978-1-64116-564-8
1. Atmospheric science. 2. Atmosphere. 3. Earth sciences. I. Dean, Ela.
QC861.3 .A86 2022
551.5--dc23

Table of Contents

Preface

The study of the Earth's atmosphere along with the processes related to it is known as atmospheric science. It is also involved in studying the effects which other systems have on it. Some of the sub-disciplines which fall under atmospheric science are meteorology, climatology and aeronomy. Meteorology deals primarily with weather forecasting using atmospheric physics and atmospheric chemistry. Climatology studies the long and short term changes in the atmosphere which define average climate of a particular geographical location. Aeronomy is concerned with the study of the higher layers of the atmosphere, focusing particularly on the processes of dissociation and ionization. This textbook provides comprehensive insights into the field of atmospheric science. It presents this complex subject in the most comprehensible and easy to understand language. The book will serve as a valuable source of reference for graduate and post graduate students.

Given below is the chapter wise description of the book:

Chapter 1- The branch of study which focuses on the Earth's atmosphere, the processes associated with it as well as the effects which other systems have on the atmosphere is called atmospheric science. The diverse aspects of atmosphere and atmospheric science have been briefly introduced in this chapter.

Chapter 2- The atmosphere is divided into several layers on the basis of temperature. Some of these are troposphere, stratosphere, mesosphere, thermosphere, ionosphere and exosphere. The topics elaborated in this chapter will help in gaining a better perspective about these layers of the atmosphere.

Chapter 3- The domain of atmospheric science which deals with the study of the chemistry of the Earth's atmosphere is referred to as atmospheric chemistry. It is involved in the study of the cycles of gases such as nitrogen, oxygen, carbon dioxide and sulfur dioxide. The diverse aspects of these gases as well as the processes related to them have been thoroughly discussed in this chapter.

Chapter 4- The branch of atmospheric science which seeks to apply physics in order to study the atmosphere is called atmospheric physics. Some of the areas of study within this discipline are solar radiation, atmospheric temperature and atmospheric humidity. The topics elaborated in this chapter will help in gaining a better perspective about these focus areas of atmospheric physics.

Chapter 5- The branch of atmospheric science which focuses on weather forecasting by using elements of atmospheric chemistry and atmospheric physics is called meteorology. The chapter closely examines the key concepts of meteorology as well as weather forecasting to provide an extensive understanding of the subject.

Chapter 6- Climate refers to the weather conditions averaged over a long period of time. The branch of science which is involved in the study of climate is called climatology. There are a number of important areas of study within this field such as climate change and global warming. The topics elaborated in this chapter will help in gaining a better perspective about these areas of study within climatology.

Indeed, my job was extremely crucial and challenging as I had to ensure that every chapter is informative and structured in a student-friendly manner. I am thankful for the support provided by my family and colleagues during the completion of this book.

Ela Dean

Chapter 1

Atmosphere and Atmospheric Science

The branch of study which focuses on the Earth's atmosphere, the processes associated with it as well as the effects which other systems have on the atmosphere is called atmospheric science. The diverse aspects of atmosphere and atmospheric science have been briefly introduced in this chapter.

Atmosphere

Atmosphere is the gas and aerosol envelope that extends from the ocean, land, and ice-covered surface of a planet outward into space. The density of the atmosphere decreases outward, because the gravitational attraction of the planet, which pulls the gases and aerosols (microscopic suspended particles of dust, soot, smoke, or chemicals) inward, is greatest close to the surface. Atmospheres of some planetary bodies, such as Mercury, are almost non-existent, as the primordial atmosphere has escaped the relatively low gravitational attraction of the planet and has been released into space. Other planets, such as Venus, Earth, Mars, and the giant outer planets of the solar system, have retained an atmosphere. In addition, Earth's atmosphere has been able to contain water in each of its three phases (solid, liquid, and gas), which has been essential for the development of life on the planet.

The evolution of Earth's current atmosphere is not completely understood. It is thought that the current atmosphere resulted from a gradual release of gases both from the planet's interior and from the metabolic activities of life-forms—as opposed to the primordial atmosphere, which developed by outgassing (venting) during the original formation of the planet. Current volcanic gaseous emissions include water vapour (H_2O), carbon dioxide (CO_2), sulfur dioxide (SO_2), hydrogen sulfide (H_2S), carbon monoxide (CO), chlorine (Cl), fluorine (F), and diatomic nitrogen (N_2; consisting of two atoms in a single molecule), as well as traces of other substances. Approximately 85 percent of volcanic emissions are in the form of water vapour. In contrast, carbon dioxide is about 10 percent of the effluent.

During the early evolution of the atmosphere on Earth, water must have been able to exist as a liquid, since the oceans have been present for at least three billion years. Given that solar output four billion years ago was only about 60 percent of what it is today, enhanced levels of carbon dioxide and perhaps ammonia (NH_3) must have been present in order to retard the loss of infrared radiation into space. The initial life-forms that evolved in this environment must have been anaerobic (i.e., surviving in the absence of oxygen). In addition, they must have been able to resist the biologically destructive ultraviolet radiation in sunlight, which was not absorbed by a layer of ozone as it is now.

Once organisms developed the capability for photosynthesis, oxygen was produced in large quantities. The build up of oxygen in the atmosphere also permitted the development of the ozone layer as O_2 molecules were dissociated into monatomic oxygen (O; consisting of single oxygen atoms) and recombined with other O_2 molecules to form triatomic ozone molecules (O_3). The capability for photosynthesis arose in primitive forms of plants between two and three billion years ago. Previous to the evolution of photosynthetic organisms, oxygen was produced in limited quantities as a by-product of the decomposition of water vapour by ultraviolet radiation.

The current molecular composition of Earth's atmosphere is diatomic nitrogen (N_2), 78.08 percent; diatomic oxygen (O_2), 20.95 percent; argon (A), 0.93 percent; water (H_2O), about 0 to 4 percent; and carbon dioxide (CO_2), 0.04 percent. Inert gases such as neon (Ne), helium (He), and krypton (Kr) and other constituents such as nitrogen oxides, compounds of sulfur, and compounds of ozone are found in lesser amounts.

Surface Budgets

Energy Budget

Earth's atmosphere is bounded at the bottom by water and land—that is, by the surface of Earth. Heating of this surface is accomplished by three physical processes—radiation, conduction, and convection—and the temperature at the interface of the atmosphere and surface is a result of this heating.

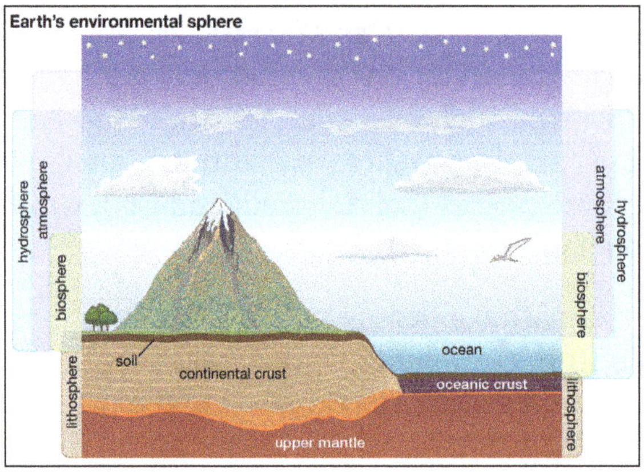

Earth's environmental spheres: Earth's environment includes the atmosphere, the hydrosphere, the lithosphere, and the biosphere.

The relative contributions of each process depend on the wind, temperature, and moisture structure in the atmosphere immediately above the surface, the intensity of solar insolation, and the physical characteristics of the surface. The temperature occurring at this interface is of critical importance in determining how suitable a location is for different forms of life.

Radiation

The temperature of the atmosphere and surface is influenced by electromagnetic radiation, and this radiation is traditionally divided into two types: insolation from the Sun and emittance from the surface and the atmosphere. Insolation is frequently referred to as shortwave radiation; it falls primarily within the ultraviolet and visible portions of the electromagnetic spectrum and consists predominantly of wavelengths of 0.39 to 0.76 micrometres (0.00002 to 0.00003 inch). Radiation emitted from Earth is called longwave radiation; it falls within the infrared portion of the spectrum and has typical wavelengths of 4 to 30 micrometres (0.0002 to 0.001 inch). Wavelengths of radiation emitted by a body depend on the temperature of the body, as specified by Planck's radiation law. The Sun, with its surface temperature of around 6,000 kelvins (K; about 5,725 °C, or 10,337 °F), emits at a much shorter wavelength than does Earth, which has lower surface and atmospheric temperatures around 250 to 300 K (−23 to 27 °C, or −9.4 to 80.6 °F).

A fraction of the incoming shortwave radiation is absorbed by atmospheric gases, including water vapour, and warms the air directly, but in the absence of clouds most of this energy reaches the surface. The scattering of a fraction of the shortwave radiation—particularly of the shortest wavelengths by air molecules in a process called Rayleigh scattering—produces Earth's blue skies.

When tall thick clouds are present, a large percentage (up to about 80 percent) of the insolation is reflected back into space. (The fraction of reflected shortwave radiation is called the cloud albedo.) Of the solar radiation reaching Earth's surface, some is reflected back into the atmosphere. Values of the surface albedo range as high as 0.95 for fresh snow to 0.10 for dark, organic soils. On land, this reflection occurs entirely at the surface. In water, however, albedo depends on the angle of the Sun's rays and the depth of the water column. If the Sun's rays strike the water surface at an oblique angle, albedo may be higher than 0.85; if these rays are more direct, only a small portion, perhaps as low as 0.02, is reflected, while the rest of the insolation is scattered within the water column and absorbed. Shortwave radiation penetrates a volume of water to significant depths (up to several hundred metres) before the insolation is completely attenuated. The heating by solar radiation in water is distributed through a depth, which results in smaller temperature changes at the surface of the water than would occur with the same insolation over an equal area of land.

The amount of solar radiation reaching the surface depends on latitude, time of year, time of day, and orientation of the land surface with respect to the Sun. In the Northern Hemisphere north of 23°30′, for example, solar insolation at local noon is less on slopes facing the north than on land oriented toward the south.

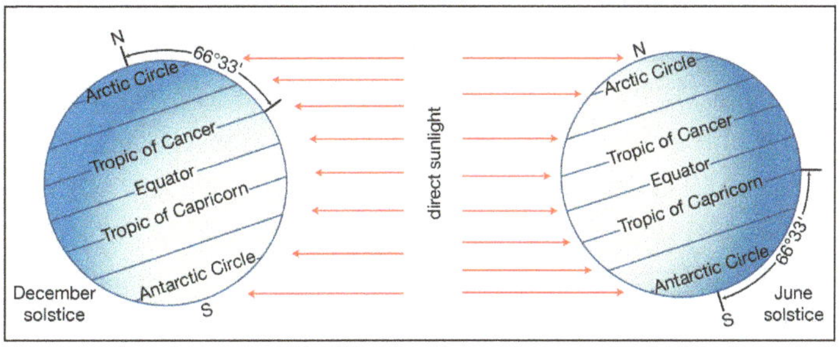

The primary cause of Earth's seasons is the change in the amount of sunlight reaching the surface at various latitudes over the course of a year. Because Earth is tilted on its axis with respect to the plane of its orbit around the Sun, different parts of its surface are in direct (overhead) sunlight at different times of the year.

Solar radiation is made up of direct and diffuse radiation. Direct shortwave radiation reaches the surface without being absorbed or scattered from its line of propagation by the intervening atmosphere. The image of the Sun's disk as a sharp and distinct object represents that portion of the solar radiation that reaches the viewer directly. Diffuse radiation, in contrast, reaches the surface after first being scattered from its line of propagation. On an overcast day, for example, the Sun's disk is not visible, and all of the shortwave radiation is diffuse.

Long-wave radiation is emitted by the atmosphere and propagates both upward and downward. According to the Stefan-Boltzmann law, the total amount of long-wave energy emitted is proportional to the fourth power of the temperature of the emitting material (e.g., the ground surface or the atmospheric layer). The magnitude of this radiation reaching the surface depends on the temperature at the height of emission and the amount of absorption that takes place between the height of emission and the surface. A larger fraction of the long-wave radiation is absorbed when the intervening atmosphere holds large amounts of water vapour and carbon dioxide. Clouds with liquid water concentrations near 2.5 grams per cubic metre absorb almost 100 percent of the long-wave radiation within a depth of 12 metres (40 feet) into the cloud. Clouds with lower liquid water concentrations require greater depths before complete absorption is attained (e.g., a cloud with a water content of 0.05 gram per cubic metre requires about 600 metres [about 2,000 feet] for complete absorption). Clouds that are at least this thick emit long-wave radiation from their bases downward to Earth's surface. The amount of long-wave radiation emitted corresponds to the temperature of the lowest levels of the cloud.

Conduction

The magnitude of heat flux by conduction below a surface depends on the thermal conductivity and the vertical gradient of temperature in the material beneath the surface. Soils such as dry peat, which has very low thermal conductivity (i.e., 0.06 watt per metre per K), permit little heat flux. In contrast, concrete has a thermal conductivity about 75 times as large (i.e., 4.60 watts per metre per K) and allows substantial heat flux. In water, the thermal conductivity is relatively unimportant, since, in contrast to land surfaces, insolation extends to substantial depths in the water; in addition, water can be mixed vertically.

Convection

Vertical mixing (convection) occurs in the atmosphere as well as in bodies of water. This process of mixing is also referred to as turbulence. It is a mechanism of heat flux that occurs in the atmosphere in two forms. When the surface is substantially warmer than the overlying air, mixing will spontaneously occur in order to redistribute the heat. This process, referred to as free convection, occurs when the environmental lapse rate (the rate of change of an atmospheric variable, such as temperature or density, with increasing altitude) of temperature decreases at a rate greater than 1 °C per 100 metres (approximately 1 °F per 150 feet). This rate is called the adiabatic lapse rate (the rate of temperature change occurring within a rising or descending air parcel). In the ocean, the temperature increase with depth that results in free convection is dependent on the temperature, salinity, and depth of the water. For example, if the surface has a temperature of 20 °C (68 °F) and a salinity of 34.85 parts per thousand, an increase in temperature with depth of greater than about 0.19 °C per km (0.55 °F per mile) just below in the upper layers of the ocean will result in free convection. In the atmosphere, the temperature profile with height determines whether free convection occurs or not. In the ocean, free convection depends on the temperature and salinity profile with depth. Colder and more saline conditions in a surface parcel of water, for example, make it more likely for that parcel to sink spontaneously and thus become part of the process of free convection.

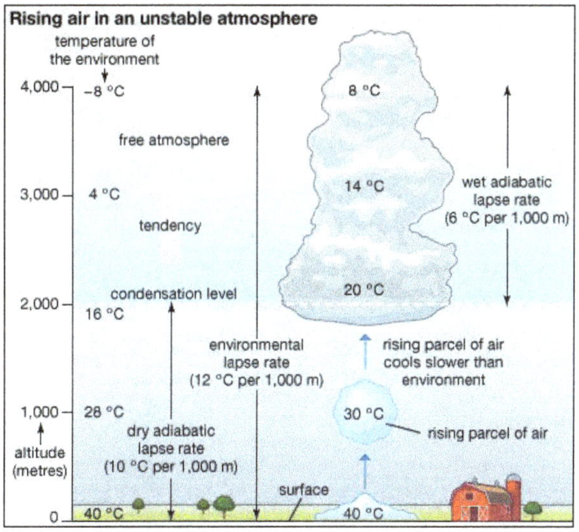

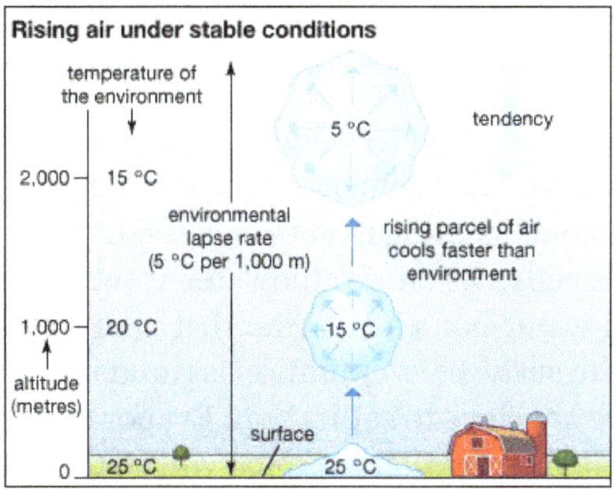

Mixing can also occur because of the shear stress of the wind on the surface. Shear stress is the pulling force of a fluid moving in one direction as it passes close to a fluid or object moving in another. As a result of surface friction, the average wind velocity at Earth's surface must be zero unless that surface is itself moving, such as in rivers or ocean currents. Winds above the surface decelerate when the vertical wind shear (the change in wind velocity at differing altitudes) becomes large enough to result in vertical mixing. The process by which heat and other atmospheric properties are mixed as a result of wind shear is called forced convection. Free and forced convection are also called convective and mechanical turbulence, respectively. This convection occurs as either sensible turbulent heat flux (heat directly transported to or from a surface) or latent turbulent heat flux (heat used to evaporate water from a surface). When this mixing does not occur, wind speeds are weak and change little with time; plumes from power-plant stacks within this layer, for example, spread very little in the vertical and remain in close proximity to the stacks.

Water Budget

The water budget at the air-surface interface is also of crucial importance in influencing atmospheric processes. The surface gains water through precipitation (rain and snow), direct condensation, and deposition (dew and frost). On land, the precipitation is often so large that some of it infiltrates into the ground or runs off into streams, rivers, lakes, and the oceans. Some of the precipitation remaining on the surface, such as in puddles or on vegetation, immediately evaporates back into the atmosphere.

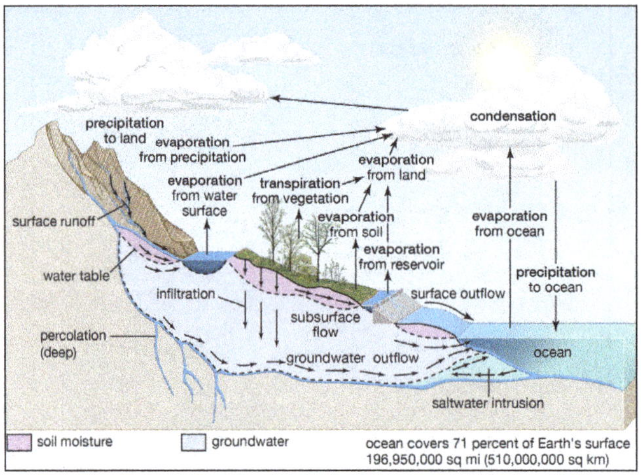

In the hydrologic cycle, water is transferred between the land surface, the ocean, and the atmosphere.

Liquid water in the soil is also converted to water vapour by transpiration from the leaves and stems of plants and by evaporation. The roots of vegetation may extract water from within the soil and emit it through stoma, or small openings, on the leaves. In addition, water may be evaporated from the surface of the soil directly, when groundwater from below is diffused upward. Evaporation occurs at the surface of water bodies at a rate that is inversely proportional to the relative humidity immediately above the surface. Evaporation is rapid in dry air but much slower when the lowest levels of the atmosphere are close to saturation. Evaporation from soils is dependent on the rate at which moisture is supplied by capillary suction within the soil, whereas transpiration from vegetation is dependent on both the water available within the root zone of plants and whether the stoma are open on the leaf surfaces. Water that evaporates and transpires into the atmosphere is often transported long distances before it precipitates out.

The input, transport, and removal of water from the atmosphere is part of the hydrologic cycle. At any one time, only a very small fraction of Earth's water is present within the atmosphere; if all the atmospheric water was condensed out, it would cover the surface of the planet only to an average of about 2.5 cm (1 inch).

Nitrogen Budget

The nitrogen budget involves the chemical transformation of diatomic nitrogen (N_2), which makes up 78 percent of the atmospheric gases, into compounds containing ammonium (NH^+), nitrite (NO_2^-), and nitrate (NO_3^-). In a process called nitrification, or nitrogen fixation, bacteria such as *Rhizobium* living within nodules on the roots of peas, clover, and other legumes convert diatomic nitrogen gas to ammonia. A small amount of nitrogen is also fixed by lightning. Ammonia may be further transformed by other bacteria into nitrites and nitrates and used by plants for growth. These compounds are eventually converted back to N_2 after the plants die or are eaten by denitrifying bacteria. These bacteria, in their consumption of plants and both the excrement and corpses of plant-eating animals, convert much of the nitrogen compounds back to N_2. Some of these compounds are also converted to N_2 by a series of chemical processes associated with ultraviolet light from the Sun. The combustion of petroleum by motor vehicles also produces oxides of nitrogen, which enhance the natural concentrations of these compounds. Smog, which occurs in many urban areas, is associated with substantially higher levels of nitrogen oxides.

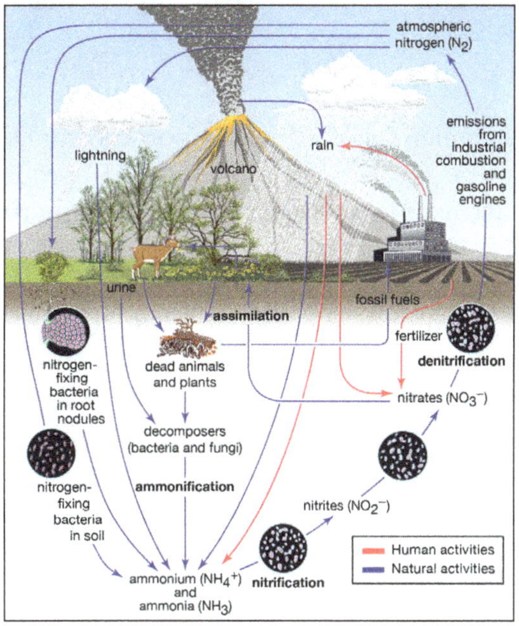

The nitrogen cycle

Sulfur Budget

The sulfur budget is also of major importance. Sulfur is put into the atmosphere as a result of weathering of sulfur-containing rocks and by intermittent volcanic emissions. Organic forms of sulfur are incorporated into living organisms and represent an important component in both the structure and the function of proteins. Sulfur also appears in the atmosphere as the gas sulfur dioxide (SO_2) and as part of particulate compounds containing sulfate (SO_4). Alone, both are directly dry-deposited or precipitated out onto Earth's surface. When wetted, these compounds are converted to caustic sulfuric acid (H_2SO_4).

Mount St. Helens volcano, viewed from the south.

Since the beginning of the Industrial Revolution, human activities have injected significant quantities of sulfur into the atmosphere through the combustion of fossil fuels. In and near regions of urbanization and heavy industrial activity, the enhanced deposition and precipitation of sulfur in the form of sulfuric acid, and of nitrogen oxides in the form of nitric acid (HNO_3), resulting from vehicular emissions, have been associated with damage to fish populations, forests, statues, and building exteriors. The conversion of sulfur and nitrogen oxides to acids such as H_2SO_4 and HNO_3 is commonly known as the acid rain problem. Sulfur and nitrogen oxides are precipitated in rain, snow, and dry deposition (deposition to the surface during dry weather).

Carbon Budget

The carbon budget in the atmosphere is of critical importance to climate and to life. Carbon appears in Earth's atmosphere primarily as carbon dioxide (CO_2) produced naturally by the respiration of living organisms, the decay of these organisms, the weathering of carbon-containing rock strata, and volcanic emissions. Plants utilize CO_2, water, and solar insolation to convert CO_2 to diatomic oxygen (O_2). This process, known as photosynthesis, can result in local reductions of CO_2 of tens of parts per million within vegetation canopies. In contrast, nighttime respiration occurring when photosynthesis is not active can increase CO_2 concentrations. These concentrations may even double within dense tropical forest canopies for short periods before sunrise. On the global scale, seasonal variations of about 1 percent occur as a result of CO_2 uptake from photosynthesis, plant respiration, and soil respiration. Atmospheric CO_2 is primarily absorbed in the Northern Hemisphere during the growing season (spring to autumn). CO_2 is also absorbed by ocean waters; the rate of exchange to the ocean is greater for colder than for warmer waters. Currently CO_2 makes up about 0.03 percent of the gaseous composition of the atmosphere.

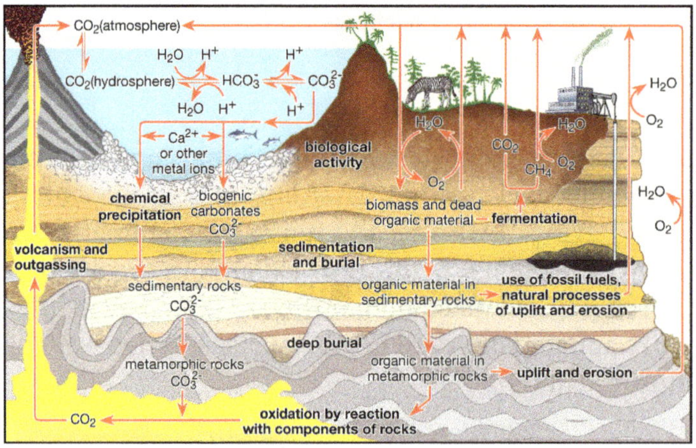

The carbon cycle

Carbon is transported in various forms through the atmosphere, the hydrosphere, and geologic formations. One of the primary pathways for the exchange of carbon dioxide (CO_2) takes place between the atmosphere and the oceans; there a fraction of the CO_2 combines with water, forming carbonic acid (H_2CO_3) that subsequently loses hydrogen ions (H^+) to form bicarbonate (HCO_3^-) and carbonate (CO_3^{2-}) ions. Mollusk shells or mineral precipitates that form by the reaction of calcium or other metal ions with carbonate may become buried in geologic strata and eventually release CO_2 through volcanic outgassing. Carbon dioxide also exchanges through photosynthesis in plants and through respiration in animals. Dead and decaying organic matter may ferment and release CO_2 or methane (CH_4) or may be incorporated into sedimentary rock, where it is converted to fossil fuels. Burning of hydrocarbon fuels returns CO_2 and water (H_2O) to the atmosphere. The biological and anthropogenic pathways are much faster than the geochemical pathways and, consequently, have a greater impact on the composition and temperature of the atmosphere.

In the geologic past, CO_2 levels have been significantly higher than they are today and have had a significant effect on both climate and ecology. During the Carboniferous Period (360 to 300 million years ago), for example, moderately warm and humid climates combined with high concentrations of CO_2 were associated with extensive lush vegetation. After these plants died and decomposed, they were converted to sedimentary rocks that eventually became the coal deposits currently used for industrial combustion.

In the atmosphere, certain wavelengths of long-wave radiation are absorbed and then reemitted by CO_2. Since the lower levels of the atmosphere are warmer than layers higher up, the absorption of upward-propagating electromagnetic radiation, and a reemission of a portion of it back downward, permits the lower atmosphere to remain warmer than it would be otherwise. The association of higher concentrations of CO_2 in the air with a warmer lower troposphere is commonly referred to as the greenhouse effect. (The name is inaccurate—an actual greenhouse is warmed primarily because solar radiation enters through the glass, which retains the heated air and prevents the mixing of cooler air into the greenhouse from above.) In recent years, there has been increasing concern that the release of CO_2 through the burning of coal and other fossil fuels will warm the lower atmosphere, a phenomenon commonly referred to as global warming. Water vapour is a more efficient greenhouse gas than carbon dioxide. However, since H_2O is ubiquitous, occurring in its three phases (solid, liquid, and gas), and since CO_2 is also a biogeochemically active gas, global temperature changes are both explained and predicted by changes in the atmospheric concentration of CO_2.

Vertical Structure of the Atmosphere

Earth's atmosphere is segmented into two major zones. The homosphere is the lower of the two and the location in which turbulent mixing dominates the molecular diffusion of gases. In this region, which occurs below 100 km (about 60 miles) or so, the composition of the atmosphere tends to be independent of height. Above 100 km, in the zone called the heterosphere, various atmospheric gases are separated by molecular mass, with the lighter gases being concentrated in the highest layers. Above 1,000 km (about 600 miles), helium and hydrogen are the dominant species. Diatomic nitrogen (N_2), a relatively heavy gas, drops off rapidly with height and exists in only trace amounts at 500 km (300 miles) and above. This decrease in the concentration of heavier gases with height is largest during periods of low Sun activity, when temperatures within the heterosphere are relatively low. The transition zone, located at a height of around 100 km between the homosphere and heterosphere, is called the turbopause.

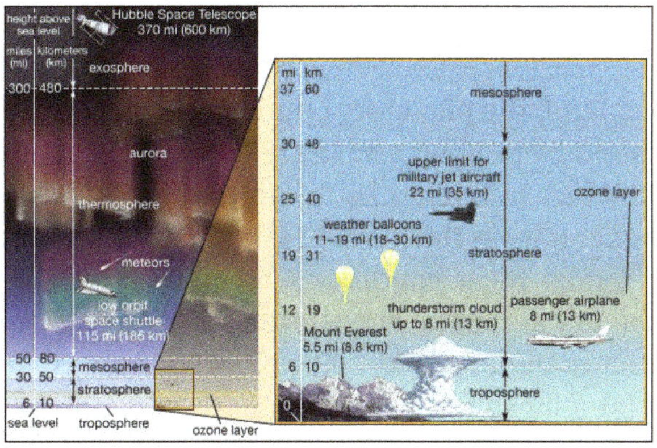

Atmosphere: vertical structure, the layers of Earth's atmosphere, showing heights of characteristic atmospheric phenomena.

The atmosphere can be further divided into several distinct layers defined by changes in air temperature with increasing height.

Troposphere

The lowest portion of the atmosphere is the troposphere, a layer where temperature generally decreases with height. This layer contains most of Earth's clouds and is the location where weather primarily occurs.

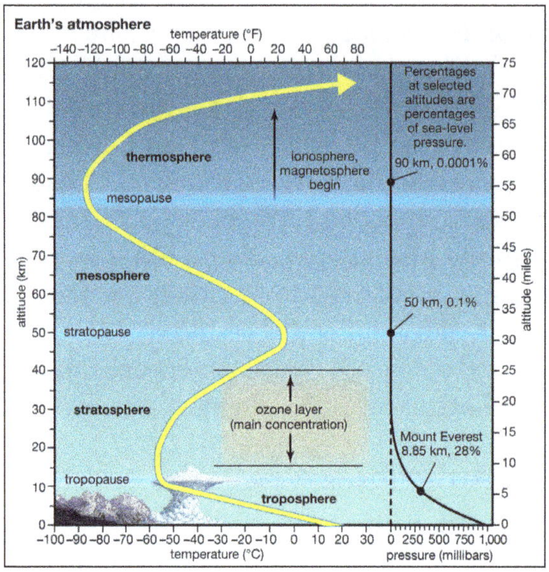

The layers of Earth's atmosphere: The yellow line shows the response of air temperature to increasing height.

Planetary Boundary Layer

The lower levels of the troposphere are usually strongly influenced by Earth's surface. This sub layer, known as the planetary boundary layer, is that region of the atmosphere in which the surface influences temperature, moisture, and wind velocity through the turbulent transfer of mass. As a result of surface friction, winds in the planetary boundary layer are usually weaker than above and tend to blow toward areas of low pressure. For this reason, the planetary boundary layer has also been called an Ekman layer, for Swedish oceanographer Vagn Walfrid Ekman, a pioneer in the study of the behaviour of wind-driven ocean currents.

Under clear, sunny skies over land, the planetary boundary layer tends to be relatively deep as a result of the heating of the ground by the Sun and the resultant generation of convective turbulence. During the summer, the planetary boundary layer can reach heights of 1 to 1.5 km (0.6 to 1 mile) above the land surface—for example, in the humid eastern United States—and up to 5 km (3 miles) in the southwestern desert. Under these conditions, when unsaturated air rises and expands, the temperature decreases at the dry adiabatic lapse rate (9.8 °C per kilometre, or roughly 23 °F per mile) throughout most of the boundary layer. Near Earth's heated surface, air temperature decreases super adiabatically (at a lapse rate greater than the dry adiabatic lapse rate). In contrast, during clear, calm nights, turbulence tends to cease, and radiational cooling (net loss of heat) from the surface results in an air temperature that increases with height above the surface.

When the rate of temperature decrease with height exceeds the adiabatic lapse rate for a region of the atmosphere, turbulence is generated. This is due to the convective overturn of the air as the warmer lower-level air rises and mixes with the cooler air aloft. In this situation, since the environmental lapse rate is greater than the adiabatic lapse rate, an ascending parcel of air remains warmer than the surrounding ambient air even though the parcel is both cooling and expanding. Evidence of this overturn is produced in the form of bubbles, or eddies, of warmer air. The larger bubbles often have sufficient buoyant energy to penetrate the top of the boundary layer. The subsequent rapid air displacement brings air from aloft into the boundary layer, thereby deepening the layer. Under these conditions of atmospheric instability, the air aloft cools according to the environmental lapse rate faster than the rising air is cooling at the adiabatic lapse rate. The air above the boundary layer replaces the rising air and undergoes compressional warming as it descends. As a result, this entrained air heats the boundary layer.

The ability of the convective bubbles to break through the top of the boundary layer depends on the environmental lapse rate aloft. The upward movement of penetrative bubbles will decrease rapidly if the parcel quickly becomes cooler than the ambient environment that surrounds it. In this situation, the air parcel will become less buoyant with additional ascent. The height that the boundary layer attains on a sunny day, therefore, is strongly influenced by the intensity of surface heating and the environmental lapse rate just above the boundary layer. The more rapidly a rising turbulent bubble cools above the boundary layer relative to the surrounding air, the lower the chance that subsequent turbulent bubbles will penetrate far above the boundary layer. The top of the daytime boundary layer is referred to as the mixed-layer inversion.

On clear, calm nights, radiational cooling results in a temperature increase with height. In this situation, known as a nocturnal inversion, turbulence is suppressed by the strong thermal stratification. Thermally stable conditions occur when warmer air overlies cooler, denser air. Over flat terrain, a nearly laminar wind flow (a pattern where winds from an upper layer easily slide past winds from a lower layer) can result. The depth of the radiationally cooled layer of air depends on a variety of factors, such as the moisture content of the air, soil and vegetation characteristics, and terrain configuration. In a desert environment, for instance, the nocturnal inversion tends to be found at greater heights than in a more humid environment. The inversion in more humid environments occurs at a lower altitude because more long-wave radiation emitted by the surface is absorbed by numerous available water molecules and reemitted back toward the surface. As a result, the lower levels of the troposphere are prevented from cooling rapidly. If the air is moist and sufficient near-surface cooling occurs, water vapour will condense into what is called "radiation fog."

Wind-generated Turbulence

During windy conditions, the mechanical production of turbulence becomes important. Turbulence eddies produced by wind shear tend to be smaller in size than the turbulence bubbles produced by the rapid convection of buoyant air. Within a few tens of metres of the surface during windy conditions, the wind speed increases dramatically with height. If the winds are sufficiently strong, the turbulence generated by wind shear can overshadow the resistance of layered, thermally stable air.

In general, there tends to be little turbulence above the boundary layer in the troposphere. Even so, there are two notable exceptions. First, turbulence is produced near jet streams, where large velocity shears exist both within and adjacent to cumuliform clouds. In these locations, buoyant turbulence occurs as a result of the release of latent heat. Second, pockets of buoyant turbulence may be found at and just above cloud tops. In these locations, the radiational cooling of the clouds destabilizes pockets of air and makes them more buoyant. Clear-air turbulence (CAT) is frequently reported when aircraft fly near one of these regions of turbulence generation.

The top of the troposphere, called the tropopause, corresponds to the level in which the pattern of decreasing temperature with height ceases. It is replaced by a layer that is essentially isothermal (of equal temperature). In the tropics and subtropics, the tropopause is high, often reaching to about 18 km (11 miles), as a result of vigorous vertical mixing of the lower atmosphere by thunderstorms. In polar regions, where such deep atmospheric turbulence is much less frequent, the tropopause is often as low as 8 km (5 miles). Temperatures at the tropopause range from as low as −80 °C (−112 °F) in the tropics to −50 °C (−58 °F) in polar regions.

Cloud Formation within the Troposphere

The region above the planetary boundary layer is commonly known as the free atmosphere. Winds at this volume are not directly retarded by surface friction. Clouds occur most frequently in this portion of the troposphere, though fog and clouds that impinge or develop over elevated terrain often occur at lower levels.

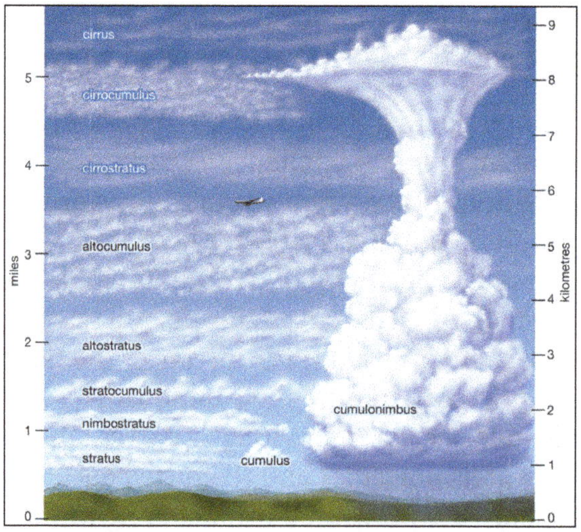

Cloud formation: Cloud formation at various heights.

There are two basic types of clouds: cumuliform and strati form. Cloud types develop when clear air ascends, cooling adiabatically as it expands until either water begins to condense or deposition occurs. Water undergoes a change of state from gas to liquid under these conditions, because cooler air can hold less water vapour than warmer air. For example, air at 20 °C (68 °F) can contain almost four times as much water vapour as at 0 °C (32 °F) before saturation takes place and water vapour condenses into liquid droplets.

Stratiform clouds occur as saturated air is mechanically forced upward and remains colder than the surrounding clear air at the same height. In the lower troposphere, such clouds are called stratus. Advection fog is a stratus cloud with a base lying at Earth's surface. In the middle troposphere, stratiform clouds are known as altostratus. In the upper troposphere, the terms cirrostratus and cirrus are used. The cirrus cloud type refers to thin, often wispy, cirrostratus clouds. Stratiform clouds that both extend through a large fraction of the troposphere and precipitate are called nimbostratus.

Cirrus fibratus are high clouds that are nearly straight or irregularly curved. They appear as fine white filaments and are generally distinct from one another.

Altocumulus radiatus, a cloud layer with laminae arranged in parallel bands.

Cumuliform clouds occur when saturated air is turbulent. Such clouds, with their bubbly turreted shapes, exhibit the small-scale up-and-down behaviour of air in the turbulent planetary boundary layer. Often such clouds are seen with bases at or near the top of the boundary layer as turbulent eddies generated near Earth's surface reach high enough for condensation to occur.

Cumulus humilis, flattened clouds characterized by only a small vertical extent.

Cumuliform clouds will form in the free atmosphere if a parcel of air, upon saturation, is warmer than the surrounding ambient atmosphere. Since this air parcel is warmer than its surroundings, it will accelerate upward, creating the saturated turbulent bubble characteristic of a cumuliform cloud. Cumuliform clouds, which reach no higher than the lower troposphere, are known as cumulus humulus when they are randomly distributed and as stratocumulus when they are organized into lines. Cumulus congestus clouds extend into the middle troposphere, while deep, precipitating cumuliform clouds that extend throughout the troposphere are called cumulonimbus. Cumulonimbus clouds are also called thunderstorms, since they usually have lightning and thunder associated with them. Cumulonimbus clouds develop from cumulus humulus and cumulus congestus clouds.

Stratosphere and Mesosphere

The stratosphere is located above the troposphere and extends up to about 50 km (30 miles). Above the tropopause and the isothermal layer in the lower stratosphere, temperature increases with height. Temperatures as high as 0 °C (32 °F) are observed near the top of the stratosphere.

The observed increase of temperature with height in the stratosphere results in strong thermodynamic stability with little turbulence and vertical mixing. The warm temperatures and very dry air result in an almost cloud-free volume. The infrequent clouds that do occur are called nacreous, or mother-of-pearl, clouds because of their striking iridescence, and they appear to be composed of both ice and supercooled water. These clouds form up to heights of 30 km (19 miles).

The pattern of temperature increase with height in the stratosphere is the result of solar heating as ultraviolet radiation in the wavelength range of 0.200 to 0.242 micrometre dissociates diatomic oxygen (O_2). The resultant attachment of single oxygen atoms to O_2 produces ozone (O_3). Natural stratospheric ozone is produced mainly in the tropical and middle latitudes. Regions of nearly complete ozone depletion, which have occurred in the Antarctic during the spring, are associated with nacreous clouds, chlorofluorocarbons (CFCs), and other pollutants from human activities. These regions are more commonly known as ozone holes. Ozone is also transported downward into the troposphere, primarily in the vicinity of the polar front.

The stratopause caps the top of the stratosphere, separating it from the mesosphere near 45–50 km (28–31 miles) in altitude and a pressure of 1 millibar (approximately equal to 0.75 mm of mercury at 0 °C, or 0.03 inch of mercury at 32 °F). In the mesosphere, temperatures again decrease with increasing altitude. Unlike the situation in the stratosphere, vertical air currents in the mesosphere are not strongly inhibited. Ice crystal clouds, called noctilucent clouds, occasionally form in the upper mesosphere. Above the mesopause, a region occurring at altitudes near 85 to 90 km (50 to 55 miles), temperature again increases with height in a layer called the thermosphere.

Thermosphere

Temperatures in the thermosphere range from near 500 K (approximately 227 °C, or 440 °F) during periods of low sunspot activity to 2,000 K (1,725 °C, or 3,137 °F) when the Sun is active. The thermopause, defined as the level of transition to a more or less isothermal temperature profile at the top of the thermosphere, occurs at heights of around 250 km (150 miles) during quiet Sun periods and almost 500 km (300 miles) when the Sun is active. Above 500 km, molecular collisions are infrequent enough that temperature is difficult to define.

The portion of the thermosphere where charged particles (ions) are abundant is called the ionosphere. These ions result from the removal of electrons from atmospheric gases by solar ultraviolet radiation. Extending from about 80 to 300 km (about 50 to 185 miles) in altitude, the ionosphere is an electrically conducting region capable of reflecting radio signals back to Earth.

Maximum ion density, a condition that makes for efficient radio transmission, occurs within two sub layers the lower E region, which exists from 90 to 120 km (about 55 to 75 miles) in altitude; and the F region, which exists from 150 to 300 km (about 90 to 185 miles) in altitude. The F region has two maxima (i.e., two periods of highest ion density) during daylight hours, called F1 and F2. Both the F1 and F2 regions possess high ion density and are strongly influenced by both solar activity and time of day. Of these, the F2 region is the more variable of the two and may reach an ion density as high as 106 electrons per cubic centimetre. Shortwave radio transmissions, capable of reaching around the world, take advantage of the ability of layers in the ionosphere to reflect certain wavelengths of electromagnetic radiation. In addition, electrical discharges from the tops of thunderstorms into the ionosphere, called transient luminous events, have been observed.

Magnetosphere and Exosphere

Above approximately 500 km (300 miles), the motion of ions is strongly constrained by the presence of Earth's magnetic field. This region of Earth's atmosphere, called the magnetosphere, is compressed by the solar wind on the daylight side of the planet and stretched outward in a long tail on the night side. The colourful auroral displays often seen in polar latitudes are associated with bursts of high-energy particles generated by the Sun. When these particles are influenced by the magnetosphere, some are subsequently injected into the lower ionosphere.

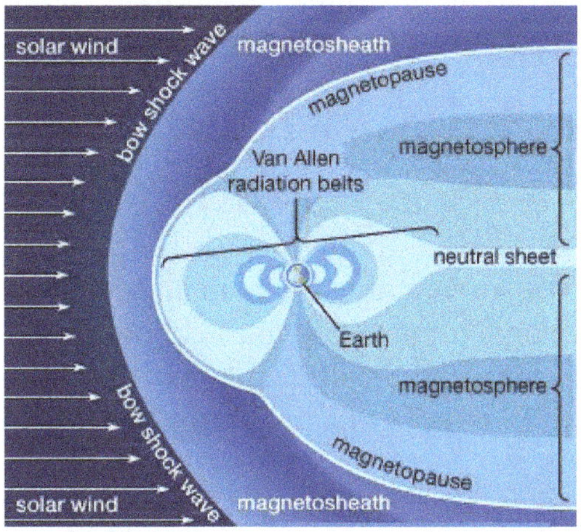

The Van Allen radiation belts contained within Earth's magnetosphere. Pressure from the solar wind is responsible for the asymmetrical shape of the magnetosphere and the belts.

The layer above 500 km is referred to as the exosphere, a region in which at least half of the upward-moving molecules do not collide with one another. In contrast, these molecules follow long ballistic trajectories and may exit the atmosphere completely if their escape velocities are high enough. The loss rate of molecules through the exosphere is critical in determining whether Earth or any other planetary body retains an atmosphere.

Horizontal Structure of the Atmosphere

Distribution of Heat from the Sun

The primary driving force for the horizontal structure of Earth's atmosphere is the amount and distribution of solar radiation that comes in contact with the planet. Earth's orbit around the Sun is an ellipse, with a perihelion (closest approach) of 147.5 million km (91.7 million miles) in early January and an aphelion (farthest distance) of 152.6 million km (94.8 million miles) in early July. As a result of Earth's elliptical orbit, the time between the autumnal equinox and the following vernal equinox (about September 22 to about March 21) is almost one week shorter than the remainder of the year in the Northern Hemisphere. This results in a shorter astronomical winter in the Northern Hemisphere than in the Southern Hemisphere.

Earth rotates once every 24 hours around an axis that is tilted at an angle of 23°30′ with respect to the plane of its orbit around the Sun. As a result of this tilt, during the summer season of either the Northern or the Southern Hemisphere, the Sun's rays are more direct at given latitude than they are during the winter season. Poleward of latitudes 66°30′ N and 66°30′ S, the tilt of the planet is such that for at least one complete day (at 66°30′) and as long as six months (at 90°), the Sun is above the horizon during the summer season and below the horizon during the winter.

As a result of this asymmetric distribution of solar heating, during the winter season the troposphere in the high latitudes becomes very cold. In contrast, during the summer at high latitudes, the troposphere warms significantly as a result of the long hours of daylight; however, owing to the oblique angle of the sunlight near the poles, the temperatures there remain relatively cool compared with middle latitudes. Equatorward of latitudes 30° N and 30° S or so, substantial radiant heating from the Sun occurs during both winter and summer seasons. The tropical troposphere, therefore, has comparatively little variation in temperature during the year.

Convection, Circulation and Deflection of Air

The region of greatest solar heating at the surface in the humid tropics corresponds to areas of deep cumulonimbus convection. Cumulonimbus clouds routinely form in the tropics where rising parcels of air are warmer than the surrounding ambient atmosphere. They transport water vapour, sensible heat, and Earth's rotational momentum to the upper portion of the troposphere. As a result of the vigorous convective mixing of the atmosphere, the tropopause in the lower latitudes is often very high, located some 17 to 18 km (10.5 to 11 miles) above the surface.

Since motion upward into the stratosphere is inhibited by very stable thermal layering, the air transported upward by convection diverges toward the poles in the upper troposphere. (This divergence aloft results in a wide strip of low atmospheric pressure at the surface in the tropics, occurring in an area called the equatorial trough). As the diverted air in the troposphere moves toward the poles, it tends to retain the angular momentum of the near-equatorial region, which is large as a result of Earth's rotation. As a result, the poleward-moving air is deflected toward the right in the Northern Hemisphere and toward the left in the Southern Hemisphere.

Upon reaching around 30° of latitude poleward of its region of origin, the upper-level air is traveling primarily toward the poles and is tending toward the east. Since motion upward is constrained by the stratosphere, the slowly cooling air must descend. The compressional warming that occurs as the air descends creates vast regions of subtropical high pressure. These regions are centred over the oceans and are characterized by strong thermodynamic stability. The sparse precipitation in these regions, a result of stability and subsidence, is associated with such great arid regions of the world as the Sahara, Atacama, Kalahari, and Sonoran deserts. The accumulation of air as a result of the convergence in the upper troposphere causes deep high-pressure systems, known as subtropical ridges, to form in these regions. Locally, these ridges are given such names as the Bermuda High, the Azores High, and the North Pacific High.

The Erg Admer, a large area of sand dunes in southern Algeria, is located within a vast region of subtropical high pressure. The arid conditions here result from the constant presence of descending air containing little moisture.

The descending air referred to above, upon reaching the lower troposphere, is forced to diverge by the presence of Earth's surface. Some air moves poleward, while the remainder moves equatorward. In either direction, the air is deflected to the right in the Northern Hemisphere and to the left in the Southern Hemisphere. Deflection occurs because, in accordance with Newton's first law of motion, a parcel moving in a certain direction will retain the same motion unless acted on by an exterior force. With respect to a rotating Earth, a moving parcel conserving its momentum (i.e., not acted on by an exterior force) will appear to be deflected with respect to fixed points on the rotating Earth. As seen from a fixed point in space, such a parcel would be moving in a straight line. This apparent force on the motion of a fluid (in this case, air) is called the Coriolis effect. As a result of the Coriolis effect, air tends to rotate counterclockwise around large-scale low-pressure systems and clockwise around large-scale high-pressure systems in the Northern Hemisphere. In the Southern Hemisphere, the flow direction is reversed.

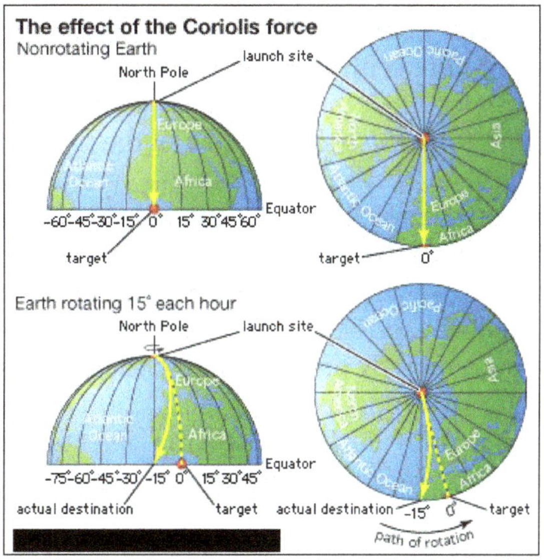

In the equatorward-moving flow, this deflection results in northeast winds north of 0° latitude and southeast winds south of that latitude. These low-level winds have been called the trade winds since 17th-century sailing vessels used them to travel to the Americas. The convergence region for lower-level northeast and southeast trade winds is called the intertropical convergence zone (ITCZ). The ITCZ corresponds to the equatorial trough and is the mechanism that helps generate the deep cumulonimbus clouds through convection. Cumulonimbus clouds are the main conduit transporting tropical heating into the upper troposphere.

The circulation pattern described above—ascent in the equatorial trough, poleward movement in the upper troposphere, descent in the subtropical ridges, and equatorward movement in the trade winds—is in effect a direct heat engine, which meteorologists call the Hadley cell. This persistent circulation mechanism transports heat from the latitudes of greatest solar insolation to the latitudes of the subtropical ridges. The geographic location of the Hadley circulation moves north and south with the seasons; however, the equatorial trough lags behind for about two months owing to the thermal inertia of Earth's surface. (For a given location on Earth's surface, the highest daily temperatures are achieved just after the period of greatest insolation, since time is required to heat the ocean surface waters and the soil).

Extratropical Cyclones

Poleward of the subtropical ridges, winds in the lower troposphere tend to be southwesterly in the Northern Hemisphere and northwesterly in the Southern Hemisphere, again owing to the Coriolis effect. Since warm air is being moved poleward at low altitudes, the wind flow is no longer associated with the direct heat engine of the Hadley cell. Instead, the continued transport of heat from the equatorial trough toward the poles is facilitated by large low-pressure eddies called extratropical cyclones. These phenomena develop along the polar front, which separates colder polar air from warmer tropical air, when sufficiently large temperature differences occur across the frontal boundary in the lower troposphere. The intensity of this temperature gradient is referred to as the baroclinicity of the front.

Extratropical cyclones have three stages of expansion: the developing stage, in which an undulating wave develops along the front; the mature stage, in which sinking cold air sweeps equatorward west of the surface low-pressure centre and ascending warm air moves poleward east of the cyclone; and the occluded stage, in which the warm air is entrained within and moved above the polar air and becomes separated from the source region of the tropical air. Cyclones that progress no farther than the developing stage are referred to as wave cyclones, while extratropical lows that reach the mature and occluded stages are called baroclinically unstable waves. Extratropical storm development is referred to as cyclogenesis. Rapid extratropical cyclone development, called explosive cyclogenesis, is often associated with major winter storms and occurs when surface pressure falls by more than about 24 millibars per day. Theoretical analysis has shown that the occurrence of baroclinically unstable waves is directly proportional to the magnitude of the temperature gradient, with maximum growth for wavelengths of 3,000 to 5,000 km (1,865 to 3,100 miles). Wavelengths that are shorter are damped by horizontal mixing. The 3,000 to 5,000 km wavelength is the typical separation between high- and low- pressure synoptic weather systems in the middle and higher latitudes.

Polar Fronts and the Jet Stream

In the troposphere, the demarcation between polar air and warmer tropical atmosphere is usually defined by the polar front. On the poleward side of the front, the air is cold and denser; equatorward of the front, the air is warmer and more buoyant. During the winter season, the polar front is generally located at lower latitudes and is more pronounced than in the summer.

Cold fronts occur at the leading edge of equatorward-moving polar air. In contrast, warm fronts are well defined at the equatorward surface position of polar air as it retreats on the eastern sides of extratropical cyclones. Equatorward-moving air behind a cold front occurs in pools of dense high pressure known as polar highs and arctic highs. The term arctic high is used to define air that originates even deeper within the high latitudes than polar highs.

When polar air neither retreats nor advances, the polar front is called a stationary front. In the occluded stage of the life cycle of an extratropical cyclone, when cold air west of the surface low-pressure centre advances more rapidly toward the east than cold air ahead of the warm front, warmer, less-dense air is forced aloft. This frontal intersection is called an occluded front. Without exception, fronts of all types follow the movement of colder air.

Clouds and often precipitation occur on the poleward sides of both warm and stationary fronts and whenever tropical air reaching the latitude of the polar front is forced upward over the colder air near the surface. Such fronts are defined as active fronts. Rain and snowfall from active fronts form a major part of the precipitation received in the middle and high latitudes. Precipitation in these areas occurs primarily during the winter months.

The position of the polar front slopes upward toward colder air. This occurs because cold air tends to undercut the warmer air of tropical origin. Since cold air is denser, atmospheric pressure decreases more rapidly with height on the poleward side of the polar front than on the warmer tropical side. This creates a large horizontal temperature contrast, which is essentially a large pressure gradient, between the polar and tropical air. In the middle and upper parts of the troposphere, this pressure gradient is responsible for the strong westerly winds occurring there. Winds created aloft circulate around a large region of upper-level low pressure near each of the poles. The centre of each low pressure region is a persistent cyclone known as the circumpolar vortex.

The region of strongest winds, which occurs at the juncture of the tropical and polar air masses, is called the jet stream. Since the temperature contrast between the tropics and the high latitudes is greatest in the winter, the jet stream is stronger during that season. In addition, since the mid-latitudes also become colder during the winter, while tropical temperatures remain relatively unchanged, the westerly jet stream approaches latitudes of 30° during the colder season. During the warmer season in both hemispheres, the jet stream moves poleward and is located between latitudes of 50° and 60°.

The jet stream reaches its greatest velocity at the tropopause. Above that level, a reversal of the horizontal temperature gradient occurs, which produces a reduction in the wind speeds of the jet stream at high latitudes. This causes a weakening of the westerlies with increasing height. At intervals ranging from 20 to 40 months, with a mean value of 26 months, westerly winds in the stratosphere reverse direction over low latitudes, so that an easterly flow develops. This feature is called the quasi-biennial oscillation (QBO). In addition, a phenomenon called sudden stratospheric warming, apparently the result of strong downward air motion, also occurs in the late winter and spring at high latitudes. Sudden stratospheric warming can significantly alter temperature-dependent chemical reactions of ozone and other reactive gases in the stratosphere and affect the development of such features as "ozone holes."

A major focus of weather forecasting in the middle and high latitudes is to forecast the movement and development of extratropical cyclones, polar and arctic highs, and the location and intensity of subtropical ridges. Spring and fall frosts, for example, are associated with the equatorward movement of polar highs behind a cold front, while droughts and heat waves in the summer are associated with unusually strong subtropical ridges.

Effect of Continents on Air Movement

Preferred geographic locations exist for subtropical ridges and for the development, movement, and decay of extratropical cyclones. During the winter months in middle and high latitudes, the lower parts of the troposphere over continents often serve as reservoirs of cold air as heat is radiated into space throughout the long nights. In contrast, the oceans lose heat less rapidly, because of the large heat capacity of water, their ability to overturn as the surfaces cool and become negatively buoyant, and the movement of ocean currents such as the Gulf Stream and the Kuroshio current. Warm currents transport heat from lower latitudes poleward and tend to occur on the western sides of oceans. The lower troposphere over these warmer oceanic areas tends to be a region of relative low pressure. As a result of this juxtaposition of cold air and warm air, the eastern sides of continents and the western fringes of oceans in middle and high latitudes are the preferred locations for extratropical storm development. Over Asia in particular, the cold high-pressure system is sufficiently permanent that a persistent offshore flow called the winter monsoon occurs.

An inverse type of flow develops in the summer as the continents heat more rapidly than their adjacent oceanic areas. Continental areas tend to become regions of relative low pressure, while high pressure in the lower troposphere becomes more prevalent offshore. As the winds travel from areas of higher pressure to areas of lower pressure, a persistent onshore flow develops over large landmasses in the lower troposphere. The result of this heating is referred to as the summer monsoon. The leading edge of this monsoon is associated with a feature called the monsoon trough, a region of low atmospheric pressure at sea level. Tropical moisture carried onshore by the summer monsoon often results in copious rainfall. The village of Cherrapunji in northeastern India, for instance, recorded over 9 metres (about 30 feet) of rain in one month (July 1861) owing to the Indian summer monsoon.

As a result of the continental effect, the subtropical ridge is segmented into surface high-pressure cells. In the summer, large landmasses in the subtropics tend to be centres of relative low pressure as a result of strong solar heating. As a consequence, persistent high-pressure cells, such as the Bermuda and Azores highs, occur over the oceans. The oval shape of these high-pressure cells creates a thermal structure on their eastern sides that differs from the thermal structure on their western sides in the lower troposphere. On the eastern side, subsidence from the Hadley circulation is enhanced by the tendency of air to preserve its angular momentum on the rotating Earth. Owing to the enhanced descent of air over the eastern parts of the oceans, landmasses adjacent to these areas (typically the western sides of continents) tend to be deserts, such as those found in northwestern and southwestern Africa and along western coastal Mexico.

Effect of Oceans on Air Movement

The arid conditions found along the western coasts of continents in subtropical latitudes are further enhanced by the influence of the equatorward surface air flow on the ocean currents. This flow exerts a shearing stress on the ocean surface, which results in the deflection of the upper layer of water above the thermocline to the right in the Northern Hemisphere and to the left in the Southern Hemisphere. (This deflection is also the result of the Coriolis effect; water from both hemispheres moves westward when displaced toward the Equator.) As warmer surface waters are carried away by this offshore ocean airflow, cold water from below the thermocline rises to the surface in a process called upwelling. Upwelling creates areas of cold coastal surface waters that stabilize the lower troposphere and reduce the chances for convection. Lower convection in turn reduces the likelihood for precipitation, although fogs and low stratus clouds are common. Upwelling regions are also associated with enriched sea life, as oxygen and organic nutrients are transported upward from the depths toward the surface of the ocean.

During periods when the intertropical convergence zone (ITCZ) is located near the Equator, trade winds from the northeast and southeast converge there. The westward-moving winds cause the displacement of surface ocean waters away from the Equator such that the deeper, colder waters move to the surface. In the central and eastern Pacific Ocean near the Equator, when this upwelling is stronger than average, the event is called La Niña. When the trade winds weaken in this region, however, warmer-than-average surface conditions occur, and upwelling is weaker than usual. This event is called El Niño. Changes in ocean surface temperatures caused by El Niño significantly affect where cumulonimbus clouds form in the ITCZ and, therefore, the geographic structure of the Hadley cell. During periods when El Niño is active, weather patterns across the entire Earth are substantially altered.

Mountain Barriers

North-south-oriented mountain barriers, such as the Rockies and the Andes, and large massifs, such as the Plateau of Tibet, also influence atmospheric flow. When the general westerly flow in the mid-latitudes reaches these barriers, air tends to be blocked. It is transported poleward west of the terrain and toward the Equator east of the obstacle. Air forced up the slopes of mountain barriers is often sufficiently moist to produce considerable precipitation on windward sides of mountains, whereas subsiding air on the lee slopes produces more-arid conditions. Essentially, the elevated terrain affects the atmosphere as if it were an anticyclone, a centre of high pressure. In addition, mountains prevent cold air from the continental interior from moving westward of the terrain. As a result, relatively mild weather occurs along the western coasts of continents with north-south mountain ranges when compared with continental interiors. For example, the West Coast of North America experiences milder winter weather than the Great Plains and Midwest, both of which occur at similar latitudes. In contrast, east-west mountain barriers, such as the Alps in Europe, offer little impediment to the general westerly flow of air. In these situations, milder maritime conditions extend much farther inland.

Cloud Processes

Condensation

The formation of cloud droplets and cloud ice crystals is associated with suspended aerosols, which are produced by natural processes as well as human activities and are ubiquitous in Earth's atmosphere. In the absence of such aerosols, the spontaneous conversion of water vapour into liquid water or ice crystals requires conditions with relative humidities much greater than 100 percent, with respect to a flat surface of H_2O. The development of clouds in such a fashion, which occurs only in a controlled laboratory environment, is referred to as homogeneous nucleation. Air containing water vapour with a relative humidity greater than 100 percent, with respect to a flat surface, is referred to as being supersaturated. In the atmosphere, aerosols serve as initiation sites for the condensation or deposition of water vapour. Since their surfaces are of discrete sizes, aerosols reduce the amount of supersaturation required for water vapour to change its phase and are referred to as cloud condensation nuclei.

The larger the aerosol and the greater its solubility, the lower the supersaturation percentage required for the aerosol to serve as a condensation surface. Condensation nuclei in the atmosphere become effective at supersaturations of around 0.1 to 1 percent (that is, levels of water vapour around 0.1 to 1 percent above the point of saturation). The concentration of cloud condensation nuclei in the lower troposphere at a supersaturation of 1 percent ranges from around 100 per cubic centimetre (approximately 1,600 per cubic inch) in size in oceanic air to 500 per cubic centimetre (8,000 per cubic inch) in the atmosphere over a continent. Higher concentrations occur in polluted air.

Aerosols that are effective for the conversion of water vapour to ice crystals are referred to as ice nuclei. In contrast to cloud condensation nuclei, the most effective ice nuclei are hydrophobic (having a low affinity for water) with molecular spacings and a crystallographic structure close to that of ice.

While cloud condensation nuclei are always readily available in the atmosphere, ice nuclei are often scarce. As a result, liquid water cooled below 0 °C (32 °F) can often remain liquid at subfreezing temperatures because of the absence of effective ice nuclei. Liquid water at temperatures less than 0 °C is referred to as supercooled water. Except for true ice crystals, which are effective at 0 °C, all other ice nuclei become effective at temperatures below freezing. In the absence of any ice nuclei, the freezing of supercooled water droplets of a few micrometres in radius, in a process called homogeneous ice nucleation, requires temperatures at or lower than −39 °C (−38 °F). While a raindrop will freeze near 0 °C, small cloud droplets have too few molecules to create an ice crystal by random chance until the molecular motion is slowed as the temperature approaches −39 °C. When ice nuclei are present, heterogeneous ice nucleation can occur at warmer temperatures.

Ice nuclei are of three types- deposition nuclei, contact nuclei, and freezing nuclei. Deposition nuclei are analogous to condensation nuclei in that water vapour directly deposits as ice crystals on the aerosol. Contact and freezing nuclei, in contrast, are associated with the conversion of supercooled water to ice. A contact nucleus converts liquid water to ice by touching a supercooled water droplet. Freezing nuclei are absorbed into the liquid water and convert the supercooled water to ice from the inside out.

Examples of cloud condensation nuclei include sodium chloride (NaCl) and ammonium sulfate ($[NH_4]_2 SO_2$), whereas the clay mineral kaolinite is an example of an ice nuclei. In addition, naturally occurring bacteria found in decayed leaf litter can serve as ice nuclei at temperatures of less than about −4 °C (24.8 °F). In a process called cloud seeding, silver iodide, with effective ice-nucleating temperatures of less than −4 °C, has been used for years in attempts to convert supercooled water to ice crystals in regions with a scarcity of natural ice nuclei.

Precipitation

Liquid Droplets

The evolution of clouds that follows the formation of liquid cloud droplets or ice crystals depends on which phase of water occurs. A cloud in which only liquid water occurs (even at temperatures less than 0 °C) is referred to as a warm cloud, and the precipitation that results is said to be due to warm-cloud processes. In such a cloud, the growth of a liquid water droplet to a raindrop begins with condensation, as additional water vapour condenses in a supersaturated atmosphere. This process continues until the droplet has attained a radius of about 10 micrometres (0.0004 inch). Above this size, since the mass of the droplet increases according to the cube of its radius, further increases by condensational growth are very slow. Subsequent growth, therefore, occurs only when the cloud droplets develop at slightly different rates. Differences in growth rates have been attributed to differences in spatial variations of the initial aerosol sizes, in solubilities, and in magnitudes of supersaturation. Cloud droplets of different sizes will fall at different velocities and will collide with droplets of different radii. If the collision is hard enough to overcome the surface tension between the two colliding droplets, coalescence will occur and result in a new and larger single droplet.

This process of cloud-droplet growth is referred to as collision-coalescence. Warm-cloud rain results when the droplets attain a sufficient size to fall to the ground. Such a raindrop (perhaps about 1 mm [0.04 inch] in radius) contains perhaps one million 10-micrometre cloud droplets. The typical radii of raindrops resulting from this type of precipitation process range up to several millimetres and have fall velocities of around 3 to 4 metres (10 to 13 feet) per second. This type of precipitation is very common from shallow cumulus clouds over tropical oceans. In these locations, the concentration of cloud condensation nuclei is so small that there is only limited competition for the available water vapour.

Precipitation of Ice

A cloud that contains ice crystals is referred to as a cold cloud, and the resulting precipitation is said to be the product of cold-cloud processes. Traditionally, this process has also been referred to as the Bergeron-Findeisen mechanism, for Swedish meteorologists Tor Bergeron and Walter Findeisen, who introduced it in the 1930s. In this type of cloud, ice crystals can grow directly from the deposition of water vapour. This water vapour may be supersaturated with respect to ice, or it may be the result of evaporation of supercooled water and subsequent deposition onto an ice crystal. Since the saturation vapour pressure of liquid water is always greater than or equal to the saturation vapour pressure of ice, ice crystals will grow at the expense of the liquid water. For example, saturated air with respect to liquid water becomes supersaturated with respect to ice by 10 percent at −10 °C (14 °F) and by 21 percent at −20 °C (-4 °F). This results in a rapid conversion of liquid water to ice. This substantial and rapid change of phase permits large ice crystals in a cloud surrounded by a large number of supercooled cloud droplets to grow quickly (often in less than 15 minutes) from tiny ice crystals to snowflakes. These snowflakes are large enough to fall by depositional growth alone. Fall velocities of snow range up to about 2 metres per second (6.5 feet per second). Ice crystals that grow by deposition have much lower densities than solid ice because of the air pockets occurring within the volume of the crystal. This lower density differentiates snow from ice. Clouds that are completely converted to ice crystals are referred to as glaciated clouds.

The specific form the ice crystals take depends on the temperature and the degree of supersaturation with respect to ice. At −14 °C (7 °F) and a relatively large supersaturation with respect to liquid water, for example, ice crystals with dendritic (treelike branching) patterns form. This type of ice crystal, the one usually used to represent snowflakes in photographs and drawings, experiences growth at the end of radial arms on one or more planes of the crystal. At −40 °C (−40 °F) and a supersaturation with respect to liquid water of close to 0 percent, hollow ice columns form.

Ice crystals can also grow large enough to precipitate either by aggregation or by riming. Aggregation occurs when the arms of the ice crystals interlock and form a clump. This collection of intermingled ice crystals can occasionally reach several centimetres in diameter. Ice crystals can also grow when supercooled water freezes directly onto the crystal to form rime. With greater accumulation of dense ice on the crystal, its fall velocity increases. When the riming is substantial enough, the crystal form of the snowflake is lost and replaced by a more or less spherical particle called graupel. Smaller-sized graupels are generally referred to as snow grains. In cumulonimbus clouds during conditions where graupels are repeatedly wetted and then injected back toward high altitudes by strong updrafts, very large graupels called hail result. Hail has been observed on the ground at sizes larger than grapefruits.

Frozen precipitation, falling to levels of the atmosphere that are much warmer than 0 °C, often melts and reaches the ground as rain. Such cold-cloud rain at the ground is usually distinguished from warm-cloud rain by its larger size. Melted hailstones, in particular, make a large-radius impact when they strike the ground. Cold-cloud rain occasionally will refreeze if a layer of subfreezing air exists near Earth's surface. When this freezing occurs in the free atmosphere, the frozen raindrops are referred to as sleet or ice pellets. When this freezing occurs only upon the impact of the raindrop with the ground, the precipitation is known as freezing rain. During ice storms, freezing rain can produce accumulations heavy enough to snap large trees and electrical lines.

Lightning and Optical Phenomena

The repeated collision of ice crystals and graupel in clouds is associated with the buildup of electrical charge. This electrification is particularly large in cumulonimbus clouds as a result of vigorous vertical mixing and collisions. On average, positive charges accumulate in the upper regions, while negative charges are concentrated lower down. In response to the negative charge near the cloud base, and as negatively charged rain falls toward the ground, a pocket of positive charge develops on the ground. When the difference in electric potential between positive and negative charges becomes large enough, a sudden electrical discharge (lightning) will occur. Lightning can occur between different regions of the cloud, as in intracloud lightning, and between the cloud and the positively charged ground, as in cloud-to-ground lightning. The passage of the lightning through the air heats it to above 30,000 K (29,725 °C, or 53,540 °F), causing a large increase in pressure. This produces a powerful shock wave that is heard as thunder.

Sunlight that propagates through clouds and precipitation often produces fascinating optical images. Rainbows are produced when sunlight is diffracted into its component colours by water droplets. In addition, halos are produced by the refraction and reflection of sunlight or moonlight by ice crystals, while coronas are formed when sunlight or moonlight passes through water droplets.

Sunlight: Rays of sunlight shining through clouds.

Cloud Research

The presence of cloud condensation and ice nuclei in air parcels is tested by using cloud chambers in which controlled temperatures and relative humidities are specified. In the upper troposphere and lower stratosphere, aircraft fly through clouds collecting droplets and ice on collection plates or photographing their presence in the airstream. In the past, identification of the different sizes of droplets and of the various types of ice crystals was performed by a researcher in a tedious and subjective procedure. Today this analysis can be automated by computerized image assessment. On the ground, rainfall impaction molds and snow crystal impressions are made. Hailstones are also collected, since an analysis of their structure often helps define the ambient environment in which they formed. Chemical analyses of the cloud droplets, ice crystals, and precipitation are also frequently performed in order to identify natural and man-made pollutants within the different forms of water.

Measurement Systems

Methods to monitor the atmosphere are of two types—in situ measurements and remote sensing observations. In situ measurements require that the instrumentation be located directly at the point of interest and in contact with the subject of interest. In contrast, remote sensors are located some distance away from the subject of interest. Remote sensors include passive systems (instruments that receive information naturally emitted from a region in the atmosphere) and active systems (instruments that emit either acoustic or electromagnetic energy and record the characteristics of this energy after it reflects off an object or surface and returns back to the sensor).

A weather balloon is released at a weather station at the South Pole.

Within the planetary boundary layer, in situ instrumentation includes towers, tethered balloons, and surface data collection platforms. A wide range of meteorological measurements are made from this equipment, including temperature, dew-point temperature, pressure, wind velocity, long-wave and shortwave radiative fluxes, and air chemistry. Active remote-sensing observations are made, using Doppler and non-Doppler radars, lidars (a type of laser that measures backscattered light), and acoustic sounders. Radars measure the backscattering of electromagnetic microwave radiation with wavelengths on the order of 3 to 10 cm (1 to 4 inches). The non-Doppler radars provide estimates of precipitation intensity, while Doppler radars can also provide estimates of wind speed and direction by detecting a shift in the frequency of an echo produced by a moving target. Shorter-wavelength Doppler radars are often able to measure winds even in clear air. Carbon dioxide lidars provide estimates of wind structure and turbulence within a few tens of kilometres of the instrument. Acoustic sounders are used primarily to monitor boundary layer depth and structure, using echo-return characteristics. Passive instrumentation includes the pyranometer, which measures direct and diffuse solar radiation, and the pyrheliometer, which samples only direct radiation from the Sun.

Above the boundary layer, but within the troposphere, the primary standard observation platform is the radiosonde. Tethered to helium balloons, radiosondes are released twice daily (simultaneously at 0000 hours and 1200 hours Greenwich Mean Time) around the world. As a result of their use, a long-period data archive of the status of the atmosphere has been achieved. Meteorological observations from radiosondes are also applied to benchmark the numerical weather prediction models used to forecast day-to-day weather. Radiosondes measure temperature, dew-point temperature, and pressure. The position of the radiosonde can be monitored by radar tracking so that wind speed and direction as a function of height are routinely available—for this reason radiosondes are also referred to as rawinsondes. Since the 1990s, the global positioning system (GPS) has been used to track the balloons and calculate wind speed and direction. The radiosondes are designed to have a rise rate of about 200 metres (650 feet) per minute.

Remote-sensing systems called profilers have been developed to provide almost continuous measurements of wind and, somewhat less accurately, of moisture and temperature throughout the lowest 10 km (6 miles) of the atmosphere. Winds are estimated by using an upward-looking Doppler radar, while temperature and moisture profiles are evaluated by using a vertically pointing radiometer that measures electromagnetic emissions of selected wavelengths at various heights in the troposphere. Used in conjunction with Earth-orbiting satellite-based passive temperature and moisture radiometric soundings, as well as active lidar wind measurements, profilers complement the data collected from radiosondes.

Aircraft also provide detailed information concerning the structure of the atmosphere. Airplanes used in field experiments, such as the Lockheed P-3 aircraft employed by the National Oceanic and Atmospheric Administration (NOAA) in the United States, are heavily instrumented and often carry Doppler radar, turbulence sensors, and in situ measurement devices for cloud water, cloud ice content, and structure. The NOAA P-3 has been used to fly through hurricanes and other types of deep precipitating cloud systems. Commercial aircraft are used routinely to collect atmospheric data temperature and wind data. This information is communicated to weather forecasters and used in the preparation of weather map analyses.

Lightning occurrences are monitored by using ground-based detectors. Such systems measure time, location, flash polarity, and stroke count of lightning strikes. When the observations from systems at different locations are combined, distribution maps of lightning strikes, and hence thunderstorm occurrences, can be made.

Above the routine maximum height of the radiosonde data (above levels where atmospheric pressure drops below 100 millibars, at about 17 km [10.5 miles]), rocketsondes, rocket-borne grenades, and falling sphere experiments have been used to monitor the thermal structure of the upper atmosphere. Since these measurements occur much less frequently than radiosonde observations, however, less is known about the meteorology above the tropopause. Satellite radiometric soundings have also been used to provide temperature readings in layers in the atmosphere from near the surface up to about 25 km (16 miles) or so, although these measurements offer less vertical and spatial resolution than in situ measurements. Similarly, ground-based radar and lidars have been used to measure atmospheric characteristics in the upper atmosphere.

The Atmospheres of other Planets

Astronomical bodies retain an atmosphere when their escape velocity is significantly larger than the average molecular velocity of the gases present in the atmosphere. There are 8 planets and over 160 moons in the solar system. Of these, the planets Venus, Earth, Mars, Jupiter, Saturn, Uranus, and Neptune have significant atmospheres. Pluto (a dwarf planet) may have an appreciable atmosphere, but perhaps only when its highly elliptical orbit is closest to the Sun. Of the moons, only Titan, a moon of Saturn, is known to have a thick atmosphere. Much of what is known of these planets and their moons has resulted from the Pioneer, Viking, Mariner, Voyager, and Venera space probes.

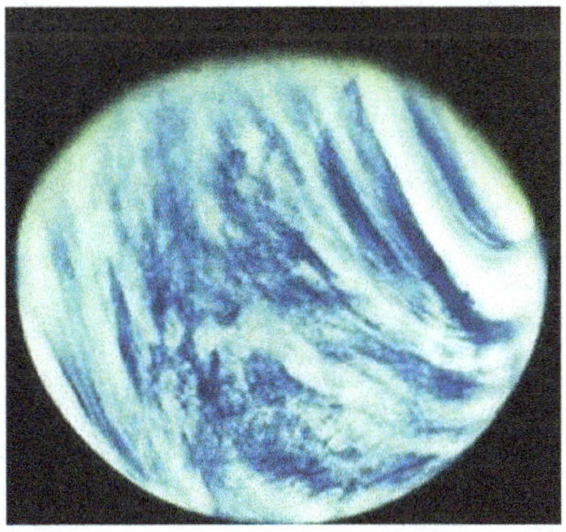

Bands of dense clouds swirl around Venus, shown in a photograph taken by the Mariner 10 spacecraft.

The atmosphere of Venus is about 96 percent carbon dioxide, with surface temperatures around 737 K (464 °C, or 867 °F). Clouds on Venus are made of sulfuric acid (H_2SO_4) and move in an easterly circulation of about 100 metres per second (224 miles per hour). Venus itself rotates only once every 243 Earth days. Surface pressures on Venus are around 95,000 millibars. (By contrast, Earth has a sea-level pressure of around 1,000 millibars).

Mars, in contrast, has a thin atmosphere composed of about 95 percent carbon dioxide, with the remainder being mostly diatomic nitrogen. Traces of water vapour also occur. Mars has a mean surface air temperature estimated at 210 K (−63 °C, or −82 °F), and surface pressures hover near 6 millibars. Both water and carbon dioxide clouds are observed on Mars, and it has well-defined seasons. In addition to periodic regional and global dust storms, cyclonic storms and clouds, associated with the boundary between cold air (from the polar cap) and warm air (from the mid-latitudes), have been observed on the planet. The rotation rate of Mars is close to the rotation rate of Earth. Evidence of river channels on Mars indicates that liquid water was present and atmospheric density was much higher in the planet's geologic past.

Along with Earth, Venus and Mars have atmospheres that were primarily formed as a result of volcanic gas emissions, although the evolution of these gases on each planet has been very different. On Mars, for example, temperatures are currently so low that most of the water vapour emitted by volcanoes has apparently been deposited as ice within the crustal soils. The closer proximity of Venus to the Sun, and the resultant higher temperatures, may have led to the loss of most of the water from that planet—most likely through the dissolution of water into hydrogen and oxygen. Hydrogen gas was lost to space; oxygen was combined with other elements through oxidation; and carbon dioxide (produced by volcanic emissions) accumulated to high concentrations. In contrast, much of the carbon dioxide in Earth's early atmosphere became part of the crustal materials, and the buildup of oxygen in Earth's atmosphere is a result of photosynthesis by plants. The development of Earth's habitable atmosphere, as contrasted with the torrid climate of Venus, appears to be directly related to Earth's distance from the Sun. Current analysis suggests that Earth's atmosphere would have evolved to the form found on Venus if the planet had been only 5 percent closer during the evolution of the atmosphere.

On the remainder of the planets, the atmospheres appear to have retained the primordial nature associated with their formation. The air on Jupiter and Saturn, for example, is made up of nearly 100 percent diatomic hydrogen (H_2) and helium (He), with small contributions of methane (CH_4) and other chemical compounds. Much less is known regarding the atmospheres of the somewhat smaller Jovian planets Uranus and Neptune, although both are thought to be similar to those of Jupiter and Saturn.

Jupiter's Great Red Spot and its surroundings. Included are the white ovals, observed and immense areas of turbulence to the left of the Great Red Spot.

On both Jupiter and Saturn, colourful cloud bands and other regional phenomena that are located at different altitudes and latitudes circulate at speeds up to several hundreds of metres per second relative to each other. The large velocity shears associated with this motion create turbulent eddies on these planets—most notably Jupiter's Great Red Spot. The bright zones on these planets correspond to the tops of upwelling clouds in the cold upper atmosphere, whereas the more colourful bands correspond to the relatively warm lower atmosphere and may be associated with the occurrence of sulfur and phosphorus compounds. Both aurora displays and intense lightning have been observed on Jupiter and Saturn.

Atmospheric Science

Atmospheric science is the interdisciplinary study which combines the different components of chemistry and physics that focuses on the dynamics and structure of the atmosphere of Earth. Atmospheric science includes the study of composition, circulation, and the chemical and physical processes of the atmosphere. Atmospheric science focuses on atmosphere, the atmospheric processes, the effects of numerous systems on the atmosphere and the effects the atmosphere has on these systems. Atmospheric science extends to planetary science and studying the atmospheres of the different planets in the solar system.

Meteorology

Meteorology is the study of the world's atmospheric which deals with weather forecasting and processes. Although this science dates back to over 1000 years ago, no significant progress occurred until the 18th century. Prior attempts in meteorology depended on historical data. The 19th century saw modest growth in meteorology after the development of weather observation networks on different corners of the world. Majority of the observed weather which helps predict an event on earth is on the troposphere. Meteorological Phenomena are the observable weather events explained by meteorology. The meteorological phenomena like acid rain, clouds, and hurricane among other, is quantified and described by numerous variables including mass flow, water vapor, temperature, atmosphere and air pressure plus the interactions and variations of these variables and the changes they go through in time. These phenomena are described and predicted using different spatial scales.

Climatology

Climatology is the study of weather conditions averaged over a specified period. Climate represents composite weather report over a specified period. Climatology is a branch of atmospheric science and a sub-field of physical geography. Fundamental climate knowledge helps in weather forecasting for a shorter period using numerous techniques like Northern Annular mode. Climatologists use different climate models for an array of purposes ranging from projecting of future climatic changes to studying the dynamics of climate and weather systems. Weather is the atmospheric conditions over a short period while climate deals with the weather conditions over an extended to an indefinite term. The climate shifts after a certain period and Shen Kuo, a Chinese scientist, noted this phenomenon after he observed petrified bamboos growing underground close to Yanzhou, a dry place which cannot support the growth of bamboos.

Paleoclimatology

Paleoclimatology is the study of ancient climatic changes. Since going back in time to observe the climatic changes is impossible, scientists use numerous climate imprints created in the past, referred to as proxies, to interpret the paleoclimate. Some of the most reliable proxies include microfossils, shells, rocks, corals, ice sheets and tree rings among others. Scientists reconstruct the ancient climate using a combination of different categories of proxy records. The proxy records are incorporated with the observations of the current climate and then uploaded into a computer model which infers the ancient climate while predicting future climatic changes. Studies of ancient environmental changes and biodiversity always reflect on the present situation especially the impact of the climatic changes on biotic recovery and mass extinctions. Paleoclimatology started in the early 19th century when numerous discoveries about the glaciations and natural changes in the ancient climate helped scientists comprehend the greenhouse effect. The first observations with reliable scientific basis were the one observed in New Zealand by John Hardcastle in 1880s.

Hardcastle discovered that the loess deposited at Timaru helped record climatic changes. Hardcastle referred to the loess as "climate registers".

Atmospheric Chemistry

Atmospheric chemistry is the field of atmospheric science which studies the chemistry of the atmosphere of earth and the other planets. Atmospheric chemistry is a multi-disciplinary approach to research which draws from volcanology, geology, environmental chemistry, meteorology, oceanography and computer modeling. The atmospheric chemistry and composition are crucial for numerous reasons, one of them being the interactions between the all the living organisms and the atmosphere. Multiple natural processes including lighting and volcano emission change the composition of the atmosphere. Atmospheric chemistry has addressed numerous problems including acid rain, global warming, photochemical smog, ozone depletion, and greenhouse gases. The atmospheric chemist tries to understand the causes of these issues and obtain a theoretical understanding of the problem which helps them create a solution which is tested and implemented.

Atmospheric Physics

Atmospheric physics is the use of physics when studying the atmosphere. Atmospheric physicists try to model the atmosphere of the earth among other planets using numerous fluid flow equations, radiation budgets, energy transfers, and chemical models. Atmospheric physics is closely related to climatology and meteorology, plus it covers the construction and design of the instruments used in studying the atmosphere and interpretation of the collected data. The atmospheric physicists use the elements of scattering theory, cloud physics, spatial statistics, and wave propagation models including the remote sensing instruments to model the weather systems. The introduction of the sounding rockets saw aeronomy become a sub-discipline dealing with the top layer of the atmosphere.

Paleotempestology

Paleotempestology refers to the study of ancient tropical cyclone activities using numerous geological proxies and documented historical records. Some of the most efficient paleotempestology methods include sedimentary proxy records, makers in coral, historical records, and tree rings and speleothems.The sedimentary proxy records method uses the over-wash deposits conserved on the sediments of marshes, microfossils and coastal lakes. The scientists adopted the use of over-wash deposits from the earlier studies of numerous paleotsunami deposits. The first study of a cyclone occurred in South Pacific and Australia from the late 1970s to early 1980s. The studies examined many parallel coral shingle ridges and marine shells and confirmed that cyclones deposit over 50 ridges on the site, and each represents an ancient severe cyclone which occurred thousands of years ago. Rocks have some natural isotopes of elements referred to as natural tracers, which help describe the state under which the rock formed. Studying the calcium carbonate present in the coral rocks helps reveal the hurricane information and surface temperature of when it developed. The heavy oxygen isotopes decrease faster as compared to lighter oxygen isotopes during the heavy rainfall periods. Since the hurricanes were the primary sources of heavy rainfall in tropical oceans, scientists can date the ancient storms by looking at the decreased lighter oxygen isotope in the coral rocks.

Chapter 2

Atmospheric Layers

The atmosphere is divided into several layers on the basis of temperature. Some of these are troposphere, stratosphere, mesosphere, thermosphere, ionosphere and exosphere. The topics elaborated in this chapter will help in gaining a better perspective about these layers of the atmosphere.

The atmosphere is comprised of layers based on temperature. These layers are the troposphere, stratosphere, mesosphere, thermosphere, ionosphere and exosphere.

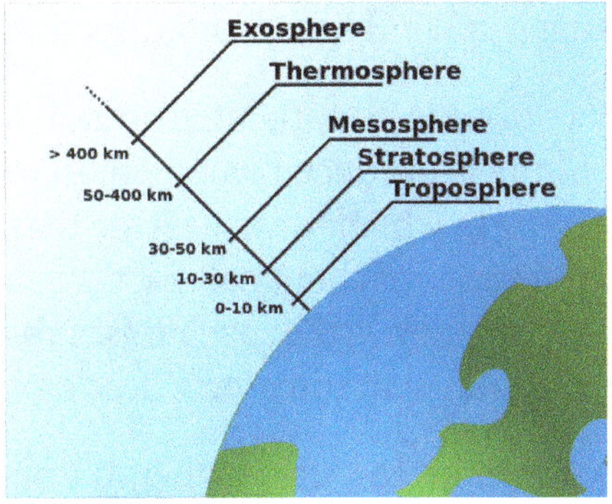

Layers of the atmosphere.

Troposphere

The troposphere is the lowest layer of Earth's atmosphere and site of all weather on Earth. The troposphere is bonded on the top by a layer of air called the tropopause, which separates the troposphere from the stratosphere and on bottom by the surface of the Earth. The troposphere is wider at the equator (10mi) than at the poles (5mi).

The troposphere contains 75 percent of atmosphere's mass- on an average day the weight of the molecules in the air is14.7 lb (sq. in.) and most of the atmosphere's water vapor. Water vapor concentration varies from trace amounts in Polar Regions to nearly 4 percent in the tropics. Most prevalent gases are nitrogen (78 percent) and oxygen (21 percent), with the remaining 1- percent consisting of argon, (.9 percent) and traces of hydrogen ozone (a form of oxygen), and other constituents. Temperature and water vapor content in the troposphere decrease rapidly with altitude. Water vapor plays a major role in regulating air temperature because it absorbs solar energy and thermal radiation from the planet's surface.

The troposphere contains 99% of the water vapor in the atmosphere. Water vapor concentrations vary with latitudinal position (north to south). They are greatest above the tropics, where they might be as high as 3% and decrease toward the polar region.

Carbon dioxide is present in small amounts, but its concentration has nearly doubled since 1900. Like water vapor, carbon dioxide is a greenhouse gas which traps some of the Earth's heat close to the surface and prevents its release into space. Scientists fear that the increasing amounts of carbon dioxide could raise the Earth's surface temperature during the next century, bringing significant changes to worldwide weather patterns. Such changes may include a shift in climatic zones and the melting of the polar ice caps, which could raise the level of the world's oceans.

The uneven heating of the regions of the troposphere by the sun (the sun warms the air at the equator more than the air at the poles) causes convection currents, large-scale patterns of winds that move heat and moisture around the globe. In the Northern and Southern hemispheres, air rises along the equator and subpolar (latitude about 50 to about 70 north and south) climatic regions and sinks in the polar and subtropical regions. Air is deflected by the Earth's rotation as it moves between the poles and equator, creating belts of surface winds moving from east to west (easterly winds) in tropical and polar regions, the winds moving from west to east (westerly winds) in the middle latitudes. This global circulation is disrupted by the circular wind patterns of migrating high and low air pressure areas, plus locally abrupt changes in wind speed and direction known as turbulence.

A common feature of the troposphere of densely populated areas is smog, which restricts visibility and is irritating to the eyes and throat. Smog is produced when pollutants accumulate close to the surface beneath an inversion layer (a layer of air in which the usual rule that temperature of air decreases with altitude doesn't apply), and undergo a series of chemical reactions in the presence pollutants from escaping into the upper atmosphere. Convection is the mechanism responsible for the vertical transport of heat in the troposphere while horizontal heat transfer is accomplished through advection.

The exchange and movement of water between the earth and atmosphere is called the water cycle. The cycle, which occurs in the troposphere, begins as the sun evaporates large amounts of water from the earth's surface and the moisture is transported to other regions by the wind. As air rises, expands, and cools, water vapor condenses and clouds develop. Clouds cover large portions of the earth at any given time and vary from fair weather cirrus to towering cumulus clouds. When liquid or solid water particles grow large enough in size, they fall toward the earth as precipitation. The type of precipitation that reaches the ground, be it rain, snow, sleet, or freezing rain, depends upon the temperature of the air through which it falls.

As sunlight enters the atmosphere, a portion is immediately reflected back to space, but the rest penetrates the atmosphere and is absorbed by the earth's surface. This energy is then remitted by the earth back into atmosphere as long-wave radiation. Carbon dioxide and water molecules absorb this energy and emit much of it back towards the earth again. This delicate exchange of energy between the earth's surface and atmosphere keeps the average global temperature from changing drastically from year to year.

Stratosphere

The stratosphere is a layer of Earth's atmosphere. It is the second layer of the atmosphere as you go upward. The troposphere, the lowest layer, is right below the stratosphere. The next higher layer above the stratosphere is the mesosphere.

The bottom of the stratosphere is around 10 km (6.2 miles or about 33,000 feet) above the ground at middle latitudes. The top of the stratosphere occurs at an altitude of 50 km (31 miles). The height of the bottom of the stratosphere varies with latitude and with the seasons. The lower boundary of the stratosphere can be as high as 20 km (12 miles or 65,000 feet) near the equator and as low as 7 km (4 miles or 23,000 feet) at the poles in winter. The lower boundary of the stratosphere is called the tropopause; the upper boundary is called the stratopause.

Ozone, an unusual type of oxygen molecule that is relatively abundant in the stratosphere, heats this layer as it absorbs energy from incoming ultraviolet radiation from the Sun. Temperatures rise as one moves upward through the stratosphere. This is exactly the opposite of the behavior in the troposphere in which we live, where temperatures drop with increasing altitude. Because of this temperature stratification, there is little convection and mixing in the stratosphere, so the layers of air there are quite stable. Commercial jet aircraft fly in the lower stratosphere to avoid the turbulence which is common in the troposphere.

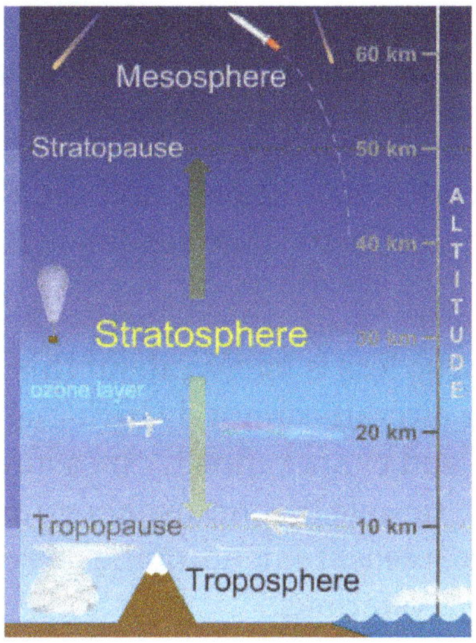

This diagram shows some of the features of the stratosphere.

The stratosphere is very dry; air there contains little water vapor. Because of this, few clouds are found in this layer; almost all clouds occur in the lower, more humid troposphere. Polar stratospheric clouds (PSCs) are the exception. PSCs appear in the lower stratosphere near the poles in winter. They are found at altitudes of 15 to 25 km (9.3 to 15.5 miles) and form only when temperatures at those heights dip below -78 °C. They appear to help cause the formation of the infamous holes in the ozone layer by "encouraging" certain chemical reactions that destroy ozone. PSCs are also called nacreous clouds.

Air is roughly a thousand times thinner at the top of the stratosphere than it is at sea level. Because of this, jet aircraft and weather balloons reach their maximum operational altitudes within the stratosphere. Due to the lack of vertical convection in the stratosphere, materials that get into the stratosphere can stay there for long times. Such is the case for the ozone-destroying chemicals called CFCs (chlorofluorocarbons). Large volcanic eruptions and major meteorite impacts can fling aerosol particles up into the stratosphere where they may linger for months or years, sometimes altering Earth's global climate. Rocket launches inject exhaust gases into the stratosphere, producing uncertain consequences.

Various types of waves and tides in the atmosphere influence the stratosphere. Some of these waves and tides carry energy from the troposphere upward into the stratosphere; others convey energy from the stratosphere up into the mesosphere. The waves and tides influence the flows of air in the stratosphere and can also cause regional heating of this layer of the atmosphere. A rare type of electrical discharge, somewhat akin to lightning, occurs in the stratosphere. These "blue jets" appear above thunderstorms, and extend from the bottom of the stratosphere up to altitudes of 40 or 50 km (25 to 31 miles).

Mesosphere

The mesosphere is the region of the atmosphere located between the stratosphere and the thermosphere, between 50 and 90 km, in which temperature decreases with height. The transition between the mesosphere and the thermosphere is called the mesopause and is the altitude at which the temperature reaches a minimum before increasing with height in the thermosphere.

The coldest temperatures in Earth's atmosphere, about -90 °C (-130 °F), are found near the top of this layer. The boundary between the mesosphere and the thermosphere above it is called the mesopause. At the bottom of the mesosphere is the stratopause, the boundary between the mesosphere and the stratosphere.

The mesosphere is difficult to study, so less is known about this layer of the atmosphere than other layers. Weather balloons and other aircraft cannot fly high enough to reach the mesosphere. Satellites orbit above the mesosphere and cannot directly measure traits of this layer. Scientists use instruments on sounding rockets to sample the mesosphere directly, but such flights are brief and infrequent. Since it is difficult to take measurements of the mesosphere directly using instruments, much about the mesosphere is still mysterious.

Most meteors vaporize in the mesosphere. Some material from meteors lingers in the mesosphere, causing this layer to have a relatively high concentration of iron and other metal atoms.

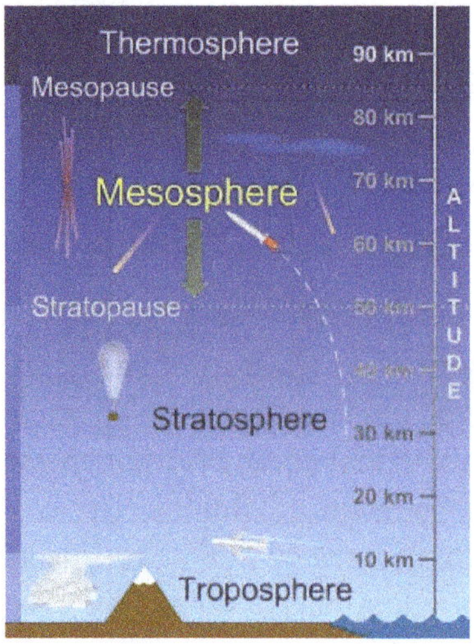

This diagram shows some of the features of the mesosphere.

Very strange, high altitude clouds called "noctilucent clouds" or "polar mesospheric clouds" sometime form in the mesosphere near the poles. These peculiar clouds form much, much higher up than other types of clouds. The mesosphere, like the stratosphere below it, is much drier than the moist troposphere we live in; making the formation of clouds in this layer a bit of a surprise. Odd electrical discharges akin to lightning, called "sprites" and "ELVES", occasionally appear in the mesosphere dozens of kilometers (miles) above thunderclouds in the troposphere.

The stratosphere and mesosphere together are sometimes referred to as the middle atmosphere. At the mesopause (the top of the mesosphere) and below, gases made of different types of atoms and molecules are thoroughly mixed together by turbulence in the atmosphere. Above the mesosphere, in the thermosphere and beyond, gas particles collide so infrequently that the gases become somewhat separated based on the types of chemical elements they contain.

Various types of waves and tides in the atmosphere influence the mesosphere. These waves and tides carry energy from the troposphere and the stratosphere upward into the mesosphere, driving most of its global circulation.

Thermosphere

The thermosphere is the atmospheric region from ~85 to ~500 km altitude, containing the ionosphere. It is characterized by high temperature and large variability, in response to changes in solar ultraviolet radiation and solar-driven geomagnetic activity.

Temperatures climb sharply in the lower thermosphere (below 200 to 300 km altitude), then level off and hold fairly steady with increasing altitude above that height. Solar activity strongly influences temperature in the thermosphere. The thermosphere is typically about 200 °C (360 °F) hotter in the daytime than at night, and roughly 500 °C (900 °F) hotter when the Sun is very active than at other times. Temperatures in the upper thermosphere can range from about 500 °C (932 °F) to 2,000 °C (3,632 °F) or higher.

The aurora (Northern Lights and Southern Lights) mostly occur in the thermosphere.

The boundary between the thermosphere and the exosphere above it is called the thermopause. At the bottom of the thermosphere is the mesopause, the boundary between the thermosphere and the mesosphere below. Although the thermosphere is considered part of Earth's atmosphere, the air density is so low in this layer that most of the thermosphere is what we normally think of as outer space. In fact, the most common definition says that space begins at an altitude of 100 km (62 miles), slightly above the mesopause at the bottom of the thermosphere. The space shuttle and the International Space Station both orbit Earth within the thermosphere.

Below the thermosphere, gases made of different types of atoms and molecules are thoroughly mixed together by turbulence in the atmosphere. Air in the lower atmosphere is mainly composed of the familiar blend of about 80% nitrogen molecules (N_2) and about 20% oxygen molecules (O_2). In the thermosphere and above, gas particles collide so infrequently that the gases become somewhat separated based on the types of chemical elements they contain. Energetic ultraviolet and X-ray photons from the Sun also break apart molecules in the thermosphere. In the upper thermosphere, atomic oxygen (O), atomic nitrogen (N), and helium (He) are the main components of air.

Much of the X-ray and UV radiation from the Sun is absorbed in the thermosphere. When the Sun is very active and emitting more high energy radiation, the thermosphere gets hotter and expands or "puffs up". Because of this, the height of the top of the thermosphere (the thermopause) varies. The thermopause is found at an altitude between 500 km and 1,000 km or higher. Since many satellites orbit within the thermosphere, changes in the density of (the very, very thin) air at orbital altitudes brought on by heating and expansion of the thermosphere generates a drag force on satellites. Engineers must take this varying drag into account when calculating orbits and satellites occasionally need to be boosted higher to offset the effects of the drag force.

High-energy solar photons also tear electrons away from gas particles in the thermosphere, creating electrically-charged ions of atoms and molecules. Earth's ionosphere, composed of several regions of such ionized particles in the atmosphere, overlaps with and shares the same space with the electrically neutral thermosphere.

Like the oceans, Earth's atmosphere has waves and tides within it. These waves and tides help move energy around within the atmosphere, including the thermosphere. Winds and the overall circulation in the thermosphere are largely driven by these tides and waves. Moving ions, dragged along by collisions with the electrically neutral gases, produce powerful electrical currents in some parts of the thermosphere.

Finally, the aurora (the Southern and Northern Lights) primarily occurs in the thermosphere. Charged particles (electrons, protons, and other ions) from space collide with atoms and molecules in the thermosphere at high latitudes, exciting them into higher energy states. Those atoms and molecules shed this excess energy by emitting photons of light, which we see as colorful auroral displays.

Ionosphere

The ionosphere is an area in the earth's atmosphere which is between the exosphere and the stratosphere. It consists of three layers, which all have the same characteristic of high concentrations of ionized particles. These particles are ionized by radiation from the sun. The ionosphere extends from approximately 50 to 250 miles (80 to 400 km) above the earth's surface.

Ions are atoms with positive or negative electrical charge. The ionosphere is important in radio communication, especially over great distances. This is because it can reflect radio waves from the earth's surface back down. Depending on the frequency of microwaves it will either allow them to pass through or reflect them away. The threshold for this is approximately 100MHz so anything above this passes through and anything below is reflected.

The Structure of Earth's Ionosphere

Due to Sun's UV radiation, Earth's upper atmosphere is partly (0.1% or less) ionized plasma at altitudes of 70-1500 km. This region, ionosphere, is coupled to both the magnetosphere and the neutral atmosphere. It is of great practical importance because of its effect on radio waves.

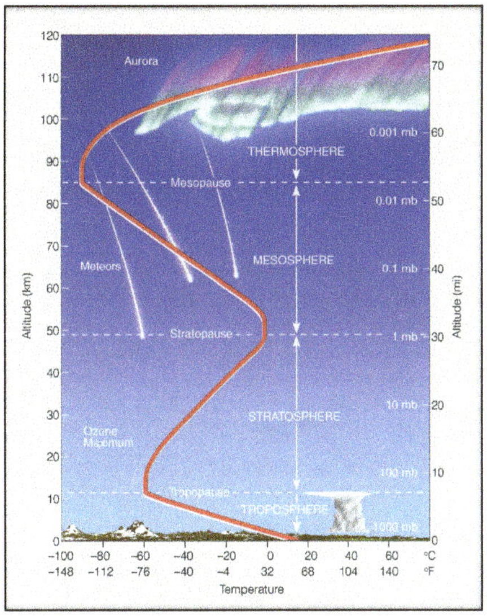

Structure of atmosphere.

The existence of a conducting layer in the upper atmosphere results also in many other interesting phenomena. Around the plasma density maximum, (F-layer) a so-called ionospheric waveguide is formed for magneto sonic waves. In addition, a so-called ionospheric Alfvén resonator (IAR) can be formed between the density maximum and an upper altitude at about 3000 km, where the Alfvén velocity has a maximum. A third natural resonator is formed between the nearly perfectly conducting terrestrial surface and the ionosphere, creating the so-called Schumann resonances.

Ionosphere Regions

The structure and dynamics of the Earth's ionosphere are subject to both large spatial and frequent temporal variations, which can be periodic as well as irregular. The changes that occur in the ionosphere are different at different altitudes because of varying relative ionization loss and transport phenomena. Consequently, the terrestrial ionosphere is divided into regular and sporadic regions according to the thermal and chemical properties of the neutral gas and ionized components.

Modern experimental and theoretical investigations divide the ionosphere into three regions of D, E, and F. Under certain solar-terrestrial conditions these regions are split into four main layers D, E, F1, and F2 as shown in figure. In principle, the lower ionosphere (up to 100 km) is a zone in which photochemical processes are the main influence on its formation and ionization balance.

The boundary of the middle ionosphere (100–170 km) marks the limit of ionization-recombination processes together with thermal and dynamic processes. The upper ionosphere is the F region, characterized by the transfer of charged particles in plasma by ambipolar diffusion, thermospheric winds, and ionosphere-magnetosphere interactions. The real heights of the ionospheric layers vary with solar zenith angle time, time of day, seasons, solar cycles, and solar activity.

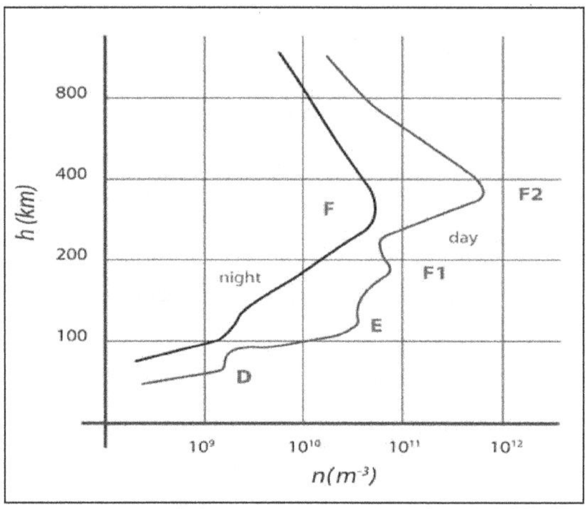

Diurnal and nocturnal electron density profiles versus altitude.

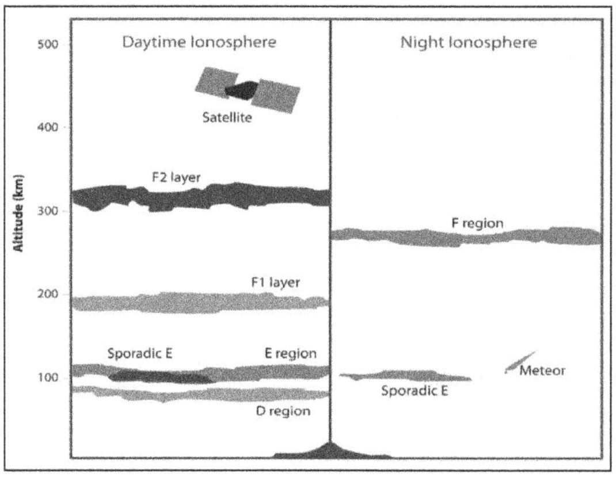

Day and Night structure of the terrestrial Ionosphere.

The first ionospheric layer found was the E layer or region at about 110 km altitude. Radio operators use it as a surface from which signals can be reflected to distant stations. It is interesting to note that this works also the other way round and, for example, the auroral kilometric radiation created by the precipitating particles high above the ionosphere does not reach the ground because of the ionospheric E layer. Above the E layer, a F layer consisting of two parts of F1 layer is at about 170 km, and F2 layer at about 250 km altitude and F layer reflects radio waves. The lowermost region of the ionosphere below 80 km altitude, D layer, however, principally absorbs radio waves. Within the auroral night time E layer plasma densities can be much higher than indicated by the figure. Densities are also very variable because of the spatial and temporal structure in the ionizing particle precipitation. The E layer plasma density profiles can also be drastically altered due to the occasional formation of so-called sporadic E layers.

In F layer altitudes one encounters such features as polar cap ionization patches and different types of troughs. Note also that there is also a clear solar cycle effect seen the average densities are higher during solar maximum years than during the minimum years. Ionospheric electric fields are the main result of the coupling between the magnetosphere and ionosphere. While at low latitudes the ionospheric plasma is co-rotating with the Earth, at higher latitudes it is convicting under the influence of the large-scale magnetospheric electric field mapped to low altitudes. The Harang discontinuity is one of the ionospheric features related to the plasma convection pattern.

Ionosphere Layers formation and their Characteristics

With the possible exception of a contribution from cosmic radiation in the formation of the lowest layer, the ionized regions in the upper atmosphere owe their existence, directly or indirectly, to the ionizing power of the solar radiation. As this radiation penetrates the earth's atmosphere, it produces ions and electrons. The nature of the ions that are produced depends upon the energy of the radiation and composition of the atmosphere at the altitude at which ionization occurs. Immediately following the production of an ion species, a number of processes occur that changes the species of ions and decrease the number of electrons and ions that exist. The total population of ions and electrons depends on the competition between ionization process, ion and electron diffusion, geomagnetic effects and mass motions. These processes superimposed upon the chemical and physical processes determine the structure of the ionosphere.

D region is the lowest lying region and hence is produced by the most penetrating of the ionizing radiation that impinges upon the earth. A secondary electron producing source is the visible and ultraviolet solar emissions, which can provide the energy necessary to release weakly held electrons from negative ions.

In the E region, the ionization of molecular species is dominant and electron-ion losses are solely due to dissociate recombination processes. The electron density in this region is determined by the photo-equilibrium equation. The unique characteristic of the F1 layer is that the principal ion produced is atomic, whereas the principal electron loss process is dissociate recombination with a molecular ion, and that the ion interchanges is also a process in this region.

Under quiet daytime conditions, below an altitude of about 200 km. all charged constituents are in or nearly in a condition of photochemical equilibrium. In fact, photochemical processes determine the structure of the ionosphere at these heights at all times of the day and night. At great heights (above 250 kin), this is not the case. At these heights the chemical loss rate of the ions is comparable to their diffusion rate though the neutral gas and transport mechanism begin to influence the ion and electron number densities. With ascending altitude, diffusion becomes more and more important, finally dominating the picture at greater heights. The competition of chemical loss (recombination) and diffusion leads to the formation of a peak in the number density profile of the ions and electrons, which is the well-known F2 peak. The geomagnetic field plays an important role in determining the ionization distribution since the transport of ionization with respect to the field is an important process in the F2 region.

The ionization density in each of these layers has a peak value at a certain altitude above and below which it decreases. The D region present only during the daytime normally has a peak density around 90 km, which may drop to around 60 km due to enhanced solar X ray flux. The E region containing both the normal E layers and at times patches of sporadic E (E), has a peak density around 120 km. Following sunset, the electron density in this region decreases by a factor of 10 or more within a very short time before attaining a nighttime equilibrium density. The F region is a combination of two different regions, namely F1 and F2. The F1 region, which disappears after sunset, has a peak density around 200 km and is more pronounced in summer than in winter and at high sunspot numbers. In the F2 region, the altitude of the peak density occurs at about 300 km in the daytime and at higher altitudes in the night.

Shortly after sunset, the absolute density near the peak of the F region often increases due to plasma transport processes before decreasing to a nighttime value and drifting downwards. The peak densities of the ionospheric layers vary with time of the day, season, latitude, and solar activity and with a variety of irregular and random disturbances. The 'topside ionosphere' is the name given to the rest of the ionosphere above the F layer peak. In a simple model, the density of the topside ionosphere decreases exponentially with height until the ionization density is below detectable levels.

Near the geomagnetic equator, the earth's magnetic field, B (in NorthSouth direction) is horizontal and there is east-west electric field due to the dynamo effect by atmospheric motions at E region heights. The electric field is eastward during the day and westward during the night. As a consequence, ionospheric plasma from equatorial F region moves upward and then diffuses downward along sloping magnetic field lines to low latitudes on both sides of equator. The electron concentration is thus depleted on the magnetic equator and enhanced in two regions, one on each side. The phenomenon is known as equatorial ionization anomaly. The distribution of ionospheric plasma is also affected by solar and magnetic disturbances like the occurrences of solar flares and magnetic storms. Varying solar activity essentially controls ionospheric parameters (like peak electron densities in different layers as well as their altitudes). There is a short (27 days) and longterm (11 years) periodicity in the solar activity. In general, it has been noticed that there is a close relationship between long-term variation in sunspot numbers and critical frequency of the F2 region. A highly significant aspect of the sunspots is their association with strong magnetic fields and they mostly occur in pairs on a line roughly parallel to the solar equator. It is believed that strong local magnetic fields are the cause of sunspots. Sunspot number is a quantitative measure of sunspot activity and the total annual sunspot number exhibits definite variations from year to year. The average time from minimum to maximum is about 5 years and from maximum to minimum 6 years with an 11-year solar cycle. Sunspot numbers are being generally used for predictions of ionospheric parameters for radio propagation and other applications. Thus, the richness in the variety of phenomena and characteristics applies to the study of radio propagation in the ionosphere as well as to the investigation of the physics of the ionosphere itself.

Mechanisms of Ionization

Photoionization

Most of the electrical activity in the ionosphere is produced by photoionization (ionization caused by light energy). Photons of short wavelength (that is, of high frequency) are absorbed by atmospheric gases. A portion of the energy is used to eject an electron, converting a neutral atom or molecule to a pair of charged species—an electron, which is negatively charged, and a companion positive ion. Ionization in the F_1 region is produced mainly by ejection of electrons from molecular oxygen (O_2), atomic oxygen (O), and molecular nitrogen (N_2). The threshold for ionization of O_2 corresponds to a wavelength of 102.7 nm (nanometres, or billionths of a metre). Thresholds for O and N_2 are at 91.1 nm and 79.6 nm, respectively.

Positive ions in turn can react with neutral gases. There is a tendency for these reactions to favour production of more-stable ions. Thus, ionized atomic oxygen, O^+, can react with O_2 and N_2, resulting in ionized molecular oxygen (O_2^+) and ionized nitric oxide (NO^+), as shown by:

$$O^+ + O_2 \rightarrow O + O_2^+$$

And,

$$O^+ + N_2 \rightarrow NO^+ + N$$

Similarly, ionized molecular nitrogen (N_2^+) can react with O and O_2 to form NO^+ and O_2^+ as follows:

$$N_2^+ + O \rightarrow NO^+ + N$$

And,

$$N_2^+ + O_2 \rightarrow N_2 + O_2^+$$

The most stable, and consequently most abundant, ions in the E and F_1 regions are O_2^+ and NO^+, the latter more so than the former. At lower altitudes, O_2^+ can react with the minor species of atomic nitrogen (N) and nitric oxide (NO) to form NO^+, as indicated by:

$$O_2^+ + N \rightarrow O + NO^+$$

And

$$O_2^+ + NO \rightarrow O_2 + NO^+$$

In the D region, NO^+ and water vapour (H_2O) can interact to form the hydronium ion, H_3O^+, and companion species such as $H_5O_2^+$ and $H_7O_4^+$. Production of hydrated ions is limited by the availability of H_2O. As a consequence, they are confined to altitudes below about 85 km (53 miles).

Recombination

The electron density in the D, E, and F_1 regions reflects for the most part a local balance between production and loss. Electrons are removed mainly by dissociative recombination, a process in which electrons attach to positively charged molecular ions and form highly energetic, unstable neutral molecules. These molecules decompose spontaneously, converting internal energy to kinetic energy possessed by the fragments. The most important processes in the ionosphere involve recombination of O_2^+ and NO^+. These reactions may be summarized by:

$$O_2^+ + e \rightarrow O + O$$

And

$$NO^+ + e \rightarrow N + O$$

A portion of the energy released in reactions above may appear as internal excitation of either nitrogen, oxygen, or both. The excited atoms can radiate, emitting faint visible light in the green and red regions of the spectrum, contributing to the phenomenon of airglow. Airglow originates mainly from altitudes above 80 km (50 miles) and is responsible for the diffuse background light that makes it possible to distinguish objects at Earth's surface on dark, moonless nights. Airglow is produced for the most part by reactions involved in the recombination of molecular oxygen. The contribution from reactions $O_2^+ + e \rightarrow O + O$ and NO+ + e → N + O is readily detectable, however, and provides a useful technique with which to observe changes in the ionosphere from the ground. Over the years, studies of airglow have contributed significantly to scientific understanding of processes in the upper atmosphere.

As indicated above, dissociative recombination provides an effective path for removal of molecular ions. There is no comparable means for removal of atomic ions. Direct recombination of ionized atomic oxygen (O^+) with an electron requires that the excess energy be radiated as light. Radiative recombination is inefficient, however, compared with dissociative recombination and plays only a small role in the removal of ionospheric electrons. The situation becomes more complicated at high altitudes where atomic oxygen (O) is the major constituent of the neutral atmosphere and where electrons are produced primarily by its photoionization. The atomic oxygen ion, O^+, may react with N_2 and O_2 to form NO^+ and O_2^+, but the abundances of N_2 and O_2 decline relative to O as a function of increasing altitude. In the absence of competing reactions, the concentration of O^+ and the density of electrons would increase steadily with altitude, paralleling the rise in the relative abundance of O. This occurs to some extent but is limited eventually by vertical transport.

Diffusion

Ions and electrons produced at high altitude are free to diffuse downward, guided by Earth's magnetic field. The lifetime of O^+ is long at high altitudes, where the densities of O_2 and N_2 are very small. As ions move downward, the densities of O_2 and N_2 increase. Eventually the time constant for reaction of O^+ with O_2 and N_2 becomes comparable to the time for diffusion, and O^+ reacts to produce either O_2^+ or NO^+ before it can move much farther. The O^+ density exhibits a maximum in this region. Competition between chemistry and transport is responsible for the formation of an electron-density maximum in the F_2 layer. The dominant positive ion is O^+.

The density of O^+ decreases with decreasing altitude below the peak, reflecting a balance between production of O by photoionization and its removal by reactions. The density of O^+ also decreases above the peak. In this case, removal of photo-ions is regulated by downward diffusion rather than by chemistry. The distribution of O^+ with altitude above the peak reflects a balance of forces—a pressure-gradient force that acts to support O^+ in opposition to gravitational and electrostatic forces that combine to pull O^+ down. The electrostatic force acts to preserve electrical charge neutrality. In its absence, the concentration of ions—which are much more massive than electrons—would tend to fall off more rapidly with altitude than electrons? The abundance of electrons would quickly exceed that of ions, and the upper atmosphere would accumulate negative charge. The electric field redresses the imbalance by drawing electrons down and providing additional upward support for positively charged ions. Though O^+ has a mass of 16 atomic units, its abundance decreases with altitude as if it had a mass of only 8 atomic units. (One atomic unit corresponds to the mass of a hydrogen atom, 1.66 10^{-24} gram.) This discrepancy occurs because the electric field exerts a force that is equivalent to that exerted by the gravitational force on a body with a mass of eight atomic units. This electrostatic force is directed upward for ions and downward for electrons, in effect buoying the ions while encouraging the electrons to sink. The concentration of electrons therefore falls off with altitude at precisely the same rate as that of O^+, preserving the balance of positive and negative charge.

Photon Absorption

Ionization at any given level depends on three factors—the availability of photons of a wavelength capable of effecting ionization, a supply of atoms and molecules necessary to intercept this radiation, and the efficiency with which the atoms and molecules are able to do so. The efficiency is relatively large for O, O_2, and N_2 from about 10 to 80 nm. This is the portion of the spectrum responsible for production of electrons and ions in the F_1 region. Photons with wavelengths between 90 and 100 nm are absorbed only by O_2. They therefore penetrate deeper and are responsible for producing about half the ionization in the E layer. The balance is derived from so-called "soft" X-rays (those of longer wavelengths), which are absorbed with relatively low efficiency in the F region and so are able to penetrate to altitudes of about 120 km (75 miles) when the Sun is high over the region. "Hard" X-rays (those of shorter wavelengths—that is, below about 5 nm) reach even deeper. This portion of the spectrum accounts for the bulk of the ionization in the D region, with an additional contribution from wavelengths longer than 102.6 nm—mainly from photons in the strong solar emission line at Lyman α at a wavelength of 121.7 nm. (The Lyman series is a related sequence of wavelengths that describe electromagnetic energy given off by energized atoms in the ultraviolet region.) Lyman α emissions are weakly absorbed by the major components of the atmosphere—O, O_2, and N_2—but they are absorbed readily by NO and have sufficient energy to ionize this relatively unstable compound. Despite the low abundance of NO, the high flux of solar radiation at Lyman α is able to provide a significant source of ionization for the D region near 90 km (55 miles).

Ionospheric variations

The ionosphere is variable in space and time. Some of the changes are chemical in origin and can be readily understood. There is a systematic variation, for example, according to the time of day. In early morning the Sun is relatively low in the sky, so that radiation must penetrate a large column of air before reaching a given level of the atmosphere. As a result, ionization rates are lower, and the location of ionized layers shifts to higher altitudes. As the Sun rises, the D, E, and F_1 layers shift in altitude. The layers are lowest and densities of electrons are highest at noon. At night, on the other hand, ionization in the D, E, and F_1 regions tends to disappear as electrons and ions recombine to form neutral gases.

The diurnal, or daily, variation of the F_2 layer is less dramatic. Ions produced at high altitudes during the day maintain a sizable density of electrons at the F_2 peak throughout the day and then diffuse downward at night. This accounts for the fact that radio reception (both in the broadcast and shortwave bands) is generally best at night. Ionization at lower altitudes—primarily those corresponding to the D region—tends to interfere with radio transmissions during the day. Interference is minimal at night because ionization in the D layer effectively disappears with the setting of the Sun.

The density of ionization varies in response to changes in the intensity and properties of radiation from the Sun. The output of solar energy is relatively constant in the visible and near-ultraviolet portions of the spectrum. It varies appreciably, however, at shorter wavelengths, reflecting changes in the temperature of the outermost regions of the solar atmosphere. The changes are particularly large, in excess of a factor of 10, at X-ray wavelengths. Variations in the D region are correspondingly large, with smaller though still significant changes in the E and F layers.

Solar activity varies on a characteristic timescale of about 11 years. It is not entirely periodic, however; successive cycles can differ significantly, and there are indications that activity can be low for centuries. The Sun was quiet for more than 200 years from about 1600 to about 1850. Solar activity was particularly intense in 1958.

The ionization above F_2 peak is removed mainly by downward diffusion of ions and electrons. Ions are constrained, however, to move along the magnetic field. The field is oriented horizontally at the magnetic equator, which is equidistant between the magnetic North Pole and the magnetic South Pole, so vertical diffusion is inhibited at low latitudes. The density of ionized atomic oxygen (O^+) and electrons at low latitudes is therefore controlled by chemistry to a larger extent than at high latitudes. The F_2 peak is correspondingly higher in altitude, and the density of electrons is elevated accordingly.

Ions and electrons formed at high altitudes and low latitudes are transported to higher latitudes by thermospheric winds. As a result, the highest density of electrons at the F_2 peak is observed at intermediate latitudes, offset from the magnetic equator by about 10 degrees.

Transport can also affect the distribution of ionization at lower altitudes. The diurnal pattern of heating in the troposphere and stratosphere excites a spectrum of waves, some of which are free to propagate vertically. The amplitude of the waves grows significantly as the disturbance enters regions of lower density. Passage of the waves is associated with strong alternating horizontal winds. Ionization can be driven up inclined magnetic field lines at one altitude, while winds blowing in an opposite direction at higher altitudes can induce simultaneous downward motion. This can lead to a bunching of ionization—a local enhancement of the electron density. The mechanism is particularly important in the E region and is responsible for the phenomenon known as sporadic E.

The buildup of ionization is normally limited by dissociative recombination of molecular ions. At D and E region altitudes, however, the ionosphere contains a small but variable concentration of atomic ions, derived from ionization of metals ablated from meteorites. The density of metallic ions—notably those of sodium (Na^+), magnesium (Mg^{++}), and potassium (K^+)—is sometimes high enough to supply a layer of ionization with a density comparable to that of the F layer. This can result in a major temporary disruption of radio communications.

Winds generated in the lower ionosphere by thermal forcing from below have characteristic periods expressed as submultiples of a day. Waves with a period of 24 hours dominate at low latitudes, whereas those with a characteristic period of 12 hours are more important at high latitudes. The origin of the waves is basically similar to that of oceanic tides caused by the pull of lunar gravity. The vertical motion that generates ionospheric waves, however, is the result of the diurnal pattern of heating and cooling rather than gravity. Additional waves can arise owing to irregular forcing, associated, for example, with thunderstorms, motion over mountain ranges, and other small-scale meteorological disturbances. These small-scale disturbances are referred to as gravity waves to distinguish them from the more regular planetary-scale motions excited by the diurnal cycle of heating and cooling. The regular response to thermal forcing is known as the atmospheric tide.

Tides and gravity waves have similar effects on ionization in the E region. They both are responsible for concentrating ionization in layers. In combination with the large-scale system of winds in the lower thermosphere, they are also effective in driving an irregular current that flows in the E and lower F regions of the ionosphere. The current owes its origin to differences in the facility with which motions of ions and electrons are constrained by the magnetic field. It is associated with an electric field and results in a modulation of the magnetic field that can be readily detected at the surface. The current is particularly intense in the equatorial region, where it is known as the electro jet. The region of strong current flow is known as the dynamo region.

Protons (H^+) and helium ions (He^+) are important components of the ionosphere above the F_2 peak. They increase in abundance relative to ionized atomic oxygen (O^+) with increasing altitude. Protons are produced by photoionization of atomic hydrogen (H):

$$hv + H \rightarrow H^+ + e$$

and by charge transfer from O^+ to H,

$$O^+ + H \rightarrow O + H^+$$

Helium ions are formed by photoionization of helium. The distribution of H^+ and He^+ with altitude reflects the influence of the polarization electric field set up to preserve charge neutrality. When O^+ is the dominant ion, the polarization field acts to lift H^+ and He^+ with a force equivalent, but in opposite direction, to that exerted by the gravitational field on a particle with a mass of eight atomic units. Protons behave as though they have an effective gravitational mass of −7 atomic units (−7 = 1 – 8). The effective mass of He^+ is −4 atomic units (−4 = 4 – 8).

The abundance of H^+ and He^+ increases with altitude. Eventually H^+ becomes the dominant component of the outermost ionosphere, which is sometimes referred to as the protonosphere. The more uniform composition of the atmosphere at this level causes a reduction in the polarization field to one equivalent to the gravitational force acting on a body with a mass of 0.5 atomic unit, directed upward for ions and downward for electrons. This field is sufficient to maintain equal densities of H^+ and electrons. The effective masses of O^+ and He^+ shift to 15.5 atomic units (15.5 = 16 – 0.5) and 3.5 atomic units (3.5 = 4 – 0.5), respectively, and the abundance of O^+, He^+, and H^+ declines with further increases in altitude.

Exosphere

The exosphere is the outermost layer of Earth's atmosphere (i.e., the upper limit of the atmosphere) and extends from the exobase, which is located at the top of the thermosphere.

Air in the exosphere is extremely thin - in many ways it is almost the same as the airless void of outer space. The layer directly below the exosphere is the thermosphere; the boundary between the two is called the thermopause. The bottom of the exosphere is sometimes also referred to as the exobase. The altitude of the lower boundary of the exosphere varies. When the Sun is active around the peak of the sunspot cycle, X-rays and ultraviolet radiation from the Sun heat and "puff up" the thermosphere - raising the altitude of the thermopause to heights around 1,000 km (620 miles) above Earth's surface. When the Sun is less active during the low point of the sunspot cycle, solar radiation is less intense and the thermopause recedes to within about 500 km (310 miles) of Earth's surface.

Not all scientists agree that the exosphere is really a part of the atmosphere. Some scientists consider the thermosphere the uppermost part of Earth's atmosphere, and think that the exosphere is really just part of space. However, other scientists do consider the exosphere part of our planet's atmosphere.

Although the International Space Station orbits Earth, it actually flies below the altitude of the exosphere.

Since the exosphere gradually fades into outer space, there is no clear upper boundary of this layer. One definition of the outermost limit of the exosphere places the uppermost edge of Earth's atmosphere around 190,000 km (120,000 miles), about halfway to the Moon. At this distance, radiation pressure from sunlight exerts more force on hydrogen atoms than does the pull of Earth's gravity. A faint glow of ultraviolet radiation scattered by hydrogen atoms in the uppermost atmosphere has been detected at heights of 100,000 km (62,000 miles) by satellites. This region of UV glow is called the geocorona.

Below the exosphere, molecules and atoms of atmospheric gases constantly collide with each other. However, air in the exosphere is so thin that such collisions are very rare. Gas atoms and molecules in the exosphere move along "ballistic trajectories", reminiscent of the arcing flight of a thrown ball (or shot cannonball!) as it gradually curves back towards Earth under the pull of gravity. Most gas particles in the exosphere zoom along curved paths without ever hitting another atom or molecule, eventually arcing back down into the lower atmosphere due to the pull of gravity. However, some of the faster-moving particles don't return to Earth - they fly off into space instead! A small portion of our atmosphere "leaks" away into space each year in this way.

Although the exosphere is technically part of Earth's atmosphere, in many ways it is part of outer space. Many satellites, including the International Space Station (ISS), orbit within the exosphere or below. For example, the average altitude of the ISS is about 330 km (205 miles), placing it in the thermosphere below the exosphere! Although the atmosphere is very, very thin in the thermosphere and exosphere, there is still enough air to cause a slight amount of drag force on satellites that orbit within these layers. This drag force gradually slows the spacecraft in their orbits, so that they eventually would fall out of orbit and burn up as they re-entered the atmosphere unless something is done to boost them back upwards. The ISS loses about 2 km (1.2 miles) in altitude each month to such "orbital decay", and must periodically be given an upward boost by rocket engines to keep it in orbit.

Ozone Layer

Ozone layer, also called ozonosphere is a region of the upper atmosphere, between roughly 15 and 35 km (9 and 22 miles) above Earth's surface, containing relatively high concentrations of ozone molecules (O_3). Approximately 90 percent of the atmosphere's ozone occurs in the stratosphere, the region extending from 10–18 km (6–11 miles) to approximately 50 km (about 30 miles) above Earth's surface. In the stratosphere the temperature of the atmosphere rises with increasing height, a phenomenon created by the absorption of solar radiation by the ozone layer. The ozone layer effectively blocks almost all solar radiation of wavelengths less than 290 nanometres from reaching Earth's surface, including certain types of ultraviolet (UV) and other forms of radiation that could injure or kill most living things.

Location in Earth's Atmosphere

In the midlatitudes the peak concentrations of ozone occur at altitudes from 20 to 25 km (about 12 to 16 miles). Peak concentrations are found at altitudes from 26 to 28 km (about 16 to 17 miles) in the tropics and from about 12 to 20 km (about 7 to 12 miles) toward the poles. The lower height of the peak-concentration region in the high latitudes largely results from pole ward and downward atmospheric transport processes that occur in the middle and high latitudes and the reduced height of the tropopause (the transition region between the troposphere and stratosphere).

Most of the remaining ozone occurs in the troposphere, the layer of the atmosphere that extends from Earth's surface up to the stratosphere. Near-surface ozone often results from interactions between certain pollutants (such as nitrogen oxides and volatile organic compounds), strong sunlight, and hot weather. It is one of the primary ingredients in photochemical smog, a phenomenon that plagues many urban and suburban areas around the world, especially during the summer months.

Ozone damage on the leaf of an English walnut.

Ozone Creation and Destruction

The production of ozone in the stratosphere results primarily from the breaking of the chemical bonds within oxygen molecules (O_2) by high-energy solar photons. This process, called photo dissociation, results in the release of single oxygen atoms, which later join with intact oxygen molecules to form ozone. Rising atmospheric oxygen concentrations some two billion years ago allowed ozone to build up in Earth's atmosphere, a process that gradually led to the formation of the stratosphere. Scientists believe that the formation of the ozone layer played an important role in the development of life on Earth by screening out lethal levels of UVB radiation (ultraviolet radiation with wavelengths between 315 and 280 nanometres) and thus facilitating the migration of life-forms from the oceans to land.

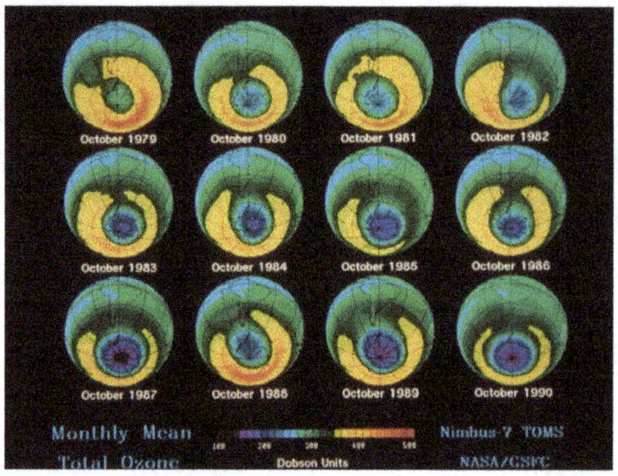

Ozone: hole changes in the size of the ozone hole.

The amount of ozone in the stratosphere varies naturally throughout the year as a result of chemical processes that create and destroy ozone molecules and as a result of winds and other transport processes that move ozone molecules around the planet. Over the course of several decades, however, human activities substantially altered the ozone layer. Ozone depletion, the global decrease in stratospheric ozone observed since the 1970s, is most pronounced in Polar Regions, and it is well correlated with the increase of chlorine and bromine in the stratosphere. Those chemicals, once freed by UV radiation from the chlorofluorocarbons (CFCs) and other halocarbons (carbon-halogen compounds) that contain them, destroy ozone by stripping away single oxygen atoms from ozone molecules. Depletion is so extensive that so-called ozone holes (regions of severely reduced ozone coverage) form over the poles during the onset of their respective spring seasons. The largest such hole—which has spanned more than 20.7 million square km (8 million square miles) on a consistent basis since 1992—appears annually over Antarctica between September and November.

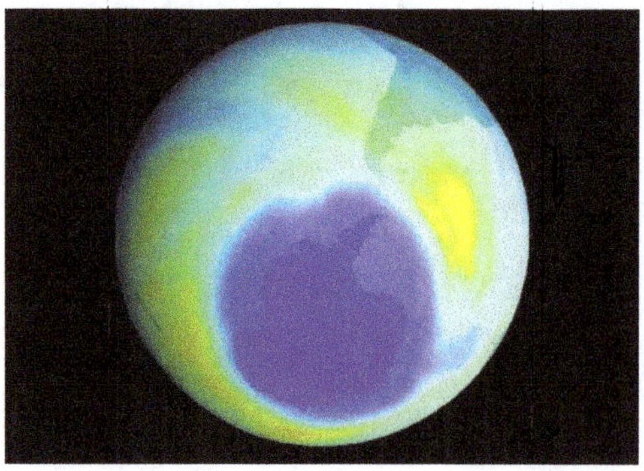

Ozone depletion: Antarctic ozone hole.

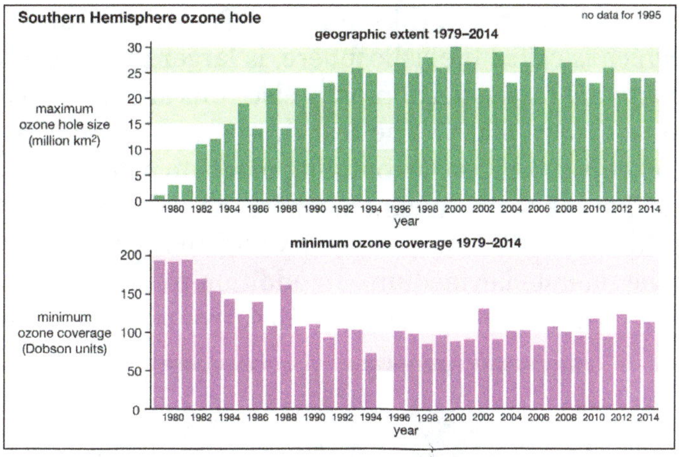

Southern Hemisphere ozone hole: Two bar graphs depicting the maximum ozone hole size and the minimum ozone coverage (in Dobson units) of the Southern Hemisphere ozone hole, 1979–2014.

As the amount of stratospheric ozone declines, more UV radiation reaches Earth's surface, and scientists worry that such increases could have significant effects on ecosystems and human health. The concern over exposure to biologically harmful levels of UV radiation has been the main driver of the creation of international treaties such as the Montreal Protocol on Substances That Deplete the Ozone Layer and its amendments, designed to protect Earth's ozone layer. Compliance with international treaties that phased out the production and delivery of many ozone-depleting chemicals, combined with upper stratospheric cooling due to increased carbon dioxide, is thought to have contributed to the shrinking of the ozone holes over the poles and to slightly higher stratospheric ozone levels overall. Continued reductions in chlorine loading are expected to result in smaller ozone holes above Antarctica after 2040. However, some scientists noted that gains in stratospheric ozone levels have only occurred in the upper stratosphere, with declines in ozone concentrations in the lower stratosphere outpacing increases in the upper stratosphere.

Ozonesonde: Researchers launching a balloon carrying an ozonesonde, an instrument that measures ozone in the atmosphere, at Amundsen-Scott South Pole Station in Antarctica.

Magnetosphere

A magnetosphere is a dynamically varying tear-drop shaped region of plasma comprising magnetic fields and charged particles surrounding a magnetized astronomical object, whether it is a planet (like the earth), a planet's moon (like Jupiter's Ganymede), or a star (like the sun). The sun's magnetosphere, which is called the heliosphere, is larger than the solar system itself and is characterized by the solar wind (a plasma of mostly electrons and protons) flowing outward from the sun and past the most distant planet. The tear drop shape of a magnetosphere arises as the magnetized astronomical object with its surrounding magnetic field and charged particles passes through ambient plasma. For the earth's magnetosphere the ambient plasma is the solar wind and the magnetosphere shields the earth from the solar wind's powerful effects. For the heliosphere the ambient plasma is the interstellar medium. In addition to the earth, the magnetized planets Mercury, Jupiter, Saturn, Uranus, and Neptune are all surrounded by a magnetosphere.

An artistic rendition of the earth, earth's magnetosphere, and the sun accurately portrays to scale the earth and its magnetosphere, while grossly distorting the relative size of the sun and its distance from the earth. An accurately scaled representation, taking the size of the earth as a scale reference, would have the sun enlarged roughly seven times and placed roughly 1160 times further away. Human knowledge of the magnetosphere dates from 1958 when the first U.S. earth-orbiting satellite, Explorer 1, detected a belt of charged particles (later named the Van Allen radiation belt), trapped by the earth's magnetic field. Subsequent studies have mapped and labeled the magnetosphere revealing that in the downwind side of the planet the long tear-drop shape of the magnetosphere can extend out to as much as 200 earth radii. Also, it is at the far end that variable openings in the magnetosphere at times permit some of the solar wind particles to flow into the magnetosphere. Some of these circulate back to the earth and may even flow along magnetic field lines at the poles to produce the auroras.

The discipline of space physics is largely the study of magnetospheres because their magnetic fields and plasmas are pervasive throughout space, except for the surfaces and interiors of the planets.

Earth's Magnetosphere

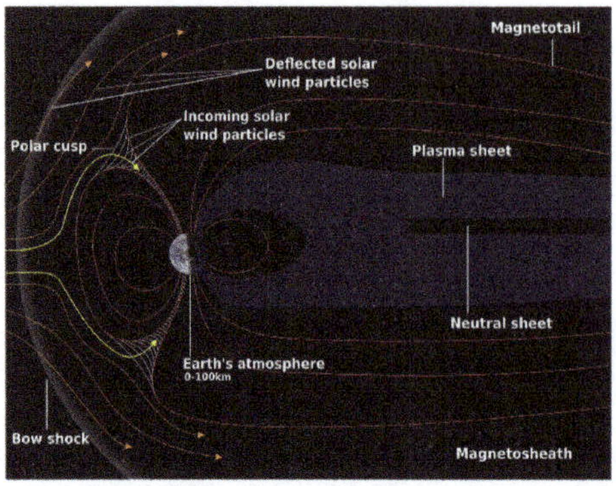

Schematic of Earth's magnetosphere: The solar wind flows from left to right.

The magnetosphere of Earth is a region in space whose shape is determined by the extent of Earth's internal magnetic field, the solar wind plasma, and the interplanetary magnetic field (IMF). In the magnetosphere, a mix of free ions and electrons from both the solar wind and the Earth's ionosphere is confined by magnetic and electric forces that are much stronger than gravity and collisions.

In spite of its name, the magnetosphere is distinctly non-spherical. On the side facing the Sun, the distance to its boundary (which varies with solar wind intensity) is about 70,000 km (10-12 Earth radii or R_E, where 1 R_E=6371 km; unless otherwise noted, all distances here are from the Earth's center). The boundary of the magnetosphere ("magnetopause") is roughly bullet shaped, about 15 R_E abreast of Earth and on the night side (in the "magnetotail" or "geotail") approaching a cylinder with a radius 20-25 R_E. The tail region stretches well past 200 R_E, and the way it ends is not well-known.

The outer neutral gas envelope of Earth, or geocorona, consists mostly of the lightest atoms, hydrogen and helium, and continues beyond 4-5 R_E, with diminishing density. The hot plasma ions of the magnetosphere acquire electrons during collisions with these atoms and create an escaping "glow" of fast atoms that have been used to image the hot plasma clouds by the IMAGE mission.

The upward extension of the ionosphere, known as the plasmasphere, also extends beyond 4-5 R_E with diminishing density, beyond which it becomes a flow of light ions called the polar wind that escapes out of the magnetosphere into the solar wind. Energy deposited in the ionosphere by auroras strongly heats the heavier atmospheric components such as oxygen and molecules of oxygen and nitrogen, which would not otherwise escape from Earth's gravity. Owing to this highly variable heating, however, a heavy atmospheric or ionospheric outflow of plasma flows during disturbed periods from the auroral zones into the magnetosphere, extending the region dominated by terrestrial material, known as the fourth or plasma geosphere, at times out to the magnetopause.

General Properties

Two factors determine the structure and behavior of the magnetosphere: (1) The internal field of the Earth, and (2) The solar wind.

1. The internal field of the Earth (its "main field") appears to be generated in the Earth's core by a dynamo process, associated with the circulation of liquid metal in the core, driven by internal heat sources. Its major part resembles the field of a bar magnet ("dipole field") inclined by about 10° to the rotation axis of Earth, but more complex parts ("higher harmonics") also exist, as first shown by Carl Friedrich Gauss. The dipole field has an intensity of about 30,000-60,000 nanoteslas (nT) at the Earth's surface, and its intensity diminishes like the inverse of the cube of the distance, i.e. at a distance of R Earth radii it only amounts to $1/R^3$ of the surface field in the same direction. Higher harmonics diminish faster, like higher powers of $1/R$, making the dipole field the only important internal source in most of the magnetosphere.

2. The solar wind is a fast outflow of hot plasma from the sun in all directions. Above the sun's equator it typically attains 400 km/s; above the sun's poles, up to twice as much. The flow is powered by the million-degree temperature of the sun's corona, for which no generally accepted explanation exists as yet. Its composition resembles that of the Sun—about 95 percent of the ions are protons, about 4 percent helium nuclei, with 1 percent of heavier

matter (C, N, O, Ne, Si, Mg up to Fe) and enough electrons to keep charge neutrality. At Earth's orbit its typical density is 6 ions/cm^3 (variable, as is the velocity), and it contains a variable interplanetary magnetic field (IMF) of (typically) 2–5 nT. The IMF is produced by stretched-out magnetic field lines originating on the Sun, referred to in what follows as simply MSPF.

Physical reasons (MSPF) make it difficult for solar wind plasma with its embedded IMF to mix with terrestrial plasma whose magnetic field has a different source. The two plasmas end up separated by a boundary, the magnetopause, and the Earth's plasma is confined to a cavity inside the flowing solar wind, the magnetosphere. The isolation is not complete, thanks to secondary processes such as magnetic reconnection (MSPF)—otherwise it would be hard for the solar wind to transmit much energy to the magnetosphere—but it still determines the overall configuration.

An additional feature is a collision-free bow shock that forms in the solar wind ahead of Earth, typically at 13.5 R_E on the sunward side. It forms because the solar velocity of the wind exceeds (typically 2–3 times) that of Alfvén waves, a family of characteristic waves with which disturbances propagate in a magnetized fluid. In the region behind the shock ("magnetosheath") the velocity drops briefly to the Alfvén velocity (and the temperature rises, absorbing lost kinetic energy), but the velocity soon rises back as plasma is dragged forward by the surrounding solar wind flow.

To understand the magnetosphere, one needs to visualize its magnetic field lines, that everywhere point in the direction of the magnetic field—for example, diverging out near the magnetic north pole (or geographic southpole), and converging again around the magnetic south pole (or the geographic northpole), where they enter the Earth. they can be visualized like wires which tie the magnetosphere together—wires that also guide the motions of trapped particles, which slide along them like beads (though other motions may also occur).

Radiation Belts

When the first scientific satellites were launched in the first half of 1958—Explorers 1 and 3 by the US, Sputnik 3 by the Soviet Union—they observed an intense (and unexpected) radiation belt around Earth, held by its magnetic field. "My God, Space is Radioactive!" exclaimed one of Van Allen's colleagues, when the meaning of those observations was realized. That was the "inner radiation belt" of protons with energies in the range 10-100 MeV (megaelectronvolts), attributed later to "albedo neutron decay," a secondary effect of the interaction of cosmic radiation with the upper atmosphere. It is centered on field lines crossing the equator about 1.5 R_E from the Earth's center.

Later a population of trapped ions and electrons was observed on field lines crossing the equator at 2.5–8 R_E. The high-energy part of that population (about 1 MeV) became known as the "outer radiation belt," but its bulk is at lower energies (peak about 65 keV) and is identified as the ring current plasma.

The trapping of charged particles in a magnetic field can be quite stable. This is particularly true in the inner belt, because the build-up of trapped protons from albedo neutrons is quite slow, requiring years to reach observed intensities. In July 1962, the United States tested an H-bomb high over the South Pacific at around 400 km in the upper atmosphere, in this region, creating an artificial belt of high-energy electrons, and some of them were still around 4–5 years later (such tests are now banned by treaty).

The outer belt and ring current are less persistent, because charge-exchange collisions with atoms of the geocorona tend to remove their particles. That suggests the existence of an effective source mechanism, continually supplying this region with fresh plasma. It turns out that the magnetic barrier can be broken down by electric forces. If plasma is pushed hard enough, it generates electric fields which allow it to move in response to the push, often (not always) deforming the magnetic field in the process.

Magnetic Tails

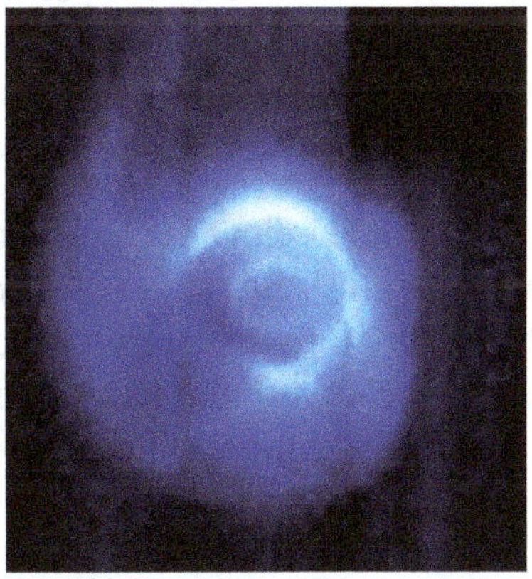

A view from the IMAGE satellite showing Earth's plasmasphere using its Extreme Ultraviolet (EUV) imager instrument.

A magnetic tail or magnetotail is formed by pressure from the solar wind on a planet's magnetosphere. The magnetotail can extend great distances away from its originating planet. Earth's magnetic tail extends at least 200 Earth radii in the anti-sunward direction well beyond the orbit of the Moon at about 60 Earth radii, while Jupiter's magnetic tail extends beyond the orbit of Saturn. On occasion Saturn is immersed inside the Jovian magnetosphere.

The extended magnetotail results from energy stored in the planet's magnetic field. At times this energy is released and the magnetic field becomes temporarily more dipole-like. As it does so that stored energy goes to energize plasma trapped on the involved magnetic field lines. Some of that plasma is driven tailward and into the distant solar wind. The rest is injected into the inner magnetosphere where it results in the aurora and the ring current plasma population. The resulting energetic plasma and electric currents can disrupt spacecraft operations, communication and navigation.

Electric Currents in Space

Magnetic fields in the magnetosphere arise from the Earth's internal magnetic field as well as from electric currents that flow in the magnetospheric plasma: The plasma acts as a kind of electromagnet. Magnetic fields from currents that circulate in the magnetospheric plasma extend the Earth's magnetism much further in space than would be predicted from the Earth's internal field alone.

Unlike in a conventional resistive electric circuit, where currents are best thought of as arising as a response to an applied voltage, currents in the magnetosphere are better seen as being caused by the structure and motion of the plasma in its associated magnetic field. For instance, electrons and positive ions trapped in the dipole-like field near the Earth tend to circulate around the magnetic axis of the dipole (the line connecting the magnetic poles) in a ring around the Earth, without gaining or losing energy (this is known as guiding center motion). Viewed from above the magnetic north pole (geographic south), ions circulate clockwise, electrons counter clockwise, producing a net circulating clockwise current, known (from its shape) as the ring current. No voltage is needed—the current arises naturally from the motion of the ions and electrons in the magnetic field.

Any such current will modify the magnetic field. The ring current, for instance, strengthens the field on its outside, helping expand the size of the magnetosphere. At the same time, it weakens the magnetic field in its interior. In a magnetic storm, plasma is added to the ring current, making it temporarily stronger, and the field at Earth is observed to weaken by up to 1-2 percent.

The deformation of the magnetic field, and the flow of electric currents in it, is intimately linked, making it often hard to label one as cause and the other as effect. Frequently (as in the magnetopause and the magnetotail) it is intuitively more useful to regard the distribution and flow of plasma as the primary effect, producing the observed magnetic structure, with the associated electric currents just one feature of those structures, more of a consistency requirement of the magnetic structure.

As noted, one exception (at least) exists, a case where voltages do drive currents. That happens with Birkeland currents, which flow from distant space into the near-polar ionosphere, continue at least some distance in the ionosphere, and then return to space. (Part of the current then detours and leaves Earth again along field lines on the morning side, flows across midnight as part of the ring current, then comes back to the ionosphere along field lines on the evening side and rejoins the pattern.) The full circuit of those currents, under various conditions, is still under debate.

Because the ionosphere is an ohmic conductor of sorts, such flow will heat it up. It will also give rise to secondary Hall currents, and accelerate magnetospheric particles—electrons in the arcs of the polar aurora, and singly-ionized oxygen ions (O^+) which contribute to the ring current.

Classification of Magnetic Fields

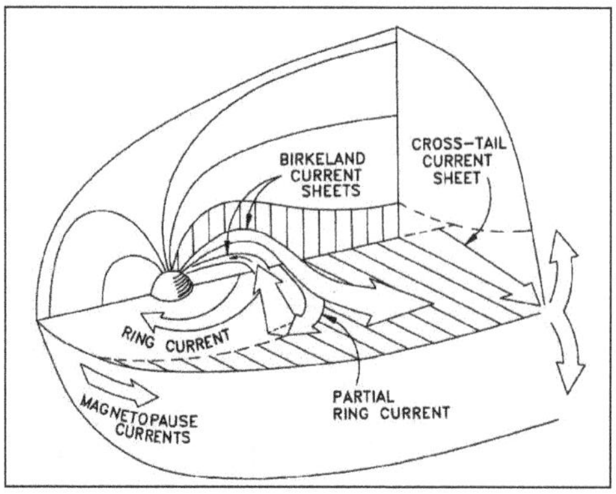

Schematic view of the different current systems which shape the Earth's magnetosphere.

Regardless of whether they are viewed as sources or consequences of the magnetospheric field structure, electric currents flow in closed circuits. That makes them useful for classifying different parts of the magnetic field of the magnetosphere, each associated with a distinct type of circuit. In this way the field of the magnetosphere is often resolved into 5 distinct parts, as follows:

1. The internal field of the Earth ("main field") arising from electric currents in the core. It is dipole-like, modified by higher harmonic contributions.

2. The ring current field, carried by plasma trapped in the dipole-like field around Earth, typically at distances 3–8 R_E (less during large storms). Its current flows (approximately) around the magnetic equator, mainly clockwise when viewed from north. (A small counter clockwise ring current flows at the inner edge of the ring, caused by the fall-off in plasma density as Earth is approached).

3. The field confining the Earth's plasma and magnetic field inside the magnetospheric cavity. The currents responsible for it flow on the magnetopause, the interface between the magnetosphere and the solar wind, described in the introduction. Their flow, again, may be viewed as arising from the geometry of the magnetic field (rather than from any driving voltage), a consequence of "Ampére's law" (embodied in Maxwell's equations) which in this case requires an electric current to flow along any interface between magnetic fields of different directions and/or intensities.

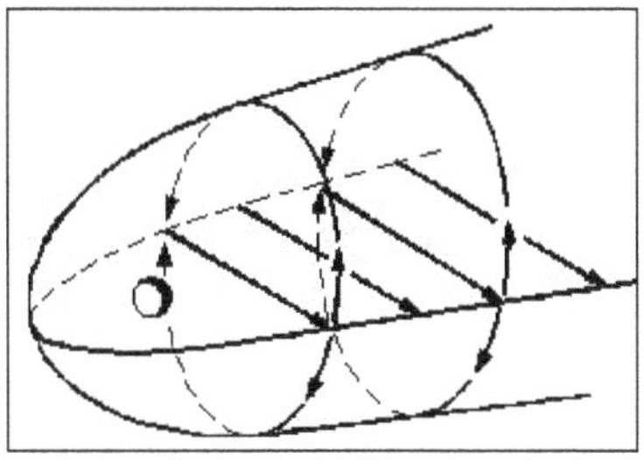

4. The system of tail currents. The magnetotail consists of twin bundles of oppositely directed magnetic field (the "tail lobes"), directed earthwards in the northern half of the tail and away from Earth in the southern half. In between the two exists a layer ("plasma sheet") of denser plasma (0.3-0.5 ions/cm^3 vs. 0.01-0.02 in the lobes), and because of the difference between the adjoining magnetic fields, by Ampére's law an electric current flows there too, directed from dawn to dusk. The flow closes (as it must) by following the tail magnetopause—part over the northern lobe, part over the southern one.

5. The Birkeland current field (and its branches in the ionosphere and ring current), a circuit is associated with the polar aurora. Unlike the 3 preceding current systems, it does require a constant input of energy, to provide the heating of its ionospheric path and the acceleration of auroral electrons and of positive ions. The energy probably comes from a dynamo process, meaning that part of the circuit threads a plasma moving relative to Earth, either in the solar wind and in "boundary layer" flows which it drives just inside the magnetopause, or by plasma moving earthward in the magnetotail, as observed during substorms.

Magnetic Substorms and Storms

Earlier it was stated that "if plasma is pushed hard enough, it generates electric fields which allow it to move in response to the push, often (not always) deforming the magnetic field in the process."

Two examples of such "pushing" are particularly important in the magnetosphere. The THEMIS mission is a NASA program to study in detail the physical processes involved in substorms.

The more common one occurs when the north-south component B_z of the interplanetary magnetic field (IMF) is appreciable and points southward. In this state field lines of the magnetosphere are relatively strongly linked to the IMF, allowing energy and plasma to enter it at relatively high rates. This swells up the magnetotail and makes it unstable. Ultimately the tail's structure changes abruptly and violently, a process known as a magnetic substorm.

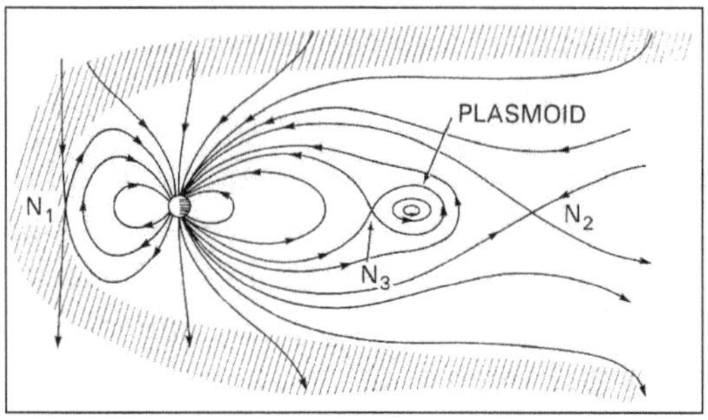

Magnetic reconnection in the near-Earth magnetotail, producing a disconnected "plasmoid".

One possible scenario (the subject is still being debated) is as follows. As the magnetotail swells, it creates a wider obstacle to the solar wind flow, causing its widening portion to be squeezed more by the solar wind. In the end, this squeezing breaks apart field lines in the plasma sheet ("magnetic reconnection"), and the distant part of the sheet, no longer attached to the Earth, is swept away as an independent magnetic structure ("plasmoid"). The near-Earth part snaps back earthwards, energizing its particles and producing Birkeland currents and bright auroras. As observed in the 1970s by the ATS satellites at 6.6 R_E, when conditions are favorable that can happen up to several times a day.

Substorms generally do not substantially add to the ring current. That happens in magnetic storms, when following an eruption on the sun (a "coronal mass ejection" or a "solar flare"—details are still being debated, see MSPF) a fast-moving plasma cloud hits the Earth. If the IMF has a southward component, this not only pushes the magnetopause boundary closer to Earth (at times to about half its usual distance), but it also produces an injection of plasma from the tail, much more vigorous than the one associated with substorms.

The plasma population of the ring current may now grow substantially, and a notable part of the addition consists of O+ oxygen ions extracted from the ionosphere as a by-product of the polar aurora. In addition, the ring current is driven earthward (which energizes its particles further), temporarily modifying the field around the Earth and thus shifting the aurora (and its current system) closer to the equator. The magnetic disturbance may decay within 1–3 days as many ions are removed by charge exchange, but the higher energies of the ring current can persist much longer.

Van Allen Radiation Belt

The Van Allen radiation belt (or Van Allen belt) is a torus of energetic charged particles (plasma) around Earth, held in place by Earth's magnetic field. Energetic electrons form two distinct radiation belts, and protons form a single belt.

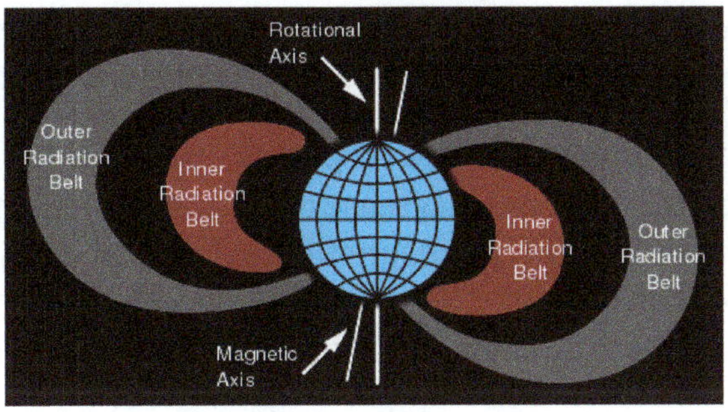

Van Allen radiation belts.

Earth's geomagnetic field is not uniformly distributed around its surface. On the Sun side, it is compressed because of the solar wind, and on the opposite side, it is elongated to around three Earth radii. This creates a cavity (called the Chapman Ferraro Cavity) in which the radiation belts reside. These belts are closely related to the polar aurora where particles strike the upper atmosphere and fluoresce. The term Van Allen belts refers specifically to the radiation belts surrounding Earth; however, similar radiation belts have been discovered around other planets.

Outer Belt

Laboratory simulation of the Van Allen belts' influence on the Solar Wind; these auroral-like Birkeland currents were created by the scientist Kristian Birkeland in his terrella, a magnetized anode globe in an evacuated chamber.

The large outer radiation belt extends from an altitude of about three to ten Earth radii (R_E) above the Earth's surface, and its greatest intensity is usually around 4-5 R_E. The outer electron radiation belt is mostly produced by the inward radial diffusion and local acceleration due to transfer of energy from whistler mode plasma waves to radiation belt electrons. Radiation belt electrons are also constantly removed by collisions with atmospheric neutrals, losses to magnetopause, and the outward radial diffusion. The outer belt consists mainly of high energy (0.1–10 MeV) electrons trapped by the Earth's magnetosphere. The gyroradii for energetic protons would be large enough to bring them into contact with the Earth's atmosphere. The electrons here have a high flux and at the outer edge (close to the magnetopause), where geomagnetic field lines open into the geomagnetic "tail," fluxes of energetic electrons can drop to the low interplanetary levels within about 100 km (a decrease by a factor of 1,000).

The trapped particle population of the outer belt is varied, containing electrons and various ions. Most of the ions are in the form of energetic protons, but a certain percentage are alpha particles and O^+ oxygen ions, similar to those in the ionosphere but much more energetic. This mixture of ions suggests that ring current particles probably come from more than one source.

The outer belt is larger than the inner belt, and its particle population fluctuates widely. Energetic (radiation) particle fluxes can increase and decrease dramatically as a consequence of geomagnetic storms, which are themselves triggered by magnetic field and plasma disturbances produced by the Sun increases are due to storm-related injections and acceleration of particles from the tail of the magnetosphere. There is debate as to whether the outer belt was discovered by the US Explorer 4 or the USSR Sputnik 2/3.

Inner Belt

The inner Van Allen Belt extends from an altitude of 700–10,000 km (0.1 to 1.5 Earth radii) above the Earth's surface, and contains high concentrations of energetic protons with energies exceeding 100 MeV and electrons in the range of hundreds of kiloelectronvolts, trapped by the strong (relative to the outer belts) magnetic fields in the region.

It is believed that protons of energies exceeding 50 MeV in the lower belts at lower altitudes are the result of the beta decay of neutrons created by cosmic ray collisions with nuclei of the upper atmosphere. The source of lower energy protons is believed to be proton diffusion due to changes in the magnetic field during geomagnetic storms. Due to the slight offset of the belts from Earth's geometric center, the inner Van Allen belt makes its closest approach to the surface at the South Atlantic Anomaly.

Impact on Space Travel

Solar cells, integrated circuits, and sensors can be damaged by radiation. In 1962, the Van Allen belts were temporarily amplified by a high-altitude nuclear explosion (the Starfish Prime test) and several satellites ceased operation. Geomagnetic storms occasionally damage electronic components on spacecraft. Miniaturization and digitization of electronics and logic circuits have made satellites more vulnerable to radiation, as incoming ions may be as large as the circuit's charge. Electronics on satellites must be hardened against radiation to operate reliably. The Hubble Space Telescope, among other satellites, often has its sensors turned off when passing through regions of intense radiation.

Missions beyond low earth orbit leave the protection of the geomagnetic field, and transit the Van Allen belts. Thus they may need to be shielded against exposure to cosmic rays, Van Allen radiation, or solar flares.

An object satellite shielded by 3 mm of aluminum in an elliptic orbit passing through the radiation belt will receive about 2,500 rem (25 Vs.) per year.

Causes

It is generally understood that the inner and outer Van Allen belts result from different processes. The inner belt, consisting mainly of energetic protons, is the product of the decay of albedo neutrons which are themselves the result of cosmic ray collisions in the upper atmosphere. The outer belt consists mainly of electrons. They are injected from the geomagnetic tail following geomagnetic storms, and are subsequently energized though wave-particle interactions. Particles are trapped in the Earth's magnetic field because it is basically a magnetic mirror. Particles gyrate around field lines and also move along field lines. As particles encounter regions of stronger magnetic field where field lines converge, their "longitudinal" velocity is slowed and can be reversed, reflecting the particle. This causes the particle to bounce back and forth between the earth's poles, where the magnetic field increases.

Simulated Van Allen Belts generated by a plasma thruster in tank 5 Electric Propulsion Laboratory at the then-called Lewis Research Center, Cleveland Ohio.

A gap between the inner and outer Van Allen belts, sometimes called safe zone or safe slot, is caused by the very low frequency (VLF) waves which scatter particles in pitch angle which results in the loss of particles to the atmosphere. Solar outbursts can pump particles into the gap but they drain again in a matter of days. The radio waves were originally thought to be generated by turbulence in the radiation belts, but recent work by James Green of the NASA Goddard Space Flight Center comparing maps of lightning activity collected by the Micro Lab 1 spacecraft with data on radio waves in the radiation-belt gap from the image spacecraft suggests that they're actually generated by lightning within Earth's atmosphere. The radio waves they generate strike the ionosphere at the right angle to pass through it only at high latitudes, where the lower ends of the gap approach the upper atmosphere. These results are still under scientific debate.

There have been nuclear tests in space that have caused artificial radiation belts. Starfish Prime, a high altitude nuclear test, created an artificial radiation belt that damaged or destroyed as many as one third of the satellites in low earth orbit at the time. Thomas Gold has argued that the outer belt is left over from the aurora while Alex Dessler has argued that the belt is a result of volcanic activity.

In another view, the belts could be considered a flow of electric current that is fed by the solar wind. With the protons being positive and the electrons being negative, the area between the belts is sometimes subjected to a current flow, which "drains" away. The belts are also thought to drive auroras, lightning and many other electrical effects.

The belts are a hazard for artificial satellites and moderately dangerous for human beings, difficult and expensive to shield against. For these reasons, the late Robert L. Forward proposed a method, called Hivolt, to drain at least the inner belt to 1 percent of its natural level within a year. The proposal involves deploying highly electrically charged tethers in orbit. The idea is that the electrons would be deflected by the large electrostatic fields and intersects the atmosphere and harmlessly dissipate.

References

- Troposphere: physics.ucsb.edu, Retrieved 2 July, 2019
- Stratosphere: ucar.edu, Retrieved 11 January, 2019
- Mesosphere, earth-and-planetary-sciences: sciencedirect.com, Retrieved 22 August, 2019
- Mesosphere: ucar.edu, Retrieved 8 February, 2019
- Thermosphere, earth-and-planetary-sciences: sciencedirect.com, Retrieved 14 January, 2019
- Thermosphere: ucar.edu, Retrieved 20 March, 2019
- Ionosphere: safeopedia.com, Retrieved 23 February, 2019
- Mechanisms-of-ionization, ionosphere-and-magnetosphere, science: britannica.com, Retrieved 27 April, 2019
- Exosphere, earth-and-planetary-sciences: sciencedirect.com, Retrieved 7 March, 2019
- Exosphere: ucar.edu, Retrieved 10 May, 2019
- Ozone-layer, science: britannica.com, Retrieved 21 April, 2019
- Magnetosphere: newworldencyclopedia.org, Retrieved 12 June, 2019
- Van-Allen-radiation-belt: newworldencyclopedia.org, Retrieved 19 May, 2019

Chapter 3

Atmospheric Chemistry

The domain of atmospheric science which deals with the study of the chemistry of the Earth's atmosphere is referred to as atmospheric chemistry. It is involved in the study of the cycles of gases such as nitrogen, oxygen, carbon dioxide and sulfur dioxide. The diverse aspects of these gases as well as the processes related to them have been thoroughly discussed in this chapter.

Atmospheric chemistry involves study of the chemistry of the atmospheres of Earth and other planets. It is a branch of atmospheric science and is a multidisciplinary field of research, drawing on environmental chemistry, meteorology, physics, computer modeling, oceanoraphy, geology, volcanology, and other disciplines. In addition, it is being increasingly associated with the field known as climatology.

Earth's atmosphere is composed of about 78 percent nitrogen, 21 percent oxygen, and small amounts of water vapor, carbon dioxide, argon, and other gases. This mixture of gases, commonly called air, protects and sustains life on Earth in a variety of ways. It provides oxygen for respiration, carbon dioxide for photosynthesis, and water vapor for the precipitation that replenishes moisture in the soil. In addition, carbon dioxide and water vapor act as "greenhouse gases" that keep the Earth sufficiently warm to maintain life. Nitrogen is used by "nitrogen-fixing" bacteria to produce compounds that are useful for plant growth. Water vapor prevents exposed living tissue from drying up. Ozone in the stratosphere absorbs ultraviolet solar radiation that could damage living tissue. In addition, higher layers of the atmosphere protect the Earth from bombardment by meteorites and charged particles in the solar wind.

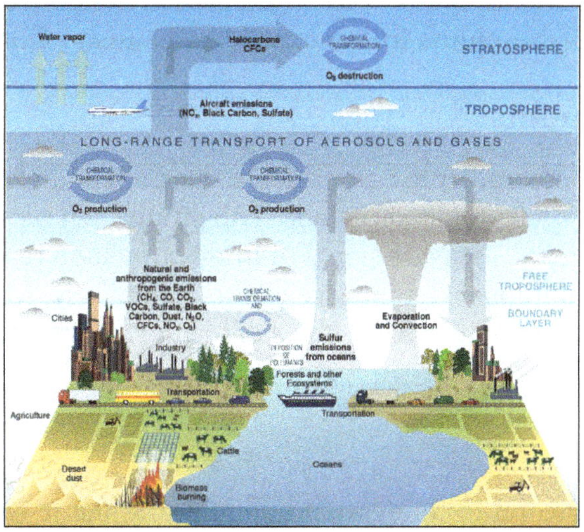

Schematic of chemical and transport processes related to atmospheric composition.

The composition of Earth's atmosphere has been altered by human activities such as fuel burning and industrial production, and a number of these changes are harmful to human health, crops, and ecosystems. Examples of problems that involve studies in atmospheric chemistry include acid rain, photochemical smog, and global warming. Researchers in the field of atmospheric chemistry seek to understand the causes of these problems and to look for possible solutions. They help inform and evaluate government policies that are related to the environment.

Methodology

Observations, laboratory measurements, and modeling are the three central elements of atmospheric chemistry. Progress in this field is often driven by interactions between these components and they form an integrated whole. For example, observations may tell us that more of a chemical compound exists than previously thought possible. This would stimulate new modeling and laboratory studies, which would increase our scientific understanding to a point where the observations can be explained.

Observations

Observations are essential to our understanding of atmospheric chemistry. Routine observations of chemical composition provide information about changes in atmospheric composition over time. One important example of this is the Keeling Curve—that show a steady rise in the concentration of carbon dioxide.

These types of observations are conducted in observatories, such as that on Mauna Loa, and on mobile platforms such as aircraft (for instance, the UK's Facility for Airborne Atmospheric Measurements), ships, and balloons. Observations of atmospheric composition are increasingly made by satellites with important instruments, such as GOME and MOPITT, giving a global picture of air pollution and chemistry. Surface observations provide long-term records at high resolution in terms of time, but they are limited in the vertical and horizontal space they provide observations from. Some surface-based instruments, such as LIDAR, can provide concentration profiles of chemical compounds and aerosols, but they are restricted in the horizontal region they can cover. Many observations are available online in Atmospheric Chemistry Observational Databases.

Laboratory Measurements

Measurements made in the laboratory are essential to our understanding of the sources and sinks of pollutants and naturally occurring compounds. Lab studies tell us which gases react with one another and how fast they react. Measurements of interest include reactions in the gas phase, on surfaces, and in water. Of additional significance is photochemistry, which quantifies how quickly molecules are split apart by sunlight and the types of products formed, plus thermodynamic data such as Henry's law coefficients.

Modeling

To synthesize and test the theoretical understanding of atmospheric chemistry, computer models are constructed. Numerical models solve the differential equations governing the concentrations of chemicals in the atmosphere. They can range from simple to highly complex.

One common trade-off in numerical models is between the number of chemical compounds and chemical reactions modeled versus the representation of transport and mixing in the atmosphere.

For example, a box model might include hundreds or even thousands of chemical reactions but will only have a very crude representation of mixing in the atmosphere. By contrast, 3D models represent many of the physical processes of the atmosphere but due to constraints on computer resources will have far fewer chemical reactions and compounds. Models can be used to interpret observations, test understanding of chemical reactions, and predict future concentrations of chemical compounds in the atmosphere. One important current trend is for atmospheric chemistry modules to become one part of Earth system models in which the links between climate, atmospheric composition, and the biosphere can be studied.

Some models are constructed by automatic code generators. In this approach, a set of constituents are chosen and the automatic code generator then selects the reactions involving those constituents from a set of reaction databases. Once the reactions have been chosen, the ordinary differential equations (ODE) that describe the changes over time can be automatically constructed.

Nitrogen and its Cycle

Nitrogen (symbol N, atomic number 7) is the chief constituent of the Earth's atmosphere and a vital element in all known forms of life. At ordinary temperatures and pressures, free nitrogen (unbound to any other element) is a colorless, odorless, and tasteless gas. As an inert gas, it reduces the amount of oxygen available for the oxidation of natural materials, thus restricting spontaneous combustion of flammable materials and the corrosion of metals. It also protects living organisms from the toxic effects of breathing pure (or highly concentrated) oxygen. The Earth's nitrogen continually cycles through the atmosphere, biosphere, and lithosphere, effected by such processes as nitrogen fixation by bacteria, metabolic processing in living things, and decomposition of dead organic matter.

In living organisms, nitrogen atoms are part of the molecular structures of such key substances as amino acids, proteins, and nucleic acids. In industry, nitrogen gas is used as an inert replacement for air in the packaging of foods and the manufacture of steel and electronic components. Liquid nitrogen is a cryogen (low-temperature refrigerant) used for freezing and transport of food and other perishable products. Ammonia, a significant compound of nitrogen, is useful for fertilizers and for the synthesis of nitric acid and other valuable compounds. Nitric acid is an oxidizing agent used in liquid-fueled rockets, potassium nitrate is used in gunpowder, and trinitrotoluene (TNT) is a significant explosive. In addition, nitrogen is a constituent element in every major class of drugs.

Nitrogen makes up 78.084 percent of the volume (and 75.5 percent of the mass) of air. Its most common isotope, nitrogen-14 (^{14}N), appears to be created through nuclear fusion processes in stars. Compounds that contain this element have been observed by astronomers, and molecular nitrogen has been detected in interstellar space by David Knauth and coworkers using the Far Ultraviolet Spectroscopic Explorer. Molecular nitrogen occurs in trace amounts in the atmospheres of various planets, but it is a major constituent of Titan, the planet Saturn's largest moon.

Nitrogen is present in all living organisms, as part of the molecular structures of proteins, nucleic acids, and other important substances. It is a large component of animal waste, usually in the form of urea, uric acid, and their derivatives.

Nitrogen Cycle

The nitrogen cycle is a repeating cycle of processes during which nitrogen moves through both living and non-living things: the atmosphere, soil, water, plants, animals and bacteria.

In order to move through the different parts of the cycle, nitrogen must change forms. In the atmosphere, nitrogen exists as a gas (N_2), but in the soils it exists as nitrogen oxide, NO, and nitrogen dioxide, NO_2, and when used as a fertilizer, can be found in other forms, such as ammonia, NH_3, which can be processed even further into a different fertilizer, ammonium nitrate, or NH_4NO_3.

Nitrogen Cycle Processes

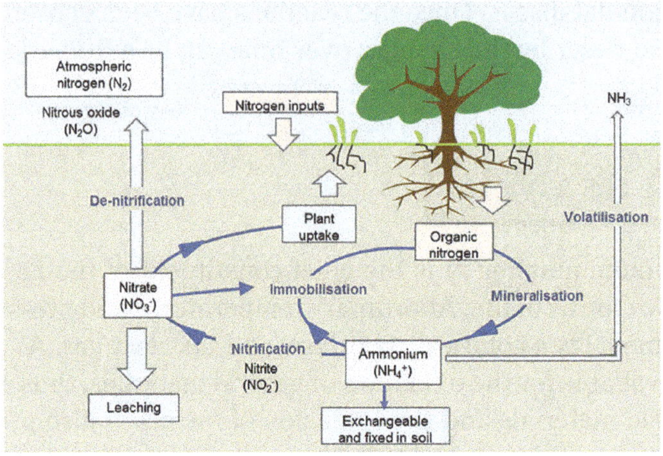

Soil Nitrogen Cycle.

Natural Soil Nitrogen Cycle

Dinitrogen is an essential element for the nitrogen cycle. The symbiotic microbes fix the dinitrogen (N_2) into volatile ammonia, NH_3, by the process of nitrogen fixation. The ammonia is further protonated to ammonium, NH_4^+, which is uptaken by plants to manufacture amino acids for growth. When plants decompose, the organic molecules present in the plant residues become a source of nitrogen and other nutrients to soil microbes. The microbes conduct mineralization (ammonification) to utilize the organic molecules as electron donors, acquiring energy and producing ammonium. In contrast, the transformation of organic nitrogen to ammonium is reversible through immobilization (assimilation) when the carbon to nitrogen ratio (C/N ratio) is high. Ammonium can also undergo nitrification to produce NO_3^- in aerobic environments. The NO_3^- can be utilized by plants or by other organisms in anaerobic environments as an electron acceptor, can leach out by dissolving in water, or be reduced to dinitrogen gas via denitrification processes in anaerobic conditions. During the denitrification and nitrification processes nitrous oxide, N_2O, is produced and escapes into atmosphere.

Anthropogenic Influences on the Soil Nitrogen Cycle

One major influence of human input is reduced biological nitrogen fixation. With excessive ammonium input, plants no longer need the symbiotic microbes to provide ammonium. As a result, the degree of symbiosis will be diminished. Furthermore, the excessive ammonium provides nutrients for nitrifiers to produce nitrate, causing an increase in the amount of leached nitrate. Another negative outcome of anthropogenic inputs is the increased emission of nitrous oxide into the atmosphere due to increased nitrification and denitrification from excessive ammonium inputs.

Nitrogen fixation

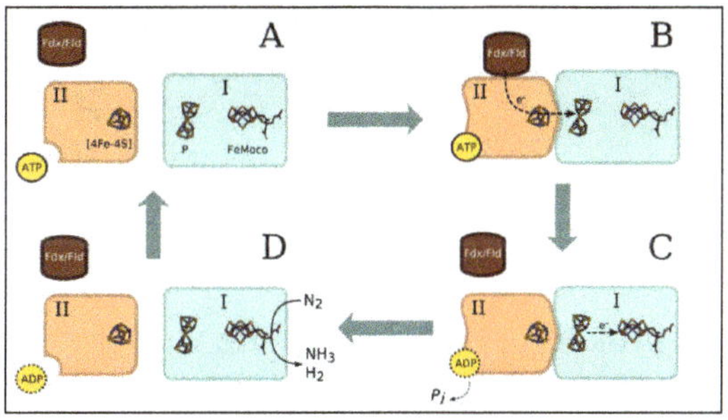

Nitrogenase Pathway.

Prokaryotes (bacteria) are responsible for the nitrogen fixation inside the soil. They mostly establish associative relationships with leguminous plants and other plant species. Yet, there are free-living bacteria like Azotobacter. All of the prokaryotes that are capable of nitrogen fixation have nitrogenase, which can fix nitrogen into ammonia. Many prokaryotes share the same enzyme complex through horizontal gene transfer via plasmids or evolutionary events.

Nitrogen Fixation Mechanism

The overall reaction catalyzed by nitrogenase is N_2 + 8H+ + $8e^-$ + 16 ATP = $2NH_3$ + H_2 + 16ADP + 16 Pi. In the process, nitrogen is used to replace a pair of hydrogen molecules on the nitrogenase complex, which has been heavily reduced along with protonation. The rest of the H+ will bond to the nitrogen and reform the structure of the molecule. Then, H+ continues to be added to the complex and binds with an NH group to form NH_3. Each complex is able to produce two NH3 molecules in each cycle of this process. Once NH_3 leaves the nitrogenase complex, it will be rapidly protonated to NH_4^+.

Mechanisms of creating Anaerobic Environment

Nitrogenase are inhibited by O_2, thus anaerobic environments are required for conducting nitrogen fixation. Free-living bacteria like *Azotobacter* have a couple of methods to overcome the aerobic condition. *Azotobacter* has a very high rate of respiratory metabolism so that it can keep the concentration of oxygen low within the cell in order to protect the nitrogenase enzyme. The second method adapted by cultured *Azotobacter* and *Rhizobium* species is to excrete large amounts of extracellular polymers that reduce the diffusion of oxygen. Leghemoglobin is a molecule in legume modules that are used to trap oxygen via chemical mechanisms that are similar to those used in hemoglobin. Other microorganisms, such as the cyanobacteria *Nostoc*, form special structures with thick cell walls that exclude oxygen and allow the cell to provide fixed nitrogen to its neighbors. Other cyanobacteria species have specific cells within their body that are only used for nitrogen fixation. In these cells, there is only photosystem I for ATP generation, whereas other cells have both photosystem I and II, which generate oxygen. Alternative nitrogenase that are not deactivated by oxygen do exist but are found in a relatively small number of organisms.

Key Microorganisms

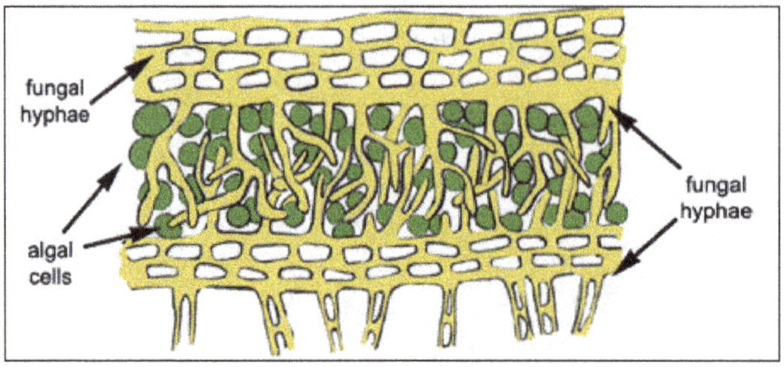

Lichen of Fungi and Algae.

The portion of nitrogen fixed by prokaryotes makes up the vast majority of all biologically active nitrogen on the planet. Diazotrophs are a category of all organisms that are capable of fixing nitrogen. Diazotrophs either coexist with plants via symbiosis or have the ability to live freely inside the soil. Azorhizobium, Bradyrhizobium, Rhizobium, Bradyrhizobium, Sinorhizobium, Mesorhizobium, and Allorhizobium are genera that have symbiotic relationships with leguminous plants. Diazotrophs modify leguminous root hairs into nodules to create an anaerobic environment and to provide an energy source in the form of organic carbon. In return, bacteria provide ammonium to the plant as a component for plant growth. There are also other genera that have different relationships with plants. For instance, Frankia has actinorrhizal symbiosis with a tree genus known as Casuarina. There are some genera, like Azotobacter, that live freely in aerobic environments and some, like Desulfovibrio, that live freely in anaerobic environments. A more unique example of a symbiotic relationship is that of lichens. Cyanobacteria, which is a photoautotrophic genus that fixes nitrogen, are protected by fungal hyphae. In return, the fungi receive a critical nutrient source in the form of ammonium.

Nitrogen Ammonification

Net nitrogen mineralization is commonly defined as the sum of concurrent ammonium production and consumption processes. However, in recent years, it has been shown that mineralization also is driven by depolymerization of nitrogen containing polymers available to use both by plants and microorganisms, and thus it is more correct to call the process ammonification or gross nitrogen mineralization.

The new paradigm regarding nitrogen cycling rethinks mineralization as the center point in the nitrogen cycle, and instead regards depolymerization of nitrogen containing polymers as what regulates the nitrogen cycle. Ammonification or gross nitrogen mineralization is the conversion of organic-nitrogen polymers to ammonium, mediated by heterotrophic microbes. The microbial degradation of amino acids, sugars, and nucleic acids is seen as a need for energy and carbon, with ammonium released as a by-product of catabolism. Since the polymers are not immediately bioavailable due to their large size, they are first cleaved by extracellular enzymes.

The extracellular enzymes are created by microorganisms, and have the capacity to depolymerize proteins, aminopolysaccharides found in cell walls, nucleic acids, and to hydrolyze urea. Proteins are broken down by a wide variety of proteinases (large proteins), proteases, and peptidases that cleave dipeptides or split individual amino acids. Some examples of isolated proteolytic enzymes are subtilism, clostripain, and thermolysin.

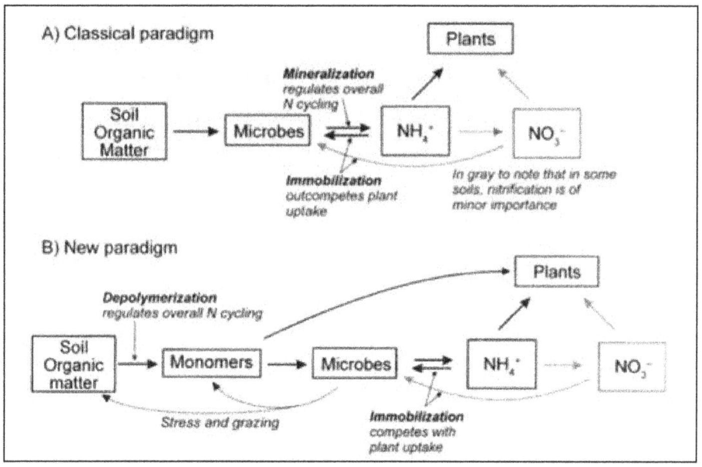

Change in Paradigm Concerning the Nitrogen Cycle.

Several common extracellular polymers degrade cell wall polymers, producing an end product of individual amino sugar monomers. In fungi which contain primarily chitin cell walls, chitinase degrades the cell walls into dimers, and then chitobiase breaks down each dimer into N-acetylglucosamine. In bacteria containing peptidoglycan cell walls, several enzymes, the most predominant of which is lysozyme, breaks the beta linkages holding the polymer together.

Nucleic acids are degraded by RNases and DNases, which hydrolyze ester ester bonds between phosphate and pentose sugars that make up the backbone of DNA and RNA, resulting in an end product of individual nucleotides. Ureases hydrolyze urea into carbon dioxide and ammonium, which plays a key role in immediate fertilization plant availability. The cofactor of the urease enzyme is nickel.

The research on extracellular enzymes and soil constituents has revealed some insight into the complexity of the soil matrix. An enzyme and substrate may bind to clay surfaces, protecting them from degradation and resulting in a slow degradation of the substrate over time. In addition, an enzyme can remain active and act on an available substrate if its catalytic site is unaffected and structure unaltered by adsorption to the clay surface. Intracellular enzymes play a critical role in nitrogen transformations. In most cases, the final product of ammonium occurs as a result of intracellular enzymes within microbial cells. There are two types of nitrogen found in amino acids: amine and amide. In the amide, the asparagine and glutamate are cleaved by aspariginase and glutaminase. In the amine, amino acid dehydrogenases and oxidases release the amino nitrogen in a process called deamination.

Amino sugars are metabolized in two steps - kinase phosphorylation followed by deamination. The degradation of nucleotides and subsequent ammonium release involves multiple steps - nucleotides are hydrolyzed to produce nucleosides and phosphate (PO_4^{3-}), the nucleosides are then further hydrolyzed to purines and pyrimidine bases and pentose sugar molecule parts, and finally the ammonium is released during the catabolism of the purines and pyrimidines with a urea intermediate.

Key Microorganisms

The groups of microorganisms responsible for ammonification are very broad and cover a wide range of species. Organoheterotrophs are an example of a group of microorganisms that mineralize nitrogen, often going after carbon in addition to the nitrogenous compounds present in the plant residue.

Nitrogen Immobilization

Immobilization is the conversion of ammonium and nitrate to organic nitrogen, primarily as a result of the assimilation of ammonium by the microbial biomass. The process requires energy for the conversion of nitrate to ammonium and subsequent incorporation of ammonium into amino acids. Microorganisms assimilate ammonium by two secondary pathways - the end result of both pathways is glutamate, which is then transferred to other carbon skeletons by transaminase reactions that form amino acids.

The first pathway occurs under a high concentration of ammonium (about 0.5 milligrams of nitrogen per kilogram of soil), in which $NADPH_2$ coenzymes adds ammonium to alpha ketoglutarate in order to form glutamate.

The second pathway occurs in lower concentrations of ammonium and uses a glutamine synthetase-glutamate synthetase (GOGAT) pathway in which ATP is used to add ammonium to glutamate to form glutamine, and in combination with alpha ketoglutarate, to subsequently remove the ammonium to form two glutamate molecules.

The Equilibrium between Ammonification and Immobilization

Ammonification and nitrogen immobilization are co-occuring processes, with ammonification occurring in the presence of a C/N ratio of less than 25.There are several factors that influence whether there is net production or consumption of nitrogen by microorganisms in soil, including factors influencing chemical pathway rates, substrate availability and composition, and the chemical composition of the soil matrix.

A very broad rule of thumb that estimates the direction of the equilibrium between ammonification and immobilization is the carbon to nitrogen ratio of the substrate. Various factors other than the C/N ratio may alter the equilibrium, such as lignin content, the moisture of the material being broken down, and the pH of the substrate and surrounding soil matrix.

These factors also alter the decomposition rate of the materials. However another broad rule of thumb is that the less carbon there is, the more quickly that the material will decompose and the nitrogenous compounds be made available.

The table below presents a few examples of C/N ratios.

Material	C/N Ratio
Wheat Straw	80:1
Corn Stover	57:1
Flowering Rye Cover Crop	37:1
Vegetative Rye Cover Crop	26:1
Mature Alfalfa Hay	25:1
Ideal Microbial Diet	24:1
Rotted Barnyard Manure	20:1
Legume Hay	17:1
Hairy Vetch Cover Crop	11:1
Young Alfalfa Hay	13:1
Soil Microbes (Average)	8:1

At C/N ratios above the ideal microbial diet (24:1) immobilization occurs due to the fact that there is a much greater amount of carbon that needs to be broken down – a process which requires more microbial biomass. Since the microbial biomass cannot retrieve the nitrogenous compounds until the carbon is broken down, the microorganisms convert inorganic nitrogen to organic nitrogen as an investment for their own growth.

Once the material is decomposed and there is an excess of microbial biomass, the microbial population begins to die off and return to equilibrium. The dead biomass, with a C/N ratio of 8:1, can be quickly utilized by the remaining biomass as a food and energy source or as a mineralization source for plant symbiosis.

C/N ratios below the ideal microbial diet cause mineralization due to the fact that there is an excess of organic nitrogenous compounds as a result of rapid substrate decomposition. The microbial biomass is not limited in growth by nitrogenous needs and organic nitrogen is converted to inorganic nitrogen for utilization for the purposes of plant growth and symbiosis.

Most organic matter is 45% carbon by mass, and thus the C/N ratio is determined mainly by the nitrogen concentration found in the residues. In agricultural systems, it is critical to properly determine and control the nitrogen concentration in the applied crop residue in order to influence the equilibrium in a manner that is conducive to the goals of the agricultural system.

The net result of the equilibrium is influenced by biotic factors as well - 30% of yearly net mineralization is directly released by grazing and activities of soil animals such as protozoa and nematodes. The same factors that influence the equilibrium in relation to microorganisms – soil moisture, pH, and the entirety of the soil matrix also have large roles in determining the activity of higher biotic agents and how much nitrogen they mineralize.

Key Microorganisms

Heterotrophic microbes and other organisms assimilate ammonium to build up biomass. As in ammonification, microorganisms responsible for immobilization encompass a very broad group due to universal need for nitrogen for life.

Nitrification

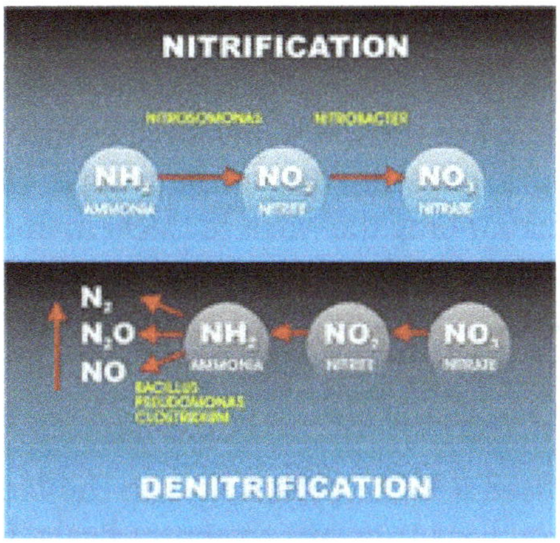

Nitrification/Denitrifcation.

Nitrification is a two-step process of ammonium (NH_4) to nitrate (NO_3) by soil bacteria. Ammonium is initially oxidized to nitrite by chemoautotrophs in the following manner:

$$NH_3 + O_2 \rightarrow NO_2^- + H^+ + H_2O$$

Chemoautotrophs are able to use carbon dioxide as a carbon source for the oxidation of ammonium. The most common bacteria that oxidizes ammonium is *Nitrosomonas*. The second step of the nitrification process is the conversion of nitrite to nitrate:

$$NO_2^- + O_2 \rightarrow 2NO_3^- + 2H^+ + 2e^-$$

The most common group of bacteria that convert nitrite to nitrate are in the genus *Nitrobacter*. This group of soil bacteria obtain their energy from the ammonium oxidation process. Nitrite can be toxic to certain plants, thus the incomplete conversion of nitrite to nitrate may prevent plant growth in those areas. It is important to ensure that nitrification occurs efficiently in food production systems as to prevent crop damage.

Denitrification

Denitrification is important in the continuation of the nitrogen cycle, as it is the step where gaseous dinitrogen is released back into the atmosphere. The natural cycle of denitrification comprises a series of enzymes such as bacterial NORs enzymes reduce nitrate to dinitrogen. As a result, reduction of nitrates to gaseous dinitrogen follows a series of respiratory biochemical reactions by microorganisms. Bacteria involved in denitrification are facultative anaerobic organoheterotrophs such as *Bacillus*. The overall process of denitrification is as follows:

$$NO_3^- \longrightarrow NO_2^- \longrightarrow N_2O \longrightarrow NO \longrightarrow N_2$$

Nitrate Nitrite⟶Nitric Oxide⟶Nitrous Oxide⟶Dinitrogen gas

Environmental Impacts

$$N_2 + O_2 \xrightarrow{\text{yields}} 2NO$$

$$2NO + O_2 \xrightarrow{\text{yields}} 2NO_2$$

$$NO_2 + hv \xrightarrow{\text{yields}} NO + O$$

$$O + O_2 \xrightarrow{\text{yields}} O_3$$

The nitrogen cycle causes many environmental issues such as global warming and pollution itself. The global increase in temperature is mainly due to the release of greenhouse gases. Greenhouse gases causes the blanket of air surrounding the earth to thicken, thus warming up the earth. The by-products of the nitrogen cycle produce greenhouse effects, specifically in agricultural fields. Farmers input high amounts of ammonium nitrate to fertilize their soil to grow their crops. Consequently this increases the production of pollutants: nitrous oxide (N_2O) and nitric oxide (NO), through denitrification. CO_2, another greenhouse gas, would also be produced from aerobic respiring microbes.

Excess nitrogen fixation has caused a lot of pollution into the biosphere. For example, nitrogen dioxide is degraded through photolysis to form oxygen radicals, which will react with molecular oxygen to form ozone in the troposphere, as shown in Figure. While ozone is beneficial in the stratosphere, where it can block UV radiation, ozone in the troposphere is a pollutant and can cause oxidative stress to humans and other organisms.

Mineralized nitrogen compounds in soil can be nitrified into nitrates, which can be leached into groundwater and cause contamination. Nitrate is soluble in water due to its negative charge; this polarization prevents adsorption onto the negatively charged soil surface and allows the soluble nitrate to leach into groundwater. Nitrate pollution is a major concern because of its many adverse effects such as methemoglobinemia, cancer, and eutrophication. Nitrification not only produces nitrates but also produces nitric acid, which will acidify bodies of water and harm sensitive marine life. Excessive nitrogen has also negatively affected terrestrial life by increasing the prevalence of pathogens, such as bacteria, viruses, and fungi. These pathogens were once limited by the lack of nitrogen but now have become more abundant.

Strategies to Lessen Environmental Impact

Some strategies to reduce mobile nitrogen in the biosphere are bioremediation, reduction of emissions, and sequestration. Bioremediation can reduce the amount of nitrate from the environment by converting it into atmospheric nitrogen and nitric oxide. However, this has its own concerns because nitric oxide production must be limited to lessen effects of global warming. Reduction of emissions reduces the amount of nitrogen entering the biosphere, thus lessening the amount of mobile nitrogen available. Sequestration of nitrogen prevents nitrogen from becoming mobile and can be increased by selecting for crops with higher nitrogen uptake efficiency that are also able to form symbiotic relationships with nitrogen sequestering soil microorganisms. There has been renewed interest in studying these relationships in the attempt to increase nitrogen sequestration in soil.

Oxygen and its Cycle

Oxygen (chemical symbol O, atomic number 8) is the second most common element on Earth and the third most common element in the universe. At ordinary temperatures and pressures, free oxygen (unbound to any other element) is a colorless, odorless, tasteless gas that makes up about 21% (by volume) of air. In combination with other elements, oxygen forms a variety of compounds, the most important of which is water. The Earth's oxygen continually cycles through the atmosphere, biosphere, and lithosphere, effected by such processes as photosynthesis and surface weathering.

Oxygen is essential for the respiratory function of humans, animals, plants, and some types of bacteria. If the human body is deprived of oxygen for more than a few minutes, the person's brain, heart, and other organs will suffer damage, leading to unconsciousness and death. On the other hand, relatively high concentrations of oxygen, if breathed at relatively high pressures or for prolonged periods, can have toxic effects.

Oxygen is the most common component of the Earth's crust (46.6% by mass), the second most common component of the Earth as a whole (28.2% by mass), and the second most common component of the Earth's atmosphere (20.947% by volume). Most of the oxygen is bonded to other elements.

Unbound oxygen (called molecular oxygen or molecular dioxygen O_2) is thought to have first appeared in significant quantities on our planet during the Paleoproterozoic era (2500–1600 million years ago), produced by the metabolic action of early single-celled organisms classified as archaea and bacteria. This new presence of large amounts of free oxygen drove most of the organisms then living to extinction. The atmospheric abundance of free oxygen in later geological epochs up to the present has been driven largely by photosynthetic organisms—roughly three quarters by phytoplankton and algae in the oceans and one quarter by terrestrial plants.

Oxygen Cycle

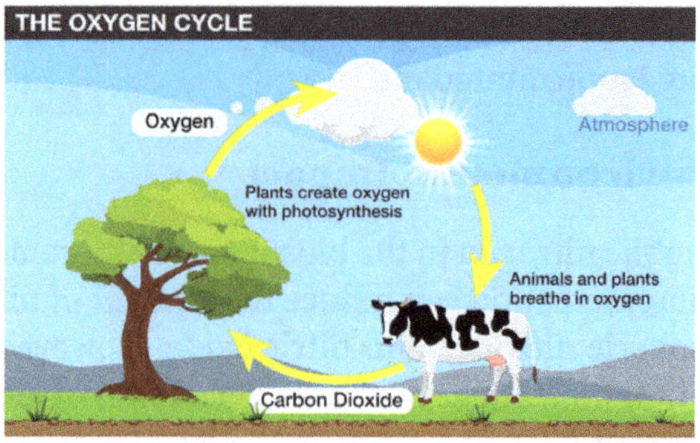

A Diagrammatic Representation of the Oxygen Cycle.

Oxygen cycle, along with the carbon cycle and nitrogen cycle plays an essential role in the existence of life on the earth. The oxygen cycle is a biological process which helps in maintaining the oxygen level by moving through three main spheres of the earth which are:

- Atmosphere
- Lithosphere
- Biosphere.

This biogeochemical cycle explains the movement of oxygen gas within the atmosphere, the ecosystem, biosphere and the lithosphere. The oxygen cycle is interconnected with the carbon cycle.

The atmosphere is the layer of gases presents above the earth's surface. The sum of all Earth's ecosystem makes a biosphere. Lithosphere, which is the solid outer section along with the Earth's crust and it is the largest reservoir of oxygen.

Stages of the Oxygen Cycle

The steps involved in the oxygen cycle are:

Stage-1: All green plants during the process of photosynthesis, release oxygen back into the atmosphere as a by-product.

Stage-2: All aerobic organisms use free oxygen for respiration.

Stage-3: Animals exhale Carbon dioxide back into the atmosphere which is again used by the plants during photosynthesis. Now oxygen is balanced within the atmosphere.

Uses of Oxygen

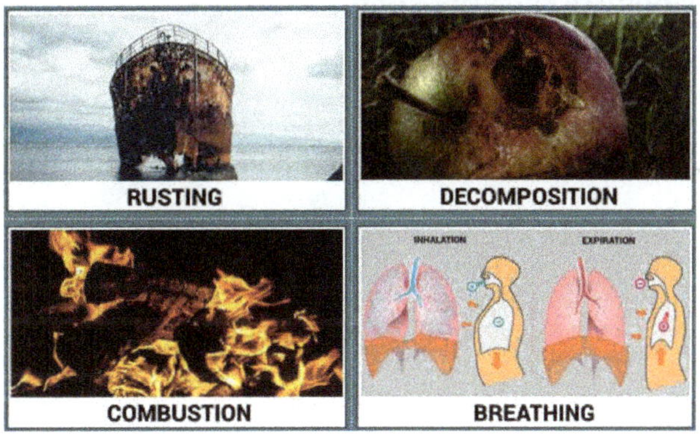

The four main processes that use Atmospheric oxygen are:

- Breathing: It is the physical process, through which all living organisms including plants, animals, and humans inhale oxygen from the outside environment into the cells of an organism and exhale carbon dioxide back into the atmosphere.
- Decomposition: It is one of the natural and most important processes in the oxygen cycle and occurs when an organism dies. The dead animal or plants decay into the ground, and the organic matter along with the carbon, oxygen, water and other components are returned back into the soil and air. This process is carried out by the invertebrates including fungi, bacteria and some insects which are collectively called as the decomposers. The entire process requires oxygen and releases carbon dioxide.
- Combustion: It is also one of the most important processes which occur when any of the organic materials including fossil fuels, plastics and wood, are burned in the presence of oxygen and releases carbon dioxide into the atmosphere.
- Rusting: This process also requires oxygen. It is the formation of oxides which is also called oxidation. In this process, metals like iron or alloy rust when they are exposed to moisture and oxygen for an extended period of time and new compounds of oxides are formed by the combination of oxygen with the metal.

Production of Oxygen

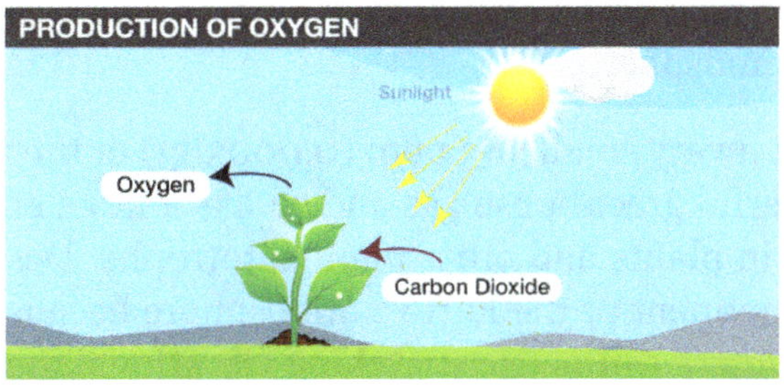

- Plants: The leading creators of oxygen are plants by the process of photosynthesis. Photosynthesis is a biological process by which all green plants synthesize their food in the presence of sunlight. During photosynthesis, plants use sunlight, water, carbon dioxide to create energy and oxygen gas is liberated as a by-product of this process.

- Sunlight: Sunlight also produces oxygen. Some oxygen gas is produced when the sunlight reacts with water vapour in the atmosphere.

Importance of Oxygen Cycle

As we all know, Oxygen is one of the most essential components of the Earth's atmosphere. It is mainly required for:

- Breathing
- Combustion
- Supports aquatic life
- Decomposition of organic waste.

Oxygen is an important element required for life, however, it can be toxic to some anaerobic bacteria (especially obligate anaerobes). The oxygen cycle is mainly involved in maintaining the level of oxygen in the atmosphere. The entire cycle can be summarized as, the oxygen cycle begins with the process of photosynthesis in the presence of sunlight, releases oxygen back into the atmosphere, which humans and animals breathe in oxygen and breathe out carbon dioxide, and again linking back to the plants. This also proves that both the oxygen and carbon cycle occur independently and are interconnected to each other.

Carbon Dioxide and its Cycle

Carbon dioxide is a chemical compound that is found as a gas in the Earth's atmosphere. It consists of simple molecules, each of which has one carbon and two oxygen atoms. Thus its chemical formula is CO_2. It is currently at a concentration of approximately 385 parts per million (ppm) by volume in the Earth's atmosphere.

It is a major component of the carbon cycle. In general, it is exhaled by animals and used for photosynthesis by growing plants. Additional carbon dioxide is created by the combustion of fossil fuels or vegetable matter, as well as other chemical processes. It is an important greenhouse gas because of its ability to absorb many infrared wavelengths of the Sun's light, and because of the length of time it stays in the Earth's atmosphere.

Carbon dioxide is present at a very small 383 ppm (0.000383) of the volume of the earth's atmosphere, but it is a very powerful greenhouse gas and so has a large effect upon climate. It is also essential to photosynthesis in plants and other photoautotrophs. Despite the low concentration, CO_2 is a very important component of the Earth's atmosphere because it absorbs infrared radiation at wavelengths of 4.26 μm (asymmetric stretching vibrational mode) and 14.99 μm (bending vibrational mode) and enhances the greenhouse effect to a great degree.

Although water vapour accounts a substantial fraction of the greenhouse effect, there is no real way to control the amount of water vapor in the Earth's climate system and it is short-lived in the atmosphere. In addition, water vapor is almost never considered a forcing, but rather almost always a feedback. On the other hand, carbon dioxide is a very powerful forcing, and it also lasts far longer in the Earth's atmosphere. With a radiative forcing of about 1.5 W/m^2, it is relatively twice as powerful as the next major forcing greenhouse gas, methane, and relatively ten times as powerful as the third, nitrous oxide. Carbon dioxide alone contributes up to 12 percent to the greenhouse effect.

The 20-year smoothed Law Dome DE02 and DE02-2 ice cores show the levels of CO_2 to have been 284 ppm in 1832. As of January 2007, the measured atmospheric CO_2 concentration at the Mauna Loa observatory was about 383 ppm. Of this 99 ppm rise in 175 years, 70 ppm of it has been in the last 47 years.

Various methods of limiting or removing the amount of carbon dioxide in the atmosphere have been suggested. Current debate on the subject mostly involves economic or political matters at a policy level.

Carbon Cycle

The carbon cycle is the cycle by which carbon moves through our Earth's various systems. The carbon cycle is influenced by living things atmospheric changes, ocean chemistry, and geologic activity is all part of this cycle. The levels of carbon are at an all-time high, largely due to human activities.

Carbon is an essential element for life as we know it because of its ability to form multiple, stable bonds with other molecules. This is why nucleotides, amino acids, sugars, and lipids all depend on carbon backbones: carbon provides a stable structure that allows the chemistry of life to happen. Without carbon, none of these molecules could exist and function in the ways that permit the chemistry of life to occur.

As a gas, carbon largely takes the form of carbon dioxide. Carbon dioxide is released by organisms as they break down by glucose. Autotrophic organisms like plants use carbon dioxide and sunlight to create glucose. However, carbon dioxide is also released by decaying organic matter, geological processes, and the burning of fossil fuels. Excess carbon dioxide is largely absorbed by the ocean, which leads to ocean acidification and may have been responsible for several mass extinctions.

Carbon Cycle Steps

Carbon in the Atmosphere

To become part of the carbon cycle, carbon atoms start out in a gaseous form. Carbon dioxide gas – CO_2 – can be produced by inorganic processes, or by the metabolisms of living things. Before Earth had life on it, carbon dioxide gas likely came from volcanic activity and asteroid impacts. Today, carbon is also released into the atmosphere through the activities of living things, such as the exhalations of animals, the actions of decomposer organisms, and the burning of wood and fossil fuels by humans. However carbon dioxide gets into the atmosphere, CO_2 gas is the starting point of the carbon cycle.

Producers absorb Carbon

"Producers" – organisms that produce food from sunlight, such as plants – absorb carbon dioxide from the atmosphere and use it to build sugars, lipids, proteins, and other essential building blocks of life.

For plants, CO_2 is absorbed through pores in their leaves called "stomata." Carbon dioxide enters the plant through the stomata and is incorporated into containing carbon compounds with the help of energy from sunlight. Plants and other producer organisms such as cyanobacteria are crucial to life on Earth because they can turn atmospheric carbon into living matter.

Producers are Eaten

"Consumers" are organisms that eat other living things. Animals are the most visible type of consumer in our ecosystems, though many types of microbes also fall into this category.

Consumers incorporate carbon compounds from plants and other food sources when they eat them. They use some of these carbon compounds from food to build their own bodies – but much of the food they eat is broken down to release energy, in a process that is almost the reverse of what producers do.

While producers use energy from sunlight to make bonds between carbon atoms – animals break these bonds to release the energy they contain, ultimately turning sugars, lipids, and other carbon compounds into single-carbon units. These are ultimately released into the atmosphere in the form of CO_2. But, what about the carbon compounds that don't get eaten, or broken down by animals.

Decomposers Release Carbon

Plants and animals that die without being eaten by other animals are broken down by other organisms, called "decomposers." Decomposers include many bacteria and some fungi. They usually only break down matter that is already dead, rather than catching and eating a living animal or plant.

Just like animals, decomposers break down the chemical bonds in their food molecules. They create many chemical products, including in some cases CO_2.

Human Activities

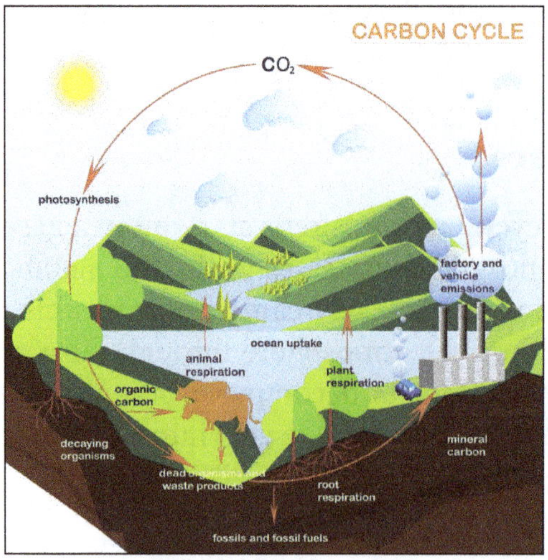

Carbon Cycle Diagram.

Recently, humans have made some big changes to the Earth's carbon cycle. By burning huge amounts of fossil fuels and cutting down roughly half of the Earth's forests, humans have decreased the Earth's ability to take carbon out of the atmosphere, while releasing large amounts of carbon into the atmosphere that had been stored in solid form as plant matter and fossil fuels.

This means more carbon dioxide in Earth's atmosphere – which is particularly dangerous since carbon dioxide is a "greenhouse gas" that plays a role in regulating the Earth's temperature and weather patterns.

The scientific community has raised alarms that by making significant changes to the Earth's carbon cycle, we may end up changing our climate or other important aspects of the ecosystem we rely upon to survive. As a result, many scientists advocate decreasing the amount of carbon burned by humans by reducing car use and electricity consumption, and advocate for investing in non-burning sources of energy such as solar power and wind power.

Carbon Cycle Examples

The carbon cycle consists of many parallel systems which can either absorb or release carbon. Together, these systems work to keep Earth's carbon cycle – and subsequently its climate and biosphere – relatively stable. Below are some examples of parts of Earth's ecosystems that can absorb carbon, turn carbon into living matter, or release carbon back into the atmosphere.

Atmosphere

One major repository of carbon is the carbon dioxide in the Earth's atmosphere. Carbon forms a stable, gaseous molecule in combination with two atoms of oxygen. In nature, this gas is released by volcanic activity, and by the respiration of animals who affix carbon molecules from the food they eat to molecules of oxygen before exhaling it.

Carbon dioxide can be removed from the atmosphere by plants, which take the atmospheric carbon and turn it into sugars, proteins, lipids, and other essential molecules for life. It can also be removed from the atmosphere by absorption into the ocean, whose water molecules can bond with carbon dioxide to form carbonic acid.

Lithosphere

The Earth's crust – called the "lithosphere" from the Greek word "litho" for "stone" and "sphere" for globe – can also release carbon dioxide into Earth's atmosphere. This gas can be created by chemical reactions in the Earth's crust and mantel.

Volcanic activity can result in natural releases of carbon dioxide. Some scientists believe that widespread volcanic activity may be to blame for the warming of the Earth that caused the Permian extinction.

While the Earth's crust can add carbon to the atmosphere, it can also remove it. Movements of the Earth's crust can bury carbon-containing chemicals such as dead plants and animals deep underground, where their carbon cannot escape back into the atmosphere. Over millions of years, these underground reservoirs of organic matter liquefy and become coal, oil, and gasoline. In recent years, humans have begun releasing much of this sequestered carbon back into the atmosphere by burning these materials to power cars, power plants, and other human equipment.

Biosphere

Among living things, some remove carbon from the atmosphere, while others release it back. The most noticeable participants in this system are plants and animals.

Plants remove carbon from the atmosphere. They don't do this as a charitable act; atmospheric carbon is actually the "food" which plants use to make sugars, proteins, lipids, and other essential molecules for life. Plants use the energy of sunlight, harvested through photosynthesis, to build these organic compounds out of carbon dioxide and other trace elements.

In a gracefully balanced set of chemical reactions, animals eat plants (and other animals), and take these synthesized molecules apart again. Animals get their fuel from the chemical energy plants have stored in the bonds between carbon atoms and other atoms during photosynthesis.

In order to do that, animal cells dissemble complex molecules such as sugars, fats, and proteins all the way down to single-carbon units – molecules of carbon dioxide, which are produced by reacting carbon-containing food molecules with oxygen from the air.

Oceans

The Earth's oceans have the ability to both absorb and release carbon dioxide. When carbon dioxide from the atmosphere comes into contact with ocean water, it can react with the water molecules to form carbonic acid – a dissolved liquid form of carbon.

When there is more carbonic acid in the ocean compared to carbon dioxide in the atmosphere, some carbonic acid may be released into the atmosphere as carbon dioxide. On the other hand, when there is more carbon dioxide in the atmosphere, more carbon dioxide will be converted to carbonic acid, and ocean acidity levels will rise.

Some scientists have raised concerns that acidity is rising in some parts of the ocean, possibly as a result of increased carbon dioxide in the atmosphere due to human activity. Although these changes in ocean acidity may sound small by human standards, many types of sea life depend on chemical reactions that need a highly specific acidity level to survive. In fact, ocean acidification is currently killing many coral reef communities.

Importance of the Carbon Cycle

The carbon cycle, under normal circumstances, works to ensure the stability of variables such as the Earth's atmosphere, the acidity of the ocean, and the availability of carbon for use by living things. Each of its components is of crucial importance to the health of all living things – especially humans, who rely on many food crops and animals to feed our large population.

Carbon dioxide in the atmosphere prevents the sun's heat from escaping into space, very much like the glass walls of a greenhouse. This isn't always a bad thing – some carbon dioxide in the atmosphere is good for keeping the Earth warm and its temperature stable.

But Earth has experienced catastrophic warming cycles in the past, such as the Permian extinction, which is thought to have been caused by a drastic increase in the atmosphere's level of greenhouse gases. No one is sure what caused the change that brought about the Permian extinction. But, greenhouse gases may have been added to an atmosphere by an asteroid impact, volcanic activity, or even massive forest fires.

Whatever the cause, during this warming episode temperatures rose drastically. Much of the Earth became desert, and over 90% of all species living at that time went extinct. This is a good example of what can happen if our planet's essential cycles experience a big change.

Water Vapour and its Cycle

Water vapor (H_2O) is found in the atmosphere in small and highly variable amounts. While it is nearly absent in most of the atmosphere, its concentration can range up to 4% in very warm, humid areas close to the surface. Despite its relative scarcity, atmospheric water probably has more of an impact on the earth than any of the major gases, aside from oxygen. Water vapor participates in the hydrologic cycle, the process that moves water between the oceans, the land surface waters, the atmosphere, and the polar ice caps. This water cycling drives erosion and rock weathering, determines the earth's weather, and sets up climate conditions that make land areas dry or wet, habitable or inhospitable. When cooled sufficiently, water vapor forms clouds by condensing to liquid water droplets, or at lower temperatures, solid ice crystals. Besides creating rain or snow, clouds affect Earth's climate by reflecting some of the energy coming from the sun, making the planet somewhat cooler. Water vapor is also an important greenhouse gas. It is concentrated near the surface and is much more prevalent near the tropics than in the polar region.

Hydrological Cycle

Hydrological cycle is the cyclic movement of water containing basic continuous processes like evaporation, precipitation and runoff as Runoff –> Evaporation –> Precipitation –> Runoff. This is a continuous cycle which starts with evaporation from the water bodies such as oceans.

Components of Hydrological Cycles

The components of hydrological cycle are:

- Runoff: It is the water flowing over the land making its way towards rivers, lakes, oceans etc. as surface or subsurface flow.
    - Surface runoff: It is the running water over the land and which ultimately discharge water to the sea.
    - Subsurface runoff: The water getting infiltrated into pervious soil mass, making its way towards rivers and lakes can be termed as subsurface runoff.
- Precipitation: It is the fall of moisture from atmosphere to the earth's surface in any form. Example: rain, hail, snow, sleet, glaze, drizzle, snowflakes.
- Evaporation: It is the conversion of natural liquids like water into gaseous form like air.
- Condensation: It is the conversion of a vapor or gas to a liquid.

- Transpiration: It is the evaporation taking place from any plant or greenery. Example, water droplet on a leaf getting evaporated into atmosphere.
- Evapotranspiration: It is the combination of evaporation and transpiration.
- Infiltration: It is the process of filtration of water to the inner layers of soil based on its structure and nature. Pervious soils go through more infiltration than impervious. Infiltration in soils like sand, gravel and coarser material is more and for finer soil particles like clay and silt, infiltration is less.
- Infiltration is inversely proportional to runoff. In a soil, if infiltration is less, then runoff is more, similarly more infiltration gives less runoff. Example: bitumen roads has more runoff than metallic red mud roads.

Process of Hydrological Cycle

Process of hydrological cycle starts with oceans. Water in oceans, gets evaporated due to heat energy provided by solar radiation and forms water vapor. This water vapor moves upwards to higher altitudes forming clouds.

Most of the clouds condense and precipitate in any form like rain, hail, snow, sleet. And a part of clouds is driven to land by winds. Precipitation, while falling to the ground, some part of it evaporates back to atmosphere. Portion of water that reaches the ground, enters the earth's surface infiltrating various strata of soil and enhancing the moisture content as well as water.

Vegetation sends a portion of water from earth's surface back to atmosphere through the process of transpiration. Once water percolates and infiltrates the earth's surface, runoff is formed over the land, flowing through the contours of land heading towards river and lakes and finally joins into oceans after many years. Some amount of water is retained as depression storage.

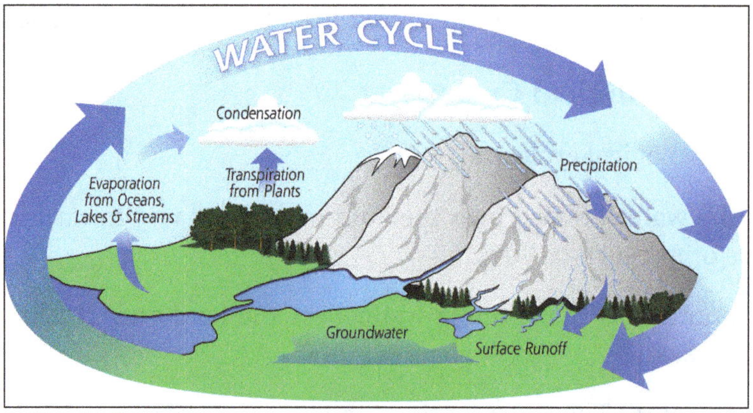

Further again the process of this hydrological cycle continues by blowing of cool air over ocean, carrying water molecules, forming into water vapor then clouds getting condensed and precipitates as rainfall. Similarly, then water gets percolate into soil, increasing water table then formation of runoff waters heading towards water bodies. Thus the cyclic process continues.

- Depression Storage: It is the part of precipitation required to fill depression zones of land.

- Interception: Part of precipitation required to wet the surface of soil, buildings and all pervious surfaces.

Water Balance Equation

Sum of inflow waters = sum of outflow waters

Out of three processes like precipitation, runoff and evaporation, Inflow is precipitation. Runoff and evaporation comes under outflow, however it can be written as:

Precipitation – runoff = Evaporation

That gives,

Precipitation (P) = Evaporation (E) + Runoff (R).

Water Reservoirs and Residence Times

Water reservoirs represent a region where part of the water participating in the water cycle is stored for a certain period of time. Oceans are the largest water reservoirs on the planet, storing nearly 97% of the waters of the hydrosphere, while the ice caps and glaciers store another 2%. Underground water reservoirs, rivers, lakes, ponds, and streams store small percentages of the total water of the hydrosphere as well, while the water content found in living organisms represents the smallest of all reservoirs. Another important term associated with the water cycle is the "residence time", which is represented by the following mathematical formula:

Residence Time = Volume of reservoir/(The rate water is added to the reservoir or leaves the reservoir)

The residence time thus represents the average time a water molecule spends in the reservoir, and can be expressed either positively or negatively depending upon whether the reservoir is experiencing net losses or net gains of water. For example, groundwater can reside for 10,000 years under the Earth's surface before leaving the underground reservoir, while atmospheric water exists for a maximum of 10 days in the atmosphere before precipitating as rain or snow.

Importance of the Water Cycle and the Impact of Human Activities

Global climates are affected by, and extremely sensitive to, changes in the water cycle's patterns, as the cycle allows for the exchange of heat and moisture between landmasses and water bodies. Evaporation of water leads to the cooling of the environment, while condensation warms the environment by releasing heat energy. The physical geography of the Earth is also highly influenced by the water cycle, as the meltdown of glaciers and runoff from rivers carves out valleys, peaks, canyons, lakes, and other landforms seen on Earth. Recently, the water cycle of the planet has intensified, and the rates of evaporation and precipitation have greatly increased. Human activities, such as the damming of rivers and streams, the extraction of surface and underground water for irrigation and other purposes, and extensive deforestation, have adversely affected the functioning of the earth's water cycle. Global warming has further impacted the hydrosphere by triggering the melting of polar ice caps, which are now losing more water by evaporation, snow-melt and runoffs than they are gaining water by precipitation. This threatens to raise the oceans' water levels and flood coastal cities around the world.

Sulphur Dioxide and its Cycle

Sulphur dioxide enters the atmosphere as a results of both natural phenomena and anthropogenic activities, such as fossil fuel combustion, oxidation of organic materials in soils, volcanic eruptions and biomass burning.

Coal burning is the single largest man-made source of sulphur dioxide, accounting for about 50% of annual global emissions, with oil burning accounting for a further 25 to 30%. Sulphur dioxide reacts on the surface of a variety of airborne solid particles (aerosols), is soluble in water and can be oxidised within airborne water droplets, producing sulphuric acid. This acidic pollution can be transported by wind over many hundreds of kilometres, and is deposited as acid rain.

Changes in the abundance of SO_2 have an impact on atmospheric chemistry and on the radiation field, and hence on the climate. Consequently, global observations of SO_2 are important for atmospheric and climate research.

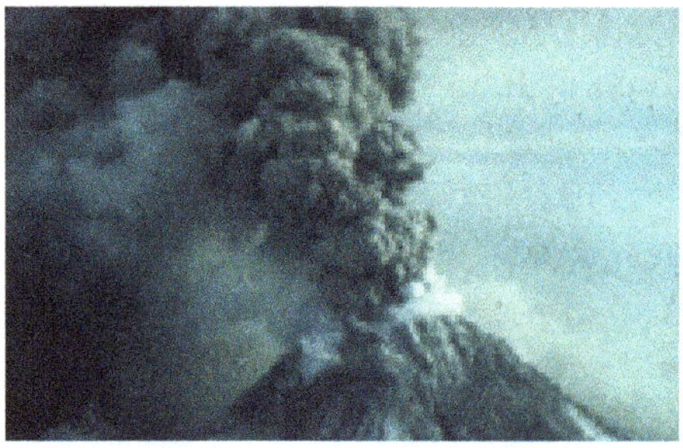

Sulphur Dioxide in Troposphere and Stratosphere

The lifetime of sulphur dioxide molecules in the troposphere is a few days. The amount is highly variable, above a low background concentration.

It is removed from the troposphere:

- In gas phase by formation of sulphuric acid (H_2SO_4), which forms condensation nuclei for aerosols and clouds and acidifies the rain;
- Directly, by way of an uptake on aerosols and clouds, which leads to dry and wet acid depositions.

Trees killed by acid rain in the Great Smoky Mountains.

Clean continental air contains less than 1 ppb of sulphur dioxide, which corresponds to a total column density less than 0.2 Dobson Units (DU) in a boundary layer of 2 km.

The lifetime of sulphur dioxide molecules in the stratosphere, on the other hand, is several weeks, during which it produces sulphate aerosols. This makes sulphur dioxide from volcanos one of the two most important sources of stratospheric aerosols.

Sulfur Cycle

Sulfur is an essential element for the macromolecules of living things, is released into the atmosphere by the burning of fossil fuels, such as coal. As a part of the amino acid cysteine, it is involved in the formation of disulfide bonds within proteins, which help to determine their 3-D folding patterns, and hence their functions. As shown in figure below, sulfur cycles between the oceans, land, and atmosphere. Atmospheric sulfur is found in the form of sulfur dioxide (SO_2) and enters the atmosphere in three ways: from the decomposition of organic molecules, from volcanic activity and geothermal vents, and from the burning of fossil fuels by humans.

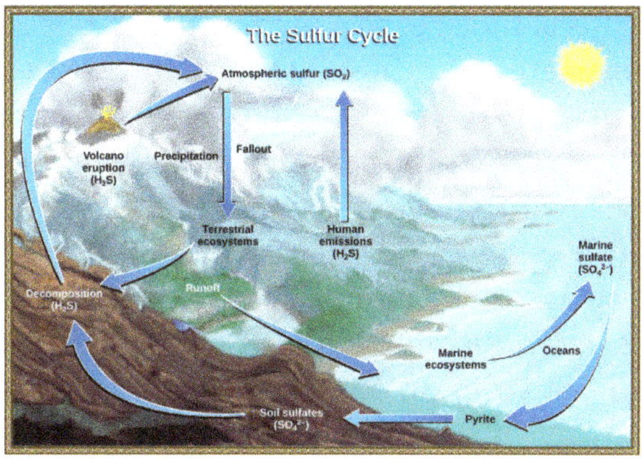

In figure, sulfur dioxide from the atmosphere becomes available to terrestrial and marine ecosystems when it is dissolved in precipitation as weak sulfuric acid or when it falls directly to the Earth as fallout. Weathering of rocks also makes sulfates available to terrestrial ecosystems. Decomposition of living organisms returns sulfates to the ocean, soil and atmosphere.

On land, sulfur is deposited in four major ways - precipitation, direct fallout from the atmosphere, rock weathering, and geothermal vents (Figure below). Atmospheric sulfur is found in the form of sulfur dioxide (SO_2), and as rain falls through the atmosphere, sulfur is dissolved in the form of weak sulfuric acid (H_2SO_4). Sulfur can also fall directly from the atmosphere in a process called fallout. Also, the weathering of sulfur-containing rocks releases sulfur into the soil. These rocks originate from ocean sediments that are moved to land by the geologic uplifting of ocean sediments. Terrestrial ecosystems can then make use of these soil sulfates (SO_4^-), and upon the death and decomposition of these organisms, release the sulfur back into the atmosphere as hydrogen sulfide (H_2S) gas.

Sulfur enters the ocean via runoff from land, from atmospheric fallout, and from underwater geothermal vents. Some ecosystems rely on chemoautotrophs using sulfur as a biological energy source. This sulfur then supports marine ecosystems in the form of sulfates.

At this sulfur vent in Lassen Volcanic National Park in northeastern California, the yellowish sulfur deposits are visible near the mouth of the vent.

Human activities have played a major role in altering the balance of the global sulfur cycle. The burning of large quantities of fossil fuels, especially from coal, releases larger amounts of hydrogen sulfide gas into the atmosphere. As rain falls through this gas, it creates the phenomenon known as acid rain. Acid rain is corrosive rain caused by rainwater falling to the ground through sulfur dioxide gas, turning it into weak sulfuric acid, which causes damage to aquatic ecosystems. Acid rain damages the natural environment by lowering the pH of lakes, which kills many of the resident fauna; it also affects the man-made environment through the chemical degradation of buildings. For example, many marble monuments, such as the Lincoln Memorial in Washington, DC, have suffered significant damage from acid rain over the years. These examples show the wide-ranging effects of human activities on our environment and the challenges that remain for our future.

References

- Atmospheric-chemistry: newworldencyclopedia.org, Retrieved 4 June, 2019
- Nitrogen: newworldencyclopedia.org, Retrieved 14 April, 2019
- Nitrogen-Cycle: kenyon.edu, Retrieved 24 July, 2019
- Oxygen: newworldencyclopedia.org, Retrieved 1 May, 2019
- Oxygen-cycle-environment, biology: byjus.com, Retrieved 11 August, 2019
- Carbon-dioxide: newworldencyclopedia.org, Retrieved 21 June, 2019
- Carbon-cycle: biologydictionary.net, Retrieved 31 January, 2019
- Atmosphere-Composition-Structure: jrank.org, Retrieved 2 July, 2019
- Hydrological-cycle-process-components, water-resources: theconstructor.org, Retrieved 22 February, 2019
- What-is-the-water-hydrologic-cycle: worldatlas.com, Retrieved 30 August, 2019
- The-sulfur-cycle, biology: lumenlearning.com, Retrieved 20 March, 2019

Chapter 4

Atmospheric Physics

The branch of atmospheric science which seeks to apply physics in order to study the atmosphere is called atmospheric physics. Some of the areas of study within this discipline are solar radiation, atmospheric temperature and atmospheric humidity. The topics elaborated in this chapter will help in gaining a better perspective about these focus areas of atmospheric physics.

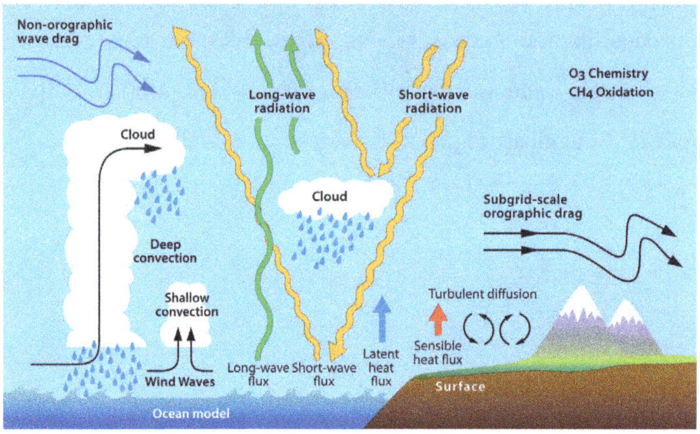

Atmospheric physics is the branch of meteorology that studies the physical regularities of processes and phenomena that occur in the atmosphere, including the processes and phenomena that determine the structure of the atmosphere. For example, atmospheric physics deals with the properties of the gases that constitute the atmosphere, the absorption and emission of radiation by the gases, the distribution of temperature and pressure, the evaporation and condensation of water vapor, the formation of clouds and precipitation, and the various forms of motion in the atmosphere.

Solar energy conversion and the thermal radiation of the atmosphere and underlying surface are studied by actinometry (in the broad sense of the term) and atmospheric optics. The latter also deals with various optical phenomena in the atmosphere, such as twilight, the colorful phenomena observed in the sky before sunrise and after sunset, halos, the color and polarization of the sky, and the visibility of objects. The study of atmospheric electricity is concerned with electrical phenomena in the atmosphere—that is, with lightning and other electrical discharges—and with the electrical properties of the atmosphere, such as conductivity, ionization, electric currents, space charges, and the charges of clouds and precipitation. Sound propagation and generation in the real atmosphere and the investigation of sound generation by acoustic methods are the subject of atmospheric acoustics. Atmospheric physics also encompasses cloud physics and the physics of the microprocesses that result in the formation of solid and liquid aerosols; these fields include the modification of atmospheric processes.

Atmospheric physics also studies the interaction of the atmosphere and the underlying surface—that is, the ocean or dry land—that occurs in the atmospheric boundary layer and results in the exchange of momentum, heat, and moisture. Turbulence in the atmosphere and hydrosphere plays a governing role in the interaction.

Processes in the upper atmosphere and the structure and dynamics of the upper atmosphere are investigated by the physics of the upper atmosphere or by aeronomy, a broader division of atmospheric physics that also studies the various chemical processes that occur in the upper atmosphere.

One of the fundamental problems of all the divisions of atmospheric physics is the development of a physical basis for the numerical modeling of various atmospheric processes. In this respect, the most important problem is parameterization, that is, the description of various small-scale processes by means of quantities that characterize average atmospheric conditions on larger scales that form the background against which the processes being studied develop. Such a description is necessary for computer simulations of atmospheric phenomena. For example, cumulus clouds, whose sizes are of the order of several kilometers, play an important role in atmospheric moisture and heat exchange, radiative transfer, and other processes. In numerical models, the effects of such clouds on radiation, heat exchange, and other atmospheric processes are parameterized, that is, expressed by means of temperature, wind, humidity, and other variables, which are given at specified points that form the three-dimensional grid of the numerical models; the distance between any two points is usually a few hundred kilometers.

Atmospheric physics also deals with the study of the atmospheres of the other planets, which contributes to a deeper understanding of the phenomena that occur in the earth's atmosphere.

Solar Radiation

Solar radiation, often called the solar resource, is a general term for the electromagnetic radiation emitted by the sun. Solar radiation can be captured and turned into useful forms of energy, such as heat and electricity, using a variety of technologies. However, the technical feasibility and economical operation of these technologies at a specific location depends on the available solar resource.

Every location on Earth receives sunlight at least part of the year. The amount of solar radiation that reaches any one spot on the Earth's surface varies according to:

- Geographic location
- Time of day
- Season
- Local landscape
- Local weather

Because the Earth is round, the sun strikes the surface at different angles, ranging from 0° (just above the horizon) to 90° (directly overhead). When the sun's rays are vertical, the Earth's surface gets all the energy possible. The more slanted the sun's rays are, the longer they travel through the atmosphere, becoming more scattered and diffuse. Because the Earth is round, the frigid polar regions never get a high sun, and because of the tilted axis of rotation, these areas receive no sun at all during part of the year.

The Earth revolves around the sun in an elliptical orbit and is closer to the sun during part of the year. When the sun is nearer the Earth, the Earth's surface receives a little more solar energy. The Earth is nearer the sun when it is summer in the southern hemisphere and winter in the northern hemisphere. However, the presence of vast oceans moderates the hotter summers and colder winters one would expect to see in the southern hemisphere as a result of this difference.

The 23.5° tilt in the Earth's axis of rotation is a more significant factor in determining the amount of sunlight striking the Earth at a particular location. Tilting results in longer days in the northern hemisphere from the spring (vernal) equinox to the fall (autumnal) equinox and longer days in the southern hemisphere during the other 6 months. Days and nights are both exactly 12 hours long on the equinoxes, which occur each year on or around March 23 and September 22.

Countries such as the United States, which lie in the middle latitudes, receive more solar energy in the summer not only because days are longer, but also because the sun is nearly overhead. The sun's rays are far more slanted during the shorter days of the winter months. Cities such as Denver, Colorado, (near 40° latitude) receive nearly three times more solar energy in June than they do in December.

The rotation of the Earth is also responsible for hourly variations in sunlight. In the early morning and late afternoon, the sun is low in the sky. Its rays travel further through the atmosphere than at noon, when the sun is at its highest point. On a clear day, the greatest amount of solar energy reaches a solar collector around solar noon.

Diffuse and Direct Solar Radiation

As sunlight passes through the atmosphere, some of it is absorbed, scattered, and reflected by:

- Air molecules
- Water vapor
- Clouds
- Dust
- Pollutants
- Forest fires
- Volcanoes

This is called diffuse solar radiation. The solar radiation that reaches the Earth's surface without being diffused is called direct beam solar radiation. The sum of the diffuse and direct solar radiation is called global solar radiation. Atmospheric conditions can reduce direct beam radiation by 10% on clear, dry days and by 100% during thick, cloudy days.

Distribution of Radiant Energy from the Sun

Nuclear fusion deep within the Sun releases a tremendous amount of energy that is slowly transferred to the solar surface, from which it is radiated into space. The planets intercept minute fractions of this energy, the amount depending on their size and distance from the Sun. A 1-square-metre (11-square-foot) area perpendicular (90°) to the rays of the Sun at the top of Earth's atmosphere, for example, receives about 1,365 watts of solar power. (This amount is comparable to the power consumption of a typical electric heater.) Because of the slight ellipticity of Earth's orbit around the Sun, the amount of solar energy intercepted by Earth steadily rises and falls by ±3.4 percent throughout the year, peaking on January 3, when Earth is closest to the Sun. Although about 31 percent of this energy is not used as it is scattered back to space, the remaining amount is sufficient to power the movement of atmospheric winds and oceanic currents and to sustain nearly all biospheric activity.

Most surfaces are not perpendicular to the Sun, and the energy they receive depends on their solar elevation angle. (The maximum solar elevation is 90° for the overhead Sun.) This angle changes systematically with latitude, the time of year, and the time of day. The noontime elevation angle reaches a maximum at all latitudes north of the Tropic of Cancer (23.5° N) around June 22 and a minimum around December 22. South of the Tropic of Capricorn (23.5° S), the opposite holds true, and between the two tropics, the maximum elevation angle (90°) occurs twice a year. When the Sun has a lower elevation angle, the solar energy is less intense because it is spread out over a larger area. Variation of solar elevation is thus one of the main factors that accounts for the dependence of climatic regime on latitude. The other main factor is the length of daylight. For latitudes poleward of 66.5° N and S, the length of day ranges from zero (winter solstice) to 24 hours (summer solstice), whereas the Equator has a constant 12-hour day throughout the year. The seasonal range of temperature consequently decreases from high latitudes to the tropics, where it becomes less than the diurnal range of temperature.

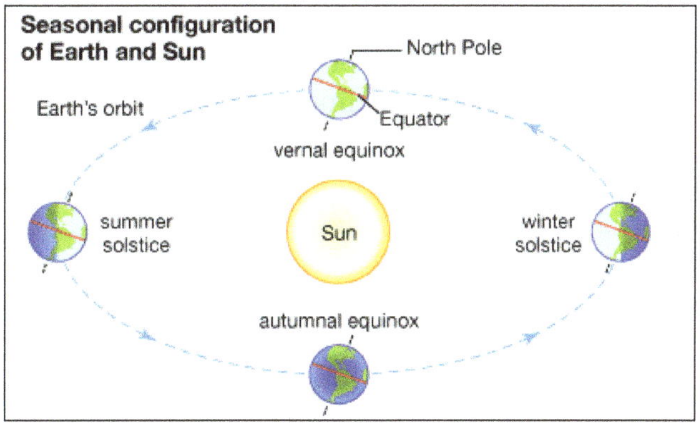

A diagram shows the position of Earth at the beginning of each season in the Northern Hemisphere.

Effects of the Atmosphere

Of the radiant energy reaching the top of the atmosphere, 46 percent is absorbed by Earth's surface on average, but this value varies significantly from place to place, depending on cloudiness, surface type, and elevation. If there is persistent cloud cover, as exists in some equatorial regions, much of the incident solar radiation is scattered back to space and very little is absorbed by Earth's surface.

Water surfaces have low reflectivity (4–10 percent), except in low solar elevations, and are the most efficient absorbers. Snow surfaces, on the other hand, have high reflectivity (40–80 percent) and so are the poorest absorbers. High-altitude desert regions consistently absorb higher-than-average amounts of solar radiation because of the reduced effect of the atmosphere above them.

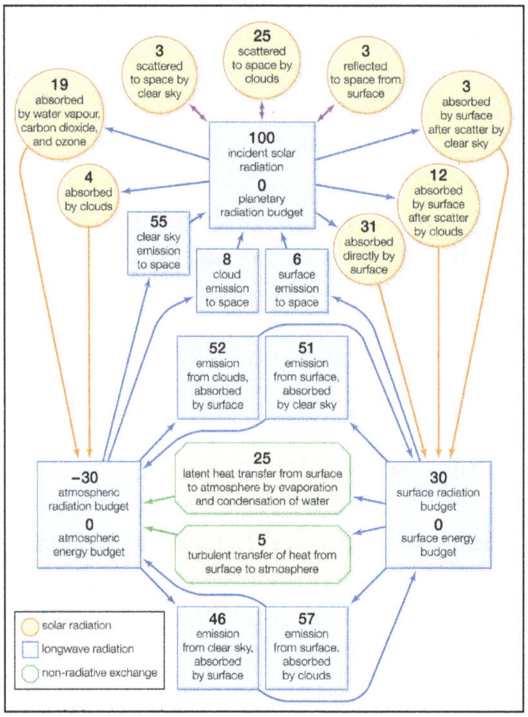

Average exchange of energy between the surface, the atmosphere, and space, as percentages of incident solar radiation (1 unit = 3.4 watts per square metre).

An additional 23 percent or so of the incident solar radiation is absorbed on average in the atmosphere, especially by water vapour and clouds at lower altitudes and by ozone (O3) in the stratosphere. Absorption of solar radiation by ozone shields the terrestrial surface from harmful ultraviolet light and warms the stratosphere, producing maximum temperatures of −15 to 10 °C (5 to 50 °F) at an altitude of 50 km (30 miles). Most atmospheric absorption takes place at ultraviolet and infrared wavelengths, so more than 90 percent of the visible portion of the solar spectrum, with wavelengths between 0.4 and 0.7 μm (0.00002 to 0.00003 inch), reaches the surface on a cloud-free day. Visible light, however, is scattered in varying degrees by cloud droplets, air molecules, and dust particles. Blue skies and red sunsets are in effect attributable to the preferential scattering of short (blue) wavelengths by air molecules and small dust particles. Cloud droplets scatter visible wavelengths impartially (hence, clouds usually appear white) but very efficiently, so the reflectivity of clouds to solar radiation is typically about 50 percent and may be as high as 80 percent for thick clouds. The constant gain of solar energy by Earth's surface is systematically returned to space in the form of thermally emitted radiation in the infrared portion of the spectrum. The emitted wavelengths are mainly between 5 and 100 μm (0.0002 and 0.004 inch), and they interact differently with the atmosphere compared with the shorter wavelengths of solar radiation. Very little of the radiation emitted by Earth's surface passes directly through the atmosphere. Most of it is absorbed by clouds, carbon dioxide, and water vapour and is then reemitted in all directions. The atmosphere thus acts as a radiative blanket over Earth's surface, hindering the loss of heat to space. The blanketing effect is greatest in the presence of low clouds and weakest for clear cold skies that contain little water vapour. Without this effect, the mean surface temperature of 15 °C (59 °F) would be some 30 °C colder. Conversely, as atmospheric concentrations of carbon dioxide, methane, chlorofluorocarbons, and other absorbing gases continue to increase, in large part owing to human activities, surface temperatures should rise because of the capacity of such gases to trap infrared radiation. The exact amount of this temperature increase, however, remains uncertain because of unpredictable changes in other atmospheric components, especially cloud cover. An extreme example of such an effect (commonly dubbed the greenhouse effect) is that produced by the dense atmosphere of the planet Venus, which results in surface temperatures of about 475 °C (887 °F). This condition exists in spite of the fact that the high reflectivity of the Venusian clouds causes the planet to absorb less solar radiation than Earth.

Average Radiation Budgets

The difference between the solar radiation absorbed and the thermal radiation emitted to space determines Earth's radiation budget. Since there is no appreciable long-term trend in planetary temperature, it may be concluded that this budget is essentially zero on a global long-term average. Latitudinally, it has been found that much more solar radiation is absorbed at low latitudes than at high latitudes. On the other hand, thermal emission does not show nearly as strong a dependence on latitude, so the planetary radiation budget decreases systematically from the Equator to the poles. It changes from being positive to negative at latitudes of about 40° N and 40° S. The atmosphere and oceans, through their general circulation, act as vast heat engines, compensating for this imbalance by providing nonradiative mechanisms for the transfer of heat from the Equator to the poles.

While Earth's surface absorbs a significant amount of thermal radiation because of the blanketing effect of the atmosphere, it loses even more through its own emission and thus experiences a net loss of long-wave radiation. This loss is only about 14 percent of the amount emitted by the surface and is less than the average gain of total absorbed solar energy. Consequently, the surface has on average a positive radiation budget.

By contrast, the atmosphere emits thermal radiation both to space and to the surface, yet it receives long-wave radiation back from only the latter. This net loss of thermal energy cannot be compensated for by the modest gain of absorbed solar energy within the atmosphere. The atmosphere thus has a negative radiation budget, equal in magnitude to the positive radiation budget of the surface but opposite in sign. Nonradiative heat transfer again compensates for the imbalance, this time largely by vertical atmospheric motions involving the evaporation and condensation of water.

Surface-energy Budgets

The rate of temperature change in any region is directly proportional to the region's energy budget and inversely proportional to its heat capacity. While the radiation budget may dominate the average energy budget of many surfaces, nonradiative energy transfer and storage also are generally important when local changes are considered.

Foremost among the cooling effects is the energy required to evaporate surface moisture, which produces atmospheric water vapour. Most of the latent heat contained in water vapour is subsequently released to the atmosphere during the formation of precipitating clouds, although a minor amount may be returned directly to the surface during dew or frost deposition. Evaporation increases with rising surface temperature, decreasing relative humidity, and increasing surface wind speed. Transpiration by plants also increases evaporation rates, which explains why the temperature in an irrigated field is usually lower than that over a nearby dry road surface.

Another important nonradiative mechanism is the exchange of heat that occurs when the temperature of the air is different from that of the surface. Depending on whether the surface is warmer or cooler than the air next to it, heat is transferred to or from the atmosphere by turbulent air motion (more loosely, by convection). This effect also increases with increasing temperature difference and with increasing surface wind speed. Direct heat transfer to the air may be an important cooling mechanism that limits the maximum temperature of hot dry surfaces. Alternatively, it may be an important warming mechanism that limits the minimum temperature of cold surfaces. Such warming is sensitive to wind speed, so calm conditions promote lower minimum temperatures.

In a similar category, whenever a temperature difference occurs between the surface and the medium beneath the surface, there is a transfer of heat to or from the medium. In the case of land surfaces, heat is transferred by conduction, a process where energy is conveyed through a material from one atom or molecule to another. In the case of water surfaces, the transfer is by convection and may consequently be affected by the horizontal transport of heat within large bodies of water.

Average values of the different terms in the energy budgets of the atmosphere and surface. The individual terms may be adjusted to suit local conditions and may be used as an aid to understanding the various temperature characteristics.

Atmospheric Temperature

Atmospheric temperature is a function of the modification of solar radiant energy by air, clouds, land, sea and other water surfaces.

The temperature of Earth's atmosphere varies with the distance from the equator (latitude) and height above the surface (altitude). It also changes with time, varying from season to season, and from day to night, as well as irregularly due to passing weather systems. If local variations are averaged out on a global basis, however, a pattern of global average temperatures emerges.

Global variation of Mean Temperature

Global variations of average surface-air temperatures are largely due to latitude, continentality, ocean currents, and prevailing winds.

The effect of latitude is evident in the large north-south gradients in average temperature that occur at middle and high latitudes in each winter hemisphere. These gradients are due mainly to the rapid decrease of available solar radiation but also in part to the higher surface reflectivity at high latitudes associated with snow and ice and low solar elevations. A broad area of the tropical ocean, by contrast, shows little temperature variation.

Continentality is a measure of the difference between continental and marine climates and is mainly the result of the increased range of temperatures that occurs over land compared with water. This difference is a consequence of the much lower effective heat capacities of land surfaces as well as of their generally reduced evaporation rates. Heating or cooling of a land surface takes place in a thin layer, the depth of which is determined by the ability of the ground to conduct heat. The greatest temperature changes occur for dry, sandy soils, because they are poor conductors with very small effective heat capacities and contain no moisture for evaporation. By far the greatest effective heat capacities are those of water surfaces, owing to both the mixing of water near the surface and the penetration of solar radiation that distributes heating to depths of several metres. In addition, about 90 percent of the radiation budget of the ocean is used for evaporation. Ocean temperatures are thus slow to change.

The effect of continentality may be moderated by proximity to the ocean, depending on the direction and strength of the prevailing winds. Contrast with ocean temperatures at the edges of each continent may be further modified by the presence of a north- or south-flowing ocean current. For most latitudes, however, continentality explains much of the variation in average temperature at fixed latitude as well as variations in the difference between January and July temperatures.

Diurnal, Seasonal and Extreme Temperatures

The diurnal range of temperature generally increases with distance from the sea and toward those places where solar radiation is strongest—in dry tropical climates and on high mountain plateaus (owing to the reduced thickness of the atmosphere to be traversed by the Sun's rays). The average difference between the day's highest and lowest temperatures is 3 °C (5 °F) in January and 5 °C (9 °F) in July in those parts of the British Isles nearest the Atlantic. The difference is 4.5 °C (8 °F) in January and 6.5 °C (12 °F) in July on the small island of Malta. At Tashkent, Uzbekistan, it is 9 °C (16 °F) in January and 15.5 °C (28 °F) in July, and at Khartoum, Sudan, the corresponding figures are 17 °C (31 °F) and 13.5 °C (24 °F). At Kandahār, Afghanistan, which lies more than 1,000 metres (about 3,300 feet) above sea level, it is 14 °C (25 °F) in January and 20 °C (36 °F) in July. There, the average difference between the day's highest and lowest temperatures exceeds 23 °C (41 °F) in September and October, when there is less cloudiness than in July. Near the ocean at Colombo, Sri L., the figures are 8 °C (14 °F) in January and 4.5 °C (8 °F) in July.

The seasonal variation of temperature and the magnitudes of the differences between the same month in different years and different epochs generally increase toward high latitudes and with distance from the ocean. Extreme temperatures observed in different parts of the world are listed in the table.

World temperature extremes			
Highest recorded air temperature			
		Temperature	
Continent or region	Place (with elevation*)	degrees C	degrees F
Africa	Kebili, Tunisia (38.1 m or 125 ft)	55	131
Antarctica	Vanda Station 77°32′ S 161°40′ E (15 m or 49 ft)	15	59
Asia	Tirat Zevi, Israel (−220 m or −722 ft)	54	129.2

Australia	Oodnadatta, South Australia (112 m or 367 ft)	50.7	123
Europe	Athens, Greece (236 m or 774 ft)	48	118.4
North America	Death Valley (Greenland Ranch), California, U.S. (−54 m or −177 ft)	56.7	134
South America	Rivadavia, Argentina (668 m or 2,192 ft)	48.9	120
Oceania	Tuguegarao, Luzon, Philippines (62 m or 203 ft)	42.2	108
Lowest recorded air temperature			
		temperature	
continent or region	place (with elevation*)	degrees C	degrees F
Africa	Ifrane, Morocco (1,635 m or 5,364 ft)	−23.9	−11
Antarctica	Vostok 77°32′ S 106°40′ E (3,420 m or 11,220 ft)	−89.2	−128.6
Asia	Verkhoyansk, Russia (107 m or 351 ft) Oymyakon, Russia (800 m or 2,624 ft)	−67.8	−90
Australia	Charlotte Pass, New South Wales (1,755 m or 5,758 ft)	−23	−9.4
Europe	Ust-Shchuger, Russia (85 m or 279 ft)	−58.1	−72.6
North America	Snag, Yukon, Canada (646 m or 2,119 ft)	−63	−81.4
South America	Sarmiento, Argentina (268 m or 879 ft)	−32.8	−27
*Above or below sea level.			

Variation with Height

There are two main levels where the atmosphere is heated—namely, at Earth's surface and at the top of the ozone layer (about 50 km, or 30 miles, up) in the stratosphere. Radiation balance shows a net gain at these levels in most cases. Prevailing temperatures tend to decrease with distance from these heating surfaces (apart from the ionosphere and the outer atmospheric layers, where other processes are at work). The world's average lapse rate of temperature (change with altitude) in the lower atmosphere is 0.6 to 0.7 °C per 100 metres (about 1.1 to 1.3 °F per 300 feet). Lower temperatures prevail with increasing height above sea level for two reasons: (1) because there is a less favourable radiation balance in the free air, and (2) because rising air—whether lifted by convection currents above a relatively warm surface or forced up over mountains—undergoes a reduction of temperature associated with its expansion as the pressure of the overlying atmosphere declines. This is the adiabatic lapse rate of temperature, which equals about 1 °C per 100 metres (about 2 °F per 300 feet) for dry air and 0.5 °C per 100 metres (about 1 °F per 300 feet) for saturated air, in which condensation (with liberation of latent heat) is produced by adiabatic cooling. The difference between these rates of change of temperature (and therefore density) of rising air currents and the state of the surrounding air determines whether the upward currents are accelerated or retarded—i.e., whether the air is unstable, so vertical convection with its characteristically attendant tall cumulus cloud and shower development is encouraged or whether it is stable and convection is damped down.

For these reasons, the air temperatures observed on hills and mountains are generally lower than on low ground, except in the case of extensive plateaus, which present a raised heating surface (and on still, sunny days, when even a mountain peak is able to warm appreciably the air that remains in contact with it).

Circulation, Currents and Ocean-atmosphere Interaction

The circulation of the ocean is a key factor in air temperature distribution. Ocean currents that have a northward or southward component, such as the warm Gulf Stream in the North Atlantic or the cold Peru (Humboldt) Current off South America, effectively exchange heat between low and high latitudes. In tropical latitudes the ocean accounts for a third or more of the poleward heat transport; at latitude 50° N, the ocean's share is about one-seventh. In the particular sectors where the currents are located, their importance is of course much greater than these figures, which represent hemispheric averages.

A good example of the effect of a warm current is that of the Gulf Stream in January, which causes a strong east-west gradient in temperatures across the eastern edge of the North American continent. The relative warmth of the Gulf Stream affects air temperatures all the way across the Atlantic, and prevailing westerlies extend the warming effect deep into northern Europe. As a result, January temperatures of Tromsø, Nor. (69°40′ N), for example, average 24 °C (43 °F) above the mean for that latitude. The Gulf Stream maintains a warming influence in July, but it is not as noticeable because of the effects of continentality.

The ocean, particularly in areas where the surface is warm, also supplies moisture to the atmosphere. This in turn contributes to the heat budget of those areas in which the water vapour is condensed into clouds, liberating latent heat in the process. This set of events occurs frequently in high latitudes and in locations remote from the ocean where the moisture was initially taken up.

The great ocean currents are themselves wind-driven—set in motion by the drag of the winds over vast areas of the sea surface, especially where the tops of waves increase the friction with the air above. At the limits of the warm currents, particularly where they abut directly upon a cold current—as at the left flank of the Gulf Stream in the neighbourhood of the Grand Banks off Newfoundland and at the subtropical and Antarctic convergences in the oceans of the Southern Hemisphere—the strong thermal gradients in the sea surface result in marked differences in the heating of the atmosphere on either side of the boundary. These temperature gradients tend to position and guide the strongest flow of the jet stream in the atmosphere above and thereby influence the development and steering of weather systems.

Interactions between the ocean and the atmosphere proceed in both directions. They also operate at different rates. Some interesting lag effects, which are of value in long-range weather forecasting, arise through the considerably slower circulation of the ocean. Thus, enhanced strength of the easterly trade winds over low latitudes of the Atlantic north and south of the Equator impels more water toward the Caribbean and the Gulf of Mexico, producing a stronger flow and greater warmth in the Gulf Stream approximately six months later. Anomalies in the position of the Gulf Stream–Labrador Current boundary, which produce a greater or lesser extent of warm water near the Grand Banks, so affect the energy supply to the atmosphere and the development and steering of weather systems from that region that they are associated with rather persistent anomalies of weather pattern over the British Isles and northern Europe. Anomalies in the equatorial Pacific and in the northern limit of the Kuroshio Current (also called the Japan Current) seem to have effects on a similar scale. Indeed, through their influence on the latitude of the jet stream and the wavelength (that is, the spacing of cold trough and warm ridge regions) in the upper westerlies, these ocean anomalies exercise an influence over the atmospheric circulation that spreads to all parts of the hemisphere.

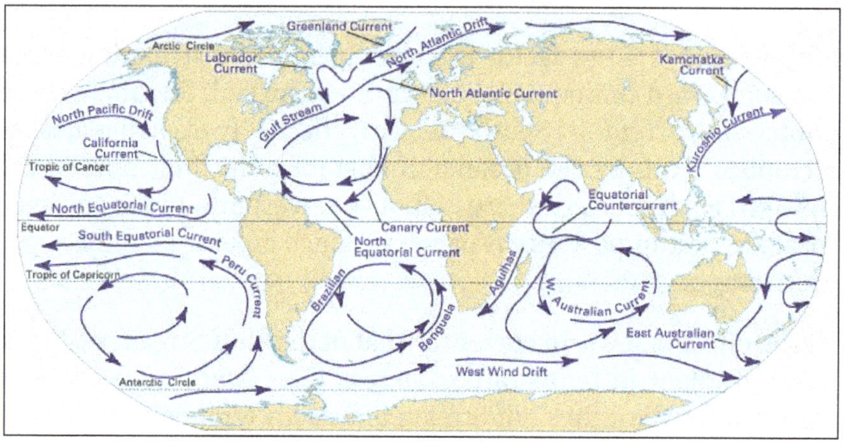

Major surface currents of the world's oceans: Subsurface currents also move vast amounts of water, but they are not known in such detail.

Sea-surface temperature anomalies that recur in the equatorial Pacific at variable intervals of two to seven years can sometimes produce major climatic perturbations. One such anomaly is known as El Niño (Spanish for "The Child"; it was so named by Peruvian fishermen who noticed its onset during the Christmas season).

During an El Niño event, warm surface water flows eastward from the equatorial Pacific, in at least partial response to weakening of the equatorial easterly winds, and replaces the normally cold upwelling surface water off the coast of Peru and Ecuador that is associated with the northward propagation of the cold Peru Current. The change in sea-surface temperature transforms the coastal climate from arid to wet. The event also affects atmospheric circulation in both hemispheres and is associated with changes in precipitation in regions of North America, Africa, and the western Pacific.

Short-term Temperature Changes

Many interesting short-term temperature fluctuations also occur, usually in connection with local weather disturbances. The rapid passage of a mid-latitude cold front, for example, can drop temperatures by 10 °C (18 °F) in a few minutes and, if followed by the sustained movement of a cold air mass, by as much as 50 °C in 24 hours, with life-threatening implications for the unwary. Temperature increases of up to 40 °C in a few hours also are possible downwind of major mountain ranges when air that has been warmed by the release of latent heat on the windward side of a range is forced to descend rapidly on the other side (such a wind is variously called chinook, foehn, or Santa Ana). Changes of this kind, however, involve a wider range of meteorological processes.

Atmospheric Humidity

The Atmospheric humidity Is the amount or volume of water vapor that is present in the atmosphere. The main source of moisture in the air comes from the surface of the oceans and seas, places where water evaporates constantly. Other sources of atmospheric moisture from lakes, glaciers and rivers, as well as the evapotranspiration processes of soil, plants and animals.

The amount of humidity found in air varies because of a number of factors. Two important factors are evaporation and condensation. At the water/atmosphere interface over our planet's oceans large amounts of liquid water are evaporated into atmospheric water vapor. This process is mainly caused by absorption of solar radiation and the subsequent generation of heat at the ocean's surface. In our atmosphere, water vapor is converted back into liquid form when air masses lose heat energy and cool. This process is responsible for the development of most clouds and also produces the rain that falls to the Earth's surface.

At 30 °C (86 °F), 4 percent of the volume of the air may be occupied by water molecules, but, where the air is colder than −40 °C (−40 °F), less than one-fifth of 1 percent of the air molecules can be water. Although the water vapour content may vary from one air parcel to another, these limits can be set because vapour capacity is determined by temperature. Temperature has profound effects upon some of the indexes of humidity, regardless of the presence or absence of vapour.

The connection between an effect of humidity and an index of humidity requires simultaneous introduction of effects and indexes. Vapour in the air is a determinant of weather, because it first absorbs the thermal radiation that leaves and cools Earth's surface and then emits thermal radiation that warms the planet. Calculation of absorption and emission requires an index of the mass of water in a volume of air. Vapour also affects the weather because, as indicated above, it condenses into clouds and falls as rain or other forms of precipitation. Tracing the moisture-bearing air masses requires a humidity index that changes only when water is removed or added.

Humidity Indexes

Absolute Humidity

Absolute humidity is the vapour concentration or density in the air. If mv is the mass of vapour in a volume of air, then absolute humidity dv is simply $dv = mv/V$, in which V is the volume and dv is expressed in grams per cubic metre. This index indicates how much vapour a beam of radiation must pass through. The ultimate standard in humidity measurement is made by weighing the amount of water gained by an absorber when a known volume of air passes through it; this measures absolute humidity, which may vary from 0 gram per cubic metre in dry air to 30 grams per cubic metre (0.03 ounce per cubic foot) when the vapour is saturated at 30 °C. The dv of a parcel of air changes, however, with temperature or pressure even though no water is added or removed, because, as the gas equation states, the volume *V* increases with the absolute, or Kelvin, temperature and decreases with the pressure.

Specific Humidity

The meteorologist requires an index of humidity that does not change with pressure or temperature. A property of this sort will identify an air mass when it is cooled or when it rises to lower pressures aloft without losing or gaining water vapour. Because all the gases will expand equally, the ratios of the weight of water to the weight of dry air, or the dry air plus vapour, will be conserved during such changes and will continue identifying the air mass.

The mixing ratio r is the dimensionless ratio $r = mv/ma$, where ma is the mass of dry air, and the specific humidity q is another dimensionless ratio $q = mv/(ma + mv)$. Because mv is less than 3 percent of ma at normal pressure and temperatures cooler than 30 °C, r and q are practically equal. These indexes are usually expressed in grams per kilogram because they are so small; the values range from 0 grams per kilogram in dry air to 28 grams per kilogram in saturated air at 30 °C. Absolute and specific humidity indexes have specialized uses, so they are not familiar to most people.

Relative Humidity

Relative humidity (U) is so commonly used that a statement of humidity, without a qualifying adjective, can be assumed to be relative humidity. U can be defined, then, in terms of the mixing ratio r that was introduced above. $U = 100r/ rw$, which is a dimensionless percentage. The divisor rw is the saturation mixing ratio, or the vapour capacity. Relative humidity is therefore the water vapour content of the air relative to its content at saturation. Because the saturation mixing ratio is a function of pressure, and especially of temperature, the relative humidity is a combined index of the environment that reflects more than water content. In many climates the relative humidity rises to about 100 percent at dawn and falls to 50 percent by noon. A relative humidity of 50 percent may reflect many different quantities of vapour per volume of air or gram of air, and it will not likely be proportional to evaporation.

An understanding of relative humidity thus requires knowledge of saturated vapour. At this point, however, the relation between U and the absorption and retention of water from the air must be considered. Small pores retain water more strongly than large pores; thus, when a porous material is set out in the air, all pores larger than a certain size (which can be calculated from the relative humidity of the air) are dried out.

The water content of a porous material at air temperature is fairly well indicated by the relative humidity. The complexity of actual pore sizes and the viscosity of the water passing through them make the relation between U and moisture in the porous material imperfect and slowly achieved. The great suction also strains the walls of the capillaries, and the consequent shrinkage is used to measure relative humidity.

The absorption of water by salt solutions is also related to relative humidity without much effect of temperature. The air above water saturated with sodium chloride is maintained at 75 to 76 percent relative humidity at a temperature between 0 and 40 °C (32 and 104 °F).

In effect, relative humidity is a widely used environmental indicator, but *U* does respond drastically to changes in temperatures as well as moisture, a response caused by the effect of temperature upon the divisor rW in *U*.

Relation between Temperature and Humidity

Tables that show the effect of temperature upon the saturation mixing ratio rw are readily available. Humidity of the air at saturation is expressed more commonly, however, as vapour pressure. Thus, it is necessary to understand vapour pressure and in particular the gaseous nature of water vapour.

The pressure of the water vapour, which contributes to the pressure of the atmosphere, can be calculated from the absolute humidity dv by the gas equation:

$$e = \frac{m_v}{V}\frac{RT}{M_w} = d_v\frac{RT}{M_w},$$

In which *R* is the gas constant, *T* the absolute temperature, M_w the molecular weight of water, and *e* the water vapour pressure in millibars (mb).

Relative humidity can be defined as the ratio of the vapour pressure of a sample of air to the saturation pressure at the existing temperature. Further, the capacity for vapour and the effect of temperature can now be presented in the usual terms of saturation vapour pressure.

Within a pool of liquid water, some molecules are continually escaping from the liquid into the space above, while more and more vapour molecules return to the liquid as the concentration of vapour rises. Finally, equal numbers are escaping and returning; the vapour is then saturated, and its pressure is known as the saturation vapour pressure, ew. If the liquid and vapour are warmed, relatively more molecules escape than return, and ew rises. There is also a saturation pressure with respect to ice. The vapour pressure curve of water has the same form as the curves for many other substances. Its location is fixed, however, by the boiling point of 100 °C (212 °F), where the saturation vapour pressure of water vapour is 1,013 mb (1 standard atmosphere), the standard pressure of the atmosphere at sea level. The decrease of the boiling point with altitude can be calculated. For example, the saturation vapour pressure at 40 °C (104 °F) is 74 mb (0.07 standard atmosphere), and the standard atmospheric pressure near 18,000 metres (59,000 feet) above sea level is also 74 mb; thus, it is where water boils at 40° C.

The everyday response of relative humidity to temperature can be easily explained. On a summer morning, the temperature might be 15 °C (59 °F) and the relative humidity 100 percent. The vapour pressure would be 17 mb (0.02 standard atmosphere) and the mixing ratio about 11 parts per thousand (11 grams of water per kilogram of air by weight). During the day the air could warm to 25 °C (77 °F), while evaporation could add little water. At 25 °C the saturation pressure is fully 32 mb (0.03 standard atmosphere). If, however, little water has been added to the air, its vapour pressure will still be about 17 mb. Thus, with no change in vapour content, the relative humidity of the air has fallen from 100 to only 53 percent, illustrating why relative humidity does not identify air masses.

The meaning of dew-point temperature can be illustrated by a sample of air with a vapour pressure of 17 mb. If an object at 15 °C is brought into the air, dew will form on the object. Hence, 15 °C is the dew-point temperature of the air—i.e., the temperature at which the vapour present in a sample of air would just cause saturation or the temperature whose saturation vapour pressure equals the present vapour pressure in a sample of air. Below freezing, this index is called the frost point. There is a one-to-one correspondence between vapour pressure and dew point. The dew point has the virtue of being easily interpreted because it is the temperature at which a blade of grass or a pane of glass will become wet with dew from the air. Ideally, it is also the temperature of fog or cloud formation.

The clear meaning of dew point suggests a means of measuring humidity. A dew-point hygrometer was invented in 1751. For this instrument, cold water was added to water in a vessel until dew formed on the vessel, and the temperature of the vessel, the dew point, provided a direct index of humidity. The greatest use of the condensation hygrometer has been to measure humidity in the upper atmosphere, where a vapour pressure of less than a thousandth millibar makes other means impractical.

Another index of humidity, the saturation deficit, can also be understood by considering air with a vapour pressure of 17 mb. At 25 °C the air has (31 − 17), or 14, mb less vapour pressure than saturated vapour at the same temperature; that is, the saturation deficit is 14 mb (0.01 standard atmosphere).

The saturation deficit has the particular utility of being proportional to the evaporation capability of the air. The saturation deficit can be expressed as,

$$e_w - e = e_w\left(1 - \frac{U}{100}\right).$$

And, because the saturation vapour pressure *ew* rises with rising temperature, the same relative humidity will correspond to a greater saturation deficit and evaporation at warm temperatures.

Humidity and Climate

The small amount of water in atmospheric vapour, relative to water on Earth, belies its importance. Compared with one unit of water in the air, the seas contain at least 100,000 units, the great glaciers 1,500, the porous earth nearly 200, and the rivers and lakes 4 or 5. The effectiveness of the vapour in the air is magnified, however, by its role in transferring water from sea to land by the media of clouds and precipitation and that in absorbing radiation.

The vapour in the air is the invisible conductor that carries water from sea to land, making terrestrial life possible. Fresh water is distilled from the salt seas and carried over land by the wind. Water evaporates from vegetation, and rain falls on the sea too, but the sea is the bigger source, and rain that falls on land is most important to humans. The invisible vapour becomes visible near the surface as fog when the air cools to the dew point. The usual nocturnal cooling will produce fog patches in cool valleys. Or the vapour may move as a tropical air mass over cold land or sea, causing widespread and persistent fog, such as occurs over the Grand Banks off Newfoundland. The delivery of water by means of fog or dew is slight, however.

When air is lifted, it is carried to a region of lower pressure, where it will expand and cool as described by the gas equation. It may rise up a mountain slope or over the front of a cooler, denser air mass. If condensation nuclei are absent, the dew point may be exceeded by the cooling air, and the water vapour becomes supersaturated. If nuclei are present or if the temperature is very low, however, cloud droplets or ice crystals form, and the vapour is no longer in the invisible guise of atmospheric humidity.

The invisible vapour has another climatic role—namely, absorbing and emitting radiation. The temperature of Earth and its daily variation are determined by the balance between incoming and outgoing radiation. The wavelength of the incoming radiation from the Sun is mostly shorter than 3 μm (0.0001 inch). It is scarcely absorbed by water vapour, and its receipt depends largely upon cloud cover. The radiation exchanged between the atmosphere and Earth's surface and the eventual loss to space is in the form of long waves. These long waves are strongly absorbed in the 3- to 8.5-μm band and in the greater than 11-μm range, where vapour is either partly or wholly opaque. As noted above, much of the radiation that is absorbed in the atmosphere is emitted back to Earth, and the surface receipt of long waves, primarily from water vapour and carbon dioxide in the atmosphere, is slightly more than twice the direct receipt of solar radiation at the surface. Thus, the invisible vapour in the atmosphere combines with clouds and the advection (horizontal movement) of air from different regions to control the surface temperature.

The world distribution of humidity can be portrayed for different uses by different indexes. To appraise the quantity of water carried by the entire atmosphere, the moisture in an air column above a given point on Earth is expressed as a depth of liquid water. It varies from 0.5 mm (0.02 inch) over the Himalayas and 2 mm (0.08 inch) over the poles in winter to 8 mm (0.3 inch) over the Sahara, 54 mm (2 inches) in the Amazon region, and 64 mm (2.5 inches) over India during the wet season. During summer the air over the United States transports 16 mm (0.6 inch) of water vapour over the Great Basin and 45 mm (1.8 inches) over Florida.

The humidity of the surface air may be mapped as vapour pressure, but a map of this variable looks much like that of temperature. Warm places are moist, and cool ones are dry; even in deserts the vapour pressure is normally 13 mb (0.01 standard atmosphere), whereas over the northern seas it is only about 4 mb (0.004 standard atmosphere). Certainly the moisture in materials in two such areas will be just the opposite, so relative humidity is a more widely useful index.

Average Relative Humidity

The average relative humidity for July reveals the humidity provinces of the Northern Hemisphere when aridity is at a maximum. At other times the relative humidity generally will be higher. The humidities over the Southern Hemisphere in July indicate the humidities that comparable regions in the Northern Hemisphere will attain in January, just as July in the Northern Hemisphere suggests the humidities in the Southern Hemisphere during January. A contrast is provided by comparing a humid cool coast to a desert. The midday humidity on the Oregon coast, for example, falls only to 80 percent, whereas in the Nevada desert it falls to 20 percent. At night the contrast is less, with averages being over 90 and about 50 percent, respectively.

Although the dramatic regular decrease of relative humidity from dawn to midday has been attributed largely to warming rather than declining vapour content, the content does vary regularly. In humid environments, daytime evaporation increases the water vapour content of the air, and the mixing ratio, which may be about 12 grams per kilogram, rises by 1 or 2 grams per kilogram in temperate places and may attain 16 grams per kilogram in a tropical rainforest. In arid environments, however, little evaporation moistens the air, and daytime turbulence tends to bring down dry air; this decreases the mixing ratio by as much as 2 grams per kilogram.

Humidity also varies regularly with altitude. On the average, fully half the water in the atmosphere lies below 0.25 km (about 0.2 mile), and satellite observations over the United States in April revealed 1 mm (0.04 inch) or less of water in all the air above 6 km (4 miles). A cross section of the atmosphere along 75° W longitude shows a decrease in humidity with height and toward the poles. The mixing ratio is 16 grams per kilogram just north of the Equator, but it decreases to 1 gram per kilogram at 50° N latitude or 8 km (5 miles) above the Equator. The transparent air surrounding mountains in fair weather is very dry indeed.

Closer to the ground, the water vapour content also changes with height in a regular pattern. When water vapour is condensing on Earth's surface at night, the content is greater aloft than at the ground; during the day the content is, in most cases, less aloft than at the ground because of evaporation.

Evaporation and Humidity

Evaporation, mostly from the sea and from vegetation, replenishes the humidity of the air. It is the change of liquid water into a gaseous state, but it may be analyzed as diffusion. The rate of diffusion, or evaporation, will be proportional to the difference between the pressure of the water vapour in the free air and the vapour that is next to, and saturated by, the evaporating liquid. If the liquid and air have the same temperature, evaporation is proportional to the saturation deficit. It is also proportional to the conductivity of the medium between the evaporator and the free air. If the evaporator is open water, the conductivity will increase with ventilation. But if the evaporator is a leaf, the diffusing water must pass through the still air within the minute pores between the water inside and the dry air outside. In this case the porosity may modify the conductivity more than ventilation.

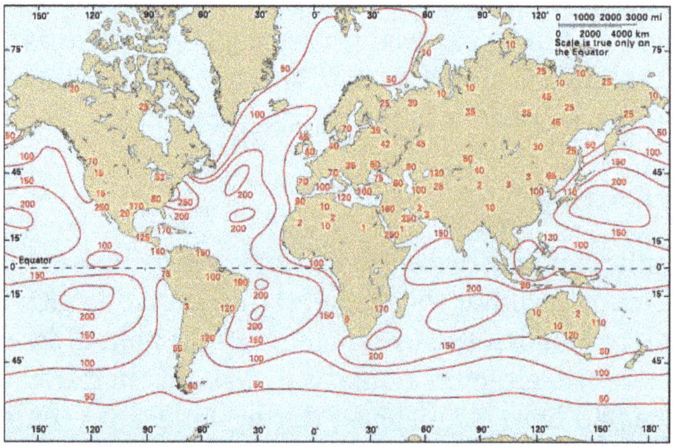

Global distribution of mean annual evaporation (in centimetres).

The temperature of the evaporator is rarely the same as the air temperature, however, because each gram of evaporation consumes about 600 calories (2,500 joules) and thus cools the evaporator. The availability of energy to heat the evaporator is therefore as important as the saturation deficit and conductivity of the air. Outdoors, some of this heat may be transferred from the surrounding air by convection, but much of it must be furnished by radiation. Evaporation is faster on sunny days than on cloudy ones not only because the sunny day may have drier air but also because the Sun warms the evaporator and thus raises the vapour pressure at the evaporator. In fact, according to the well-known Penman calculation of evaporation (an equation that considers potential evaporation as a function of humidity, wind speed, radiation, and temperature), this loss of water is essentially determined by the net radiation balance during the day.

Precipitation

Precipitation is any liquid or frozen water that forms in the atmosphere and falls back to the Earth. It comes in many forms, like rain, sleet, and snow. Along with evaporation and condensation, precipitation is one of the three major parts of the global water cycle.

Precipitation forms in the clouds when water vapor condenses into bigger and bigger droplets of water. When the drops are heavy enough, they fall to the Earth. If a cloud is colder, like it would be at higher altitudes, the water droplets may freeze to form ice. These ice crystals then fall to the Earth as snow, hail, or rain, depending on the temperature within the cloud and at the Earth's surface. Most rain actually begins as snow high in the clouds. As the snowflakes fall through warmer air, they become raindrops.

Particles of dust or smoke in the atmosphere are essential for precipitation. These particles, called "condensation nuclei," provide a surface for water vapor to condense upon. This helps water droplets gather together and become large enough to fall to the Earth.

A common misconception is that when raindrops fall, they have a teardrop shape. In fact, smaller raindrops (ones that are approximately 1 millimeter (0.039 inches) across) are almost perfectly spherical. Larger raindrops (2–3 millimeters (.078-.118 inches) across) are also round, but with a small indent on their bottom side. They look more like kidney beans when falling to the Earth. Very large rain drops (larger than 4.5 millimeters (.177 inches)) have a huge indent and look more like a parachute. These extra-large drops usually end up splitting into two smaller droplets. The indents on raindrops are caused by air resistance.

Precipitation is always fresh water, even when the water originated from the ocean. This is because sea salt does not evaporate with water. However, in some cases, pollutants in the atmosphere can contaminate water droplets before they fall to the Earth. The precipitation that results from this is called acid rain. Acid rain does not harm humans directly, but it can make lakes and streams more acidic. This harms aquatic ecosystems because plants and animals often cannot adapt to the acidity.

Precipitation is one of the three main processes (evaporation, condensation, and precipitation) that constitute the hydrologic cycle, the continual exchange of water between the atmosphere and Earth's surface. Water evaporates from ocean, land, and freshwater surfaces, is carried aloft as vapour by the air currents, condenses to form clouds, and ultimately is returned to Earth's surface as precipitation. The average global stock of water vapour in the atmosphere is equivalent to a layer of water 2.5 cm (1 inch) deep covering the whole Earth. Because Earth's average annual rainfall is about 100 cm (39 inches), the average time that the water spends in the atmosphere, between its evaporation from the surface and its return as precipitation, is about 1/40 of a year, or about nine days. Of the water vapour that is carried at all heights across a given region by the winds, only a small percentage is converted into precipitation and reaches the ground in that area. In deep and extensive cloud systems, the conversion is more efficient, but even in thunderclouds the quantities of rain and hail released amount to only some 10 percent of the total moisture entering the storm.

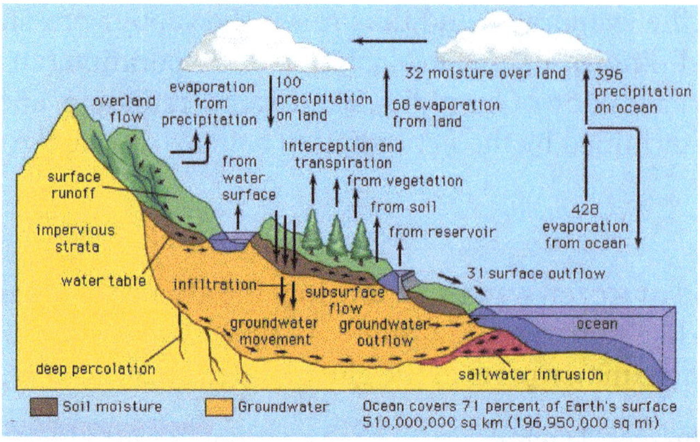

In the hydrologic cycle, water is transferred between the land surface, the ocean, and the atmosphere. The numbers on the arrows indicate relative water fluxes.

In the measurement of precipitation, it is necessary to distinguish between the amount—defined as the depth of precipitation (calculated as though it were all rain) that has fallen at a given point during a specified interval of time—and the rate or intensity, which specifies the depth of water that has fallen at a point during a particular interval of time. Persistent moderate rain, for example, might fall at an average rate of 5 mm per hour (0.2 inch per hour) and thus produce 120 mm (4.7 inches) of rain in 24 hours. A thunderstorm might produce this total quantity of rain in 20 minutes, but at its peak intensity the rate of rainfall might become much greater—perhaps 120 mm per hour (4.7 inches per hour), or 2mm (0.08 inch) per minute—for a minute or two.

The amount of precipitation falling during a fixed period is measured regularly at many thousands of places on Earth's surface by rather simple rain gauges. Measurement of precipitation intensity requires a recording rain gauge, in which water falling into a collector of known surface area is continuously recorded on a moving chart or a magnetic tape. Investigations are being carried out on the feasibility of obtaining continuous measurements of rainfall over large catchment areas by means of radar.

Apart from the trifling contributions made by dew, frost, and rime, as well as desalination plants, the sole source of fresh water for sustaining rivers, lakes, and all life on Earth is provided by precipitation from clouds. Precipitation is therefore indispensable and overwhelmingly beneficial to humankind, but extremely heavy rainfall can cause great harm: soil erosion, landslides, and flooding. Hailstorm damage to crops, buildings, and livestock can prove very costly.

Origin of Precipitation in Clouds

Cloud Formation

Clouds are formed by the lifting of damp air, which cools by expansion as it encounters the lower pressures existing at higher levels in the atmosphere. The relative humidity increases until the air has become saturated with water vapour, and then condensation occurs on any of the aerosol particles suspended in the air. A wide variety of these exist in concentrations ranging from only a few per cubic centimetre in clean maritime air to perhaps 1 million per cubic cm (16 million per cubic inch) in the highly polluted air of an industrial city. For continuous condensation leading to the formation of cloud droplets, the air must be slightly supersaturated. Among the highly efficient condensation nuclei are sea-salt particles and the particles produced by combustion (e.g., natural forest fires and man-made fires). Many of the larger condensation nuclei over land consist of ammonium sulfate. These are produced by cloud and fog droplets absorbing sulfur dioxide and ammonia from the air. Condensation onto the nuclei continues as rapidly as water vapour is made available through cooling; droplets about 10 μm (0.0004 inch) in diameter are produced in this manner. These droplets constitute a nonprecipitating cloud.

Cloud Types

The meteorologist classifies clouds mainly by their appearance, according to an international system similar to one proposed in 1803. But because the dimensions, shape, structure, and texture of clouds are influenced by the kind of air movements that result in their formation and growth and by the properties of the cloud particles, much of what was originally a purely visual classification can now be justified on physical grounds.

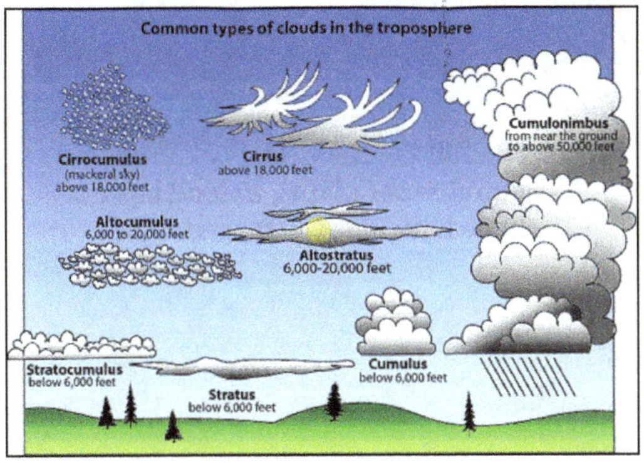

Different types of clouds form at different heights.

The first International Cloud Atlas was published in 1896. Developments in aviation during World War I stimulated interest in cloud formations and in their importance as an aid in short-range weather forecasting. This led to the publication of a more extensive atlas, the International Atlas of Clouds and States of Sky, in 1932 and to a revised edition in 1939. After World War II, the World Meteorological Organization published a new International Cloud Atlas (1956) in two volumes. It contains 224 plates, describing 10 main cloud genera (families) subdivided into 14 species based on cloud shape and structure. Nine general varieties, based on transparency and geometric arrangement, also are described. The genera, listed according to their height, are as follows:

1.High: mean heights from 5 to 13 km, or 3 to 8 miles.

Cirrus fibratus are high clouds that are nearly straight or irregularly curved.
They appear as fine white filaments and are generally distinct from one another.

a. Cirrus

b. Cirrocumulus

c. Cirrostratus

2. Middle: mean heights 2 to 7 km, or 1 to 4 miles.

Altocumulus undulatus, a cloud layer of shaded, regularly arranged rolls.

a. Altocumulus

b. Altostratus

c. Nimbostratus

3. Low: mean heights 0 to 2 km, or 0 to 1.2 miles.

a. Stratocumulus

b. Stratus

c. Cumulus

d. Cumulonimbus

Heights given are approximate averages for temperate latitudes. Clouds of each genus are generally lower in the polar regions and higher in the tropics. The definitions and descriptions of the cloud genera used in the International Cloud Atlas which illustrate some of their characteristic forms.

Four principal classes are recognized when clouds are classified according to the kind of air motions that produce them: (1) layer clouds formed by the widespread regular ascent of air, (2) layer clouds formed by widespread irregular stirring or turbulence, (3) cumuliform clouds formed by penetrative convection, and (4) orographic clouds formed by the ascent of air over hills and mountains.

The widespread layer clouds associated with cyclonic depressions, near fronts and other inclement-weather systems, are frequently composed of several layers that may extend up to 9 km (5.6 miles) or more, separated by clear zones that become filled in as rain or snow develops. These clouds are formed by the slow, prolonged ascent of a deep layer of air, in which a rise of only a few centimetres per second is maintained for several hours. In the neighbourhood of fronts, vertical velocities become more pronounced and may reach about 10 cm (4 inches) per second.

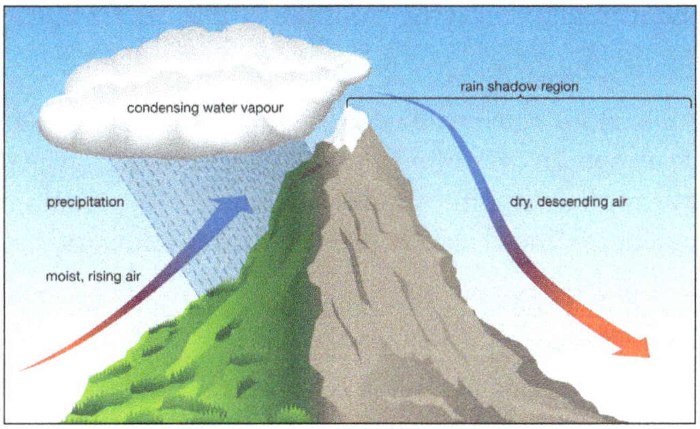

Orographic lift: Condensation, precipitation, and the rain shadow effect resulting from orographic lift.

Most of the high cirrus clouds visible from the ground lie on the fringes of cyclonic cloud systems, and, though due primarily to regular ascent, their pattern is often determined by local wave disturbances that finally trigger their formation after the air has been brought near its saturation point by the large-scale lifting.

On a cloudless night, the ground cools by radiating heat into space without heating the air adjacent to the ground. If the air were quite still, only a very thin layer would be chilled by contact with the ground. More usually, however, the lower layers of the air are stirred by motion over the rough ground, so the cooling is distributed through a much greater depth. Consequently, when the air is damp or the cooling is great, a fog a few hundred metres deep may form, rather than a dew produced by condensation on the ground.

In moderate or strong winds, the irregular stirring near the surface distributes the cooling upward, and the fog may lift from the surface to become a stratus cloud, which is not often more than 600 metres (about 2,000 feet) thick.

Radiational cooling from the upper surfaces of fogs and stratus clouds promotes an irregular convection within the cloud layer and causes the surfaces to have a waved or humped appearance. When the cloud layer is shallow, billows and clear spaces may develop; it is then described as stratocumulus instead of stratus.

Usually, cumuliform clouds appearing over land are formed by the rise of discrete masses of air from near the sunlight-warmed surface. These rising lumps of air, or thermals, may vary in diameter from a few tens to hundreds of metres as they ascend and mix with the cooler, drier air above them. Above the level of the cloud base, the release of latent heat of condensation tends to increase the buoyancy of the rising masses, which tower upward and emerge at the top of the cloud with rounded upper surfaces.

At any moment a large cloud may contain a number of active thermals and the residues of earlier ones. A new thermal rising into a residual cloud will be partially protected from having to mix with the cool, dry environment and therefore may rise farther than its predecessor. Once a thermal has emerged as a cloud turret at the summit or the flanks of the cloud, rapid evaporation of the droplets chills the cloud borders, destroys the buoyancy, and produces sinking. A cumulus thus has a characteristic pyramidal shape and, viewed from a distance, appears to have an unfolding motion, with fresh cloud masses continually emerging from the interior to form the summit and then sinking aside and evaporating.

In settled weather, cumulus clouds are well scattered and small; horizontal and vertical dimensions are only a kilometre or two. In disturbed weather, they cover a large part of the sky, and individual clouds may tower as high as 10 km (6 miles) or more, often ceasing their growth only upon reaching the stable stratosphere. These clouds produce heavy showers, hail, and thunderstorms.

At the level of the cloud base, the speed of the rising air masses is usually about 1 metre (3.3 feet) per second but may reach 5 metres (16 feet) per second, and similar values are measured inside smaller clouds. The upcurrents in thunderclouds, however, often exceed 5 metres per second and may reach 30 metres (98 feet) per second or more.

The rather special orographic clouds are produced by the ascent of air over hills and mountains. The air stream is set into oscillation when it is forced over the hill, and the clouds form in the crests of the (almost) stationary waves. There may thus be a succession of such clouds stretching downwind of the mountain, which remain stationary relative to the ground in spite of strong winds that may be blowing through the clouds. The clouds have very smooth outlines and are called lenticular (lens-shaped) or "wave" clouds. Thin wave clouds may form at great heights (up to 10 km, even over hills only a few hundred metres high) and occasionally are observed in the stratosphere (at 20 to 30 km [12 to 19 miles]) over the mountains of Norway, Scotland, Iceland, and Alaska. These atmospheric wave clouds are known as nacreous or "mother-of-pearl" clouds because of their brilliant iridescent colours.

Precipitation Release

Growing clouds are sustained by upward air currents, which may vary in strength from a few centimetres per second to several metres per second. Considerable growth of the cloud droplets (with falling speeds of only about 1 cm, or 0.4 inch, per second) is therefore necessary if they are to fall through the cloud, survive evaporation in the unsaturated air below, and reach the ground as drizzle or rain. The production of a few large particles from a large population of much smaller ones may be achieved in one of two ways. The first of these depends on the fact that cloud droplets are seldom of uniform size; droplets form on nuclei of various sizes and grow under slightly different conditions and for different lengths of time in different parts of the cloud. A droplet appreciably larger than average will fall faster than the smaller ones and so will collide and fuse (coalesce) with some of those that it overtakes. Calculations show that, in a deep cloud containing strong upward air currents and high concentrations of liquid water, such a droplet will have a sufficiently long journey among its smaller neighbours to grow to raindrop size. This coalescence mechanism is responsible for the showers that fall in tropical and subtropical regions from clouds whose tops do not reach altitudes where air temperatures are below 0 °C (32 °F) and therefore cannot contain ice crystals. Radar evidence also suggests that showers in temperate latitudes may sometimes be initiated by the coalescence of waterdrops, although the clouds may later reach heights at which ice crystals may form in their upper parts.

The second method of releasing precipitation can operate only if the cloud top reaches elevations at which air temperatures are below 0 °C and the droplets in the upper cloud regions become supercooled. At temperatures below −40 °C (−40 °F), the droplets freeze automatically or spontaneously. At higher temperatures, they can freeze only if they are infected with special minute particles called ice nuclei. The origin and nature of these nuclei are not known with certainty, but the most likely source is clay-silicate particles carried up from the ground by the wind. As the temperature falls below 0 °C, more and more ice nuclei become active, and ice crystals appear in increasing numbers among the supercooled droplets. Such a mixture of supercooled droplets and ice crystals is unstable, however. The cloudy air is usually only slightly supersaturated with water vapour with respect to the droplets and is strongly oversaturated with respect to ice crystals; the latter thus grow more rapidly than the droplets. After several minutes, the growing crystals acquire falling speeds of tens of centimetres per second, and several of them may become joined to form a snowflake. In falling into the warmer regions of the cloud, this flake may melt and hit ground as a raindrop.

The deep, extensive, multilayer cloud systems, from which precipitation of a widespread persistent character falls, are generally formed in cyclonic depressions (lows) and near fronts. Cloud systems of this type are associated with feeble upcurrents of only a few centimetres per second that last for at least several hours. Although the structure of these great rain-cloud systems is being explored by aircraft and radar, it is not yet well understood. That such systems rarely produce rain, as distinct from drizzle, unless their tops are colder than about −12 °C (10 °F) suggests that ice crystals are mainly responsible. This view is supported by the fact that the radar signals from these clouds usually take a characteristic form that has been clearly identified with the melting of snowflakes.

Showers, Thunderstorms and Hail

Precipitation from shower clouds and thunderstorms whether in the form of raindrops, pellets of soft hail, or true hailstones, is generally of great intensity and shorter duration than that from layer clouds and is usually composed of larger particles. The clouds are characterized by their large vertical depth, strong vertical air currents, and high concentrations of liquid water, all factors favouring the rapid growth of precipitation elements by the accretion of cloud droplets.

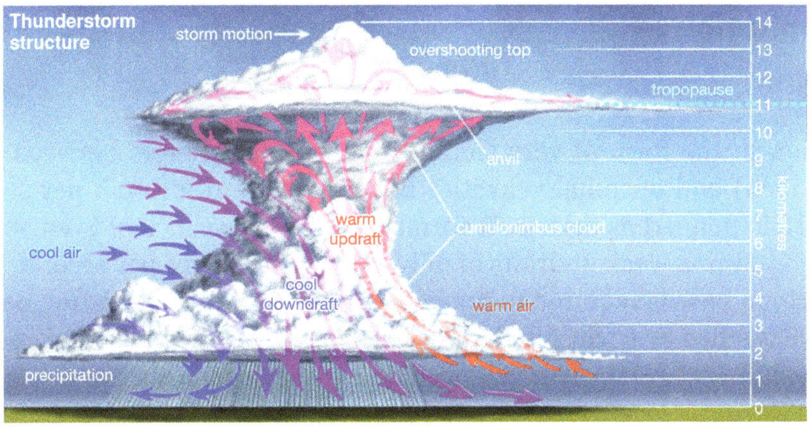

Structure of a thunderstorm.

When the atmosphere becomes unstable enough to form large, powerful updrafts and downdrafts (as indicated by the red and blue arrows), a towering thundercloud is built up. At times the updrafts are strong enough to extend the top of the cloud into the tropopause, the boundary between the troposphere (or lowest layer of the atmosphere) and the stratosphere.

In a cloud composed wholly of liquid water, raindrops may grow by coalescence. For example, a droplet being carried up from the cloud base grows as it ascends by sweeping up smaller droplets. When it becomes too heavy to be supported by the upcurrents, the droplet falls, continuing to grow by the same process on its downward journey. Finally, if the cloud is sufficiently deep, the droplet will emerge from its base as a raindrop.

In a dense, vigorous cloud several kilometres deep, the drop may attain its limiting stable diameter (about 6 mm [0.2 inch]) before reaching the cloud base and thus will break up into several large fragments. Each of these may continue to grow and attain breakup size. The number of raindrops may increase so rapidly in this manner that after a few minutes the accumulated mass of water can no longer be supported by the upcurrents and falls as a heavy shower. These conditions occur more readily in tropical regions. In temperate regions where the freezing level (0 °C) is much lower in elevation, conditions are more favourable for the ice-crystal mechanism.

The hailstones that fall from deep, vigorous clouds in warm weather consist of a core surrounded by several alternate layers of clear and opaque ice. When the growing particle traverses a region of relatively high air temperature or high concentration of liquid water, or both, the transfer of heat from the hailstone to the air cannot occur rapidly enough to allow all of the deposited water to freeze immediately. This results in the formation of a wet coating of slushy ice, which may later freeze to form a layer of compact, relatively transparent ice. If the hailstone then enters a region of lower temperature and lower water content, the impacting droplets may freeze individually to produce ice of relatively low density with air spaces between the droplets. The alternate layers are formed as the stone passes through regions in which the combination of air temperature, liquid-water content, and updraft speed allows alternately wet and dry growth.

It is held by some authorities that lightning is closely associated with the appearance of precipitation, especially in the form of soft hail, and that the charge and the strong electric fields are produced by ice crystals or cloud droplets striking and bouncing off the undersurfaces of the hail pellets.

Types of Precipitation

Drizzle

Liquid precipitation in the form of very small drops, with diameters between 0.2 and 0.5 mm (0.008 and 0.02 inch) and terminal velocities between 70 and 200 cm per second (28 and 79 inches per second), is defined as drizzle. It forms by the coalescence of even smaller droplets in low-layer clouds containing weak updrafts of only a few centimetres per second. High relative humidity below the cloud base is required to prevent the drops from evaporating before reaching the ground; drizzle is classified as slight, moderate, or thick. Slight drizzle produces negligible runoff from the roofs of buildings, and thick drizzle accumulates at a rate in excess of 1 mm per hour (0.04 inch per hour).

Rain and Freezing Rain

Liquid waterdrops with diameters greater than those of drizzle constitute rain. Raindrops rarely exceed 6 mm (0.2 inch) in diameter because they become unstable when larger than this and break up during their fall. The terminal velocities of raindrops at ground level range from 2 metres per second (7 feet per second) for the smallest to about 10 metres per second (30 feet per second) for the largest. The smaller raindrops are kept nearly spherical by surface-tension forces, but, as the diameter surpasses about 2 mm (0.08 inch), they become increasingly flattened by aerodynamic forces. When the diameter reaches 6 mm, the undersurface of the drop becomes concave because of the airstream, and the surface of the drop is sheared off to form a rapidly expanding "bubble" or "bag" attached to an annular ring containing the bulk of the water. Eventually the bag bursts into a spray of fine droplets, and the ring breaks up into a circlet of millimetre-sized drops.

A rain shaft piercing a tropical sunset as seen from Man-o'-War Bay, Tobago, Caribbean Sea.

Rain of a given intensity is composed of a spectrum of drop sizes, the average and median drop diameters being larger in rains of greater intensity. The largest drops, which have a diameter greater than 5 mm (0.2 inch), appear only in the heavy rains of intense storms.

When raindrops fall through a cold layer of air (colder than 0 °C, or 32 °F) and become supercooled, freezing rain occurs. The drops may freeze on impact with the ground to form a very slippery and dangerous "glazed" ice that is difficult to see because it is almost transparent.

Snow and Sleet

Snow in the atmosphere can be subdivided into ice crystals and snowflakes. Ice crystals generally form on ice nuclei at temperatures appreciably below the freezing point. Below −40 °C (−40 °F) water vapour can solidify without the presence of a nucleus. Snowflakes are aggregates of ice crystals that appear in an infinite variety of shapes, mainly at temperatures near the freezing point of water.

Colorado's fine, light snow attracts millions of skiers every year.

In British terminology, sleet is the term used to describe precipitation of snow and rain together or of snow melting as it falls. In the United States, it is used to denote partly frozen ice pellets.

Snow crystals generally have a hexagonal pattern, often with beautifully intricate shapes. Three- and 12-branched forms occur occasionally. The hexagonal form of the atmospheric ice crystals, their varying size and shape notwithstanding, is an outward manifestation of an internal arrangement in which the oxygen atoms form an open lattice (network) with hexagonally symmetrical structure. According to a recent internationally accepted classification, there are seven types of snow crystals: plates, stellars, columns, needles, spatial dendrites, capped columns, and irregular crystals. The size and shape of the snow crystals depend mainly on the temperature of their formation and on the amount of water vapour that is available for deposition. The two principal influences are not independent; the possible water vapour admixture of the air decreases strongly with decreasing temperature. The vapour pressure in equilibrium, or state of balance, with a level surface of pure ice is 50 times greater at −2 °C (28 °F) than at −42 °C (−44 °F), the likely limits of snow formation in the air. Crystal shape and temperature at formation are related in the table.

Ice crystal shape and temperatureat formation	
Temperature (degrees Celsius)	Form
0 to −3	thin hexagonal plates
−3 to −5	needles
−5 to −8	hollow, prismatic columns
−8 to −12	hexagonal plates
−12 to −16	dendritic crystals
−16 to −25	hexagonal plates
−25 to −50	hollow prisms

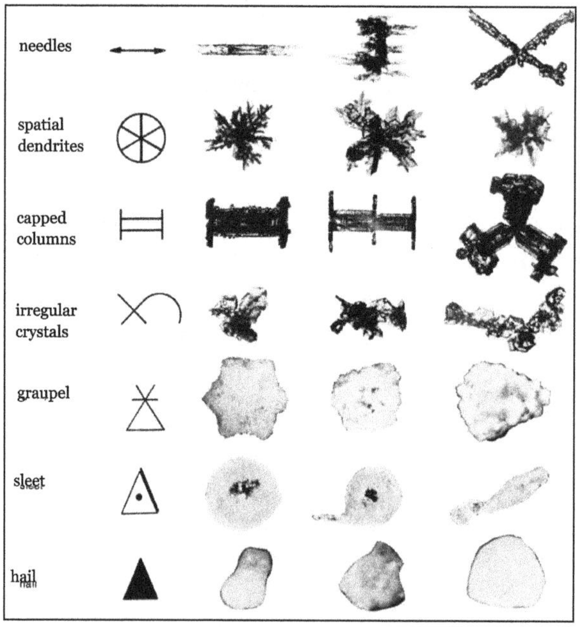

Classification of frozen precipitation.

At temperatures above about −40 °C (−40 °F), the crystals form on nuclei of very small size that float in the air (heterogeneous nucleation). The nuclei consist predominantly of silicate minerals of terrestrial origin, mainly clay minerals and micas. At still lower temperatures, ice may form directly from water vapour (homogeneous nucleation). The influence of the atmospheric water vapour depends mainly on its degree of supersaturation with respect to ice.

If the air contains a large excess of water vapour, the snow particles will grow fast, and there may be a tendency for dendritic (branching) growth. With low temperature, the excess water vapour tends to be small, and the crystals remain small. In relatively dry layers, the snow particles generally have simple forms. Complicated forms of crystals will cling together with others to form snowflakes that consist occasionally of up to 100 crystals; the diameter of such flakes may be as large as 2.5 cm (1 inch). This process will be furthered if the crystals are near the freezing point and wet, possibly by collision with undercooled water droplets. If a crystal falls into a cloud with great numbers of such drops, it will sweep some of them up. Coming into contact with ice, they freeze and form an ice cover around the crystal. Such particles are called soft hail or graupel.

Snow particles constitute the clouds of cirrus type—namely cirrus, cirrostratus, and cirrocumulus—and many clouds of alto type. Ice and snow clouds originate normally only at temperatures some degrees below the freezing point; they predominate at −20 °C (−4 °F). In temperate and low latitudes these clouds occur in the higher layers of the troposphere. In tropical regions they hardly ever occur below 4,570 metres (15,000 feet). On high mountains and particularly in polar regions, they can occur near the surface and may appear as ice fogs. If cold air near the ground is overlain by warmer air (a very common occurrence in polar regions, especially in winter), mixture at the border leads to supersaturation in the cold air. Small ice columns and needles, "diamond dust," will be formed and will float down, glittering, even from a cloudless sky. In the coldest parts of Antarctica, where temperatures near the surface are below −50 °C (−58 °F) on the average and rarely above −30 °C (−22 °F), the formation of diamond dust is a common occurrence. The floating and falling ice crystals produce in the light of the Sun and the Moon the manifold phenomena of atmospheric optics, halos, arcs, circles, mock suns, some coronas, and iridescent clouds. Most of the different optical appearances can be explained by the shapes of the crystals and their position with respect to the light source.

Most of the moderate to heavy rain in temperate latitudes depends on the presence of ice and snow particles in clouds. In the free atmosphere, droplets of fluid water can be undercooled considerably; typical ice clouds originate mainly at a temperature near −20 °C. At an identical temperature below the freezing point, the water molecules are kept more firmly in the solid than in the fluid state. The equilibrium pressure of the gaseous phase is smaller in contact with ice than with water. At −20 °C, which is the temperature of the formation of typical ice clouds (cirrus), the equilibrium pressure with respect to undercooled water (relative humidity 100 percent) is 22 percent greater than the equilibrium pressure of the water vapour in contact with ice. Hence, with an excess of water vapour beyond the equilibrium state, the ice particles tend to incorporate more water vapour and to grow more rapidly than the water droplets.

Being larger and so less retarded by friction, the ice particles fall more rapidly. In their fall they sweep up some water droplets, which on contact become frozen. Thus, a cloud layer originally consisting mainly of undercooled water with few ice crystals is transformed into an ice cloud. The development of the anvil shape at the top of a towering cumulonimbus cloud shows this transformation very clearly. The larger ice particles overcome more readily the rising tendency of the air in the cloud. Falling into lower levels they grow, aggregating with other crystals and possibly with waterdrops, melt, and form raindrops when near-surface temperatures permit.

Hail

Solid precipitation in the form of hard pellets of ice that fall from cumulonimbus clouds is called hail. It is convenient to distinguish between three types of hail particles.

The first is soft hail, or snow pellets, which are white opaque rounded or conical pellets as large as 6 mm (0.2 inch) in diameter. They are composed of small cloud droplets frozen together, have a low density, and are readily crushed.

The second is small hail (ice grains or pellets), which are transparent or translucent pellets of ice that are spherical, spheroidal, conical, or irregular in shape, with diameters of a few millimetres. They may consist of frozen raindrops, of largely melted and refrozen snowflakes, or of snow pellets encased in a thin layer of solid ice.

True hailstones, the third type, are hard pellets of ice, larger than 5 mm (0.2 inch) in diameter, that may be spherical, spheroidal, conical, discoidal, or irregular in shape and often have a structure of concentric layers of alternately clear and opaque ice. A moderately severe storm may produce stones a few centimetres in diameter, whereas a very severe storm may release stones with a maximum diameter of 10 cm (4 inches) or more. Large damaging hail falls most frequently in the continental areas of middle latitudes (e.g., in the Nebraska-Wyoming-Colorado area of the United States, in South Africa, and in northern India) but is rare in equatorial regions. Terminal velocities of hailstones range from about 5 metres (16 feet) per second for the smallest stones to perhaps 40 metres (130 feet) per second for stones 5 cm (2 inches) in diameter.

World distribution of Precipitation

Regional and Latitudinal distribution

The yearly precipitation averaged over the whole Earth is about 100 cm (39 inches), but this is distributed very unevenly. The regions of highest rainfall are found in the equatorial zone and the monsoon area of Southeast Asia. Middle latitudes receive moderate amounts of precipitation, but little falls in the desert regions of the subtropics and around the poles.

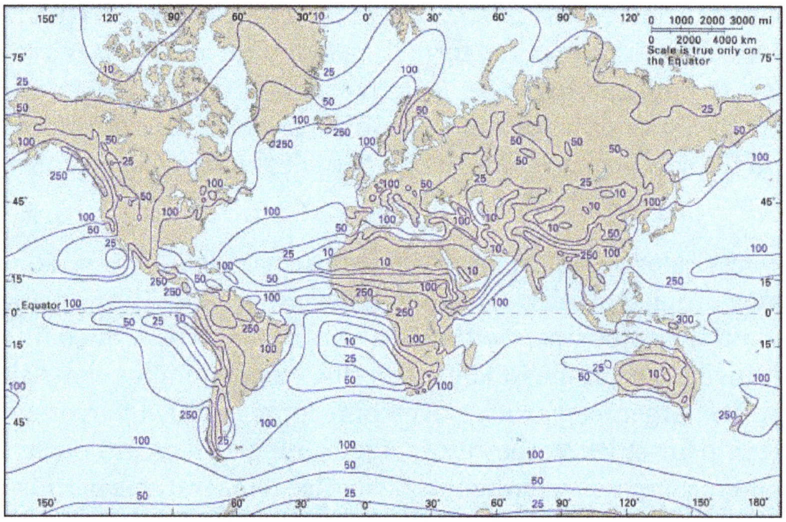

Global distribution of mean annual rainfall (in centimetres).

If Earth's surface were perfectly uniform, the long-term average rainfall would be distributed in distinct latitudinal bands, but the situation is complicated by the pattern of the global winds, the distribution of land and sea, and the presence of mountains. Because rainfall results from the ascent and cooling of moist air, the areas of heavy rain indicate regions of rising air, whereas the deserts occur in regions in which the air is warmed and dried during descent. In the subtropics, the trade winds bring plentiful rain to the east coasts of the continents, but the west coasts tend to be dry. On the other hand, in high latitudes the west coasts are generally wetter than the east coasts. Rain tends to be abundant on the windward slopes of mountain ranges but sparse on the lee sides.

In the equatorial belt, the trade winds from both hemispheres converge and give rise to a general upward motion of air, which becomes intensified locally in tropical storms that produce very heavy rains in the Caribbean, the Indian and southwest Pacific oceans, and the China Sea and in thunderstorms that are especially frequent and active over the land areas. During the annual cycle, the doldrums move toward the summer hemisphere, so outside a central region near the Equator, which has abundant rain at all seasons, there is a zone that receives much rain in summer but a good deal less in winter.

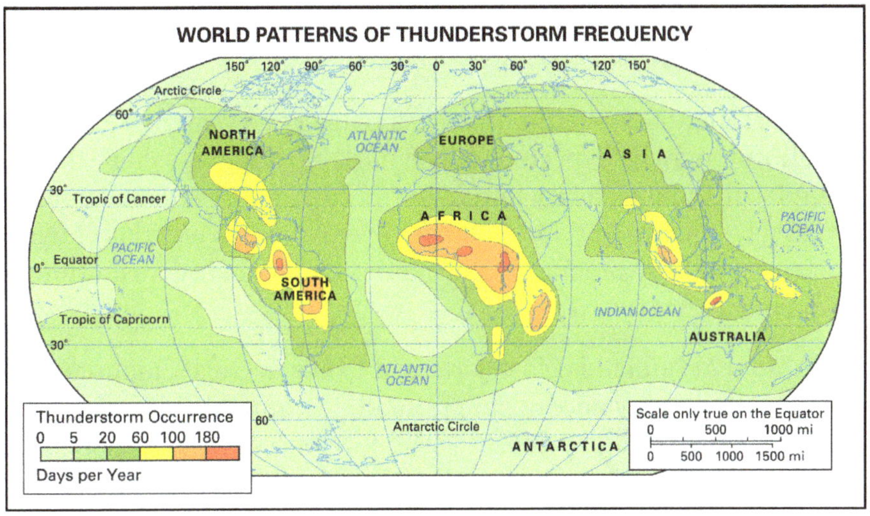

World patterns of thunderstorm frequency: Thunderstorms occur most often in the tropical latitudes over land, where the air is most likely to heat quickly and form strong updrafts.

The dry areas of the subtropics—such as the desert regions of North Africa, the Arabian Peninsula, South Africa, Australia, and central South America—are due to the presence of semipermanent subtropical anticyclones in which the air subsides and becomes warm and dry. These high-pressure belts tend to migrate with the seasons and cause summer dryness on the poleward side and winter dryness on the equatorward side of their mean positions. The easterly trade winds, having made a long passage over the warm oceans, bring plentiful rains to the east coasts of the subtropical landmasses, but the west coasts and the interiors of the continents, which are often sheltered by mountain ranges, are very dry.

In middle latitudes, weather and rainfall are dominated by traveling depressions and fronts that yield a good deal of rain in all seasons and in most places except the far interiors of the Asian and North American continents. Generally, rainfall is more abundant in summer, except on the western coasts of North America, Europe, and North Africa, where it is higher during the winter.

At high latitudes and especially in the polar regions, the low precipitation is caused partly by subsidence of air in the high-pressure belts and partly by the low temperatures. Snow or rain occur at times, but evaporation from the cold sea and land surfaces is slow, and the cold air has little capacity for moisture.

The influence of oceans and continents on rainfall is particularly striking in the case of the Indian monsoon. During the Northern Hemisphere winter, cool dry air from the interior of the continent flows southward and produces little rain over the land areas. After the air has traveled some distance over the warm tropical ocean, however, it releases heavy shower rains over the East Indies. During the northern summer, when the monsoon blows from the southwest, rainfall is heavy over India and Southeast Asia. These rains are intensified where the air is forced to ascend over the windward slopes of the Western Ghats and the Himalayas.

The combined effects of land, sea, mountains, and prevailing winds show up in South America. There the desert in southern Argentina is sheltered by the Andes from the westerly winds blowing from the Pacific Ocean, and the west-coast desert not only is situated under the South Pacific subtropical anticyclone but is also protected by the Andes against rain-bearing winds from the Atlantic.

Amounts and Variability

The long-term average amounts of precipitation for a season or a year give little information on the regularity with which rain may be expected, particularly for regions where the average amounts are small. For example, at Iquique, a city in northern Chile, four years once passed without rain, whereas the fifth year gave 15 mm (0.6 inch); the five-year average was therefore 3 mm (0.1 inch). Clearly, such averages are of little practical value, and the frequency distribution or the variability of precipitation also must be known.

The variability of the annual rainfall is closely related to the average amounts. For example, over the British Isles, which have a very dependable rainfall, the annual amount varies by less than 10 percent above the long-term average value. A variability of less than 15 percent is typical of the mid-latitude cyclonic belts of the Pacific and Atlantic oceans and of much of the wet equatorial regions. In the interiors of the desert areas of Africa, Arabia, and Central Asia, however, the rainfall in a particular year may deviate from the normal long-term average by more than 40 percent. The variability for individual seasons or months may differ considerably from that for the year as a whole, but again the variability tends to be higher where the average amounts are low.

The heaviest annual rainfall in the world was recorded at the village of Cherrapunji, India, where 26,470 mm (1,042 inches) fell between August 1860 and July 1861. The heaviest rainfall in a period of 24 hours was 1,870 mm (74 inches), recorded at the village of Cilaos, Réunion, in the Indian Ocean on March 15–16, 1952. The lowest recorded rainfall in the world occurred at Arica, a port city in northern Chile. An annual average, taken over a 43-year period, was only 0.5 mm (0.02 inch).

Although past records give some guide, it is not possible to estimate very precisely the maximum possible precipitation that may fall in a given locality during a specified interval of time. Much will depend on a favourable combination of several factors, including the properties of the storm and the effects of local topography. Thus, it is possible only to make estimates that are based on analyses of past storms or on theoretical calculations that attempt to maximize statistically the various factors or the most effective combination of factors that are known to control the duration and intensity of the precipitation. For many important planning and design problems, however, estimates of the greatest precipitation to be expected at a given location within a specified number of years are required.

In the designing of a dam, the highest 24-hour rainfall to be expected once in 30 years over the whole catchment area might be relevant. For dealing with such problems, a great deal of work has been devoted to determining from past records the frequency with which rainfalls of given intensity and total amount may be expected to reoccur at particular locations and also to determining the statistics of rainfall for a specific area from measurements made at only a few points.

Effects of Precipitation

Raindrop Impact and Soil Erosion

Large raindrops, up to 6 mm (0.2 inch) in diameter, have terminal velocities of about 10 metres (30 feet) per second and so may cause considerable compaction and erosion of the soil by their force of impact. The formation of a compacted crust makes it more difficult for air and water to reach the roots of plants and encourages the water to run off the surface and carry away the topsoil with it. In hilly and mountainous areas, heavy rain may turn the soil into mud and slurry, which may produce enormous erosion by mudflow generation. Rainwater running off hard impervious surfaces or waterlogged soil may cause local flooding.

Surface Runoff

The rainwater that is not evaporated or stored in the soil eventually runs off the surface and finds its way into rivers, streams, and lakes or percolates through the rocks and becomes stored in natural underground reservoirs. A given catchment area must achieve an overall balance such that precipitation (P) less evaporation of moisture from the surface (E) will equal storage in the ground (S) and runoff (R). This may be expressed: $P - E = S + R$. The runoff may be determined by measuring the flow of water in the rivers with stream gauges, and the precipitation may be measured by a network of rain gauges, but storage and evaporation are more difficult to estimate.

Of all the water that falls on Earth's surface, the relative amounts that run off, evaporate, or seep into the ground vary so much for different areas that no firm figures can be given for Earth as a whole. It has been estimated, however, that in the United States 10 to 50 percent of the rainfall runs off at once, 10 to 30 percent evaporates, and 40 to 60 percent is absorbed by the soil. Of the entire rainfall, 15 to 30 percent is estimated to be used by plants, either to form plant tissue or in transpiration.

Atmospheric Pressure

The air around you has weight, and it presses against everything it touches. That pressure is called atmospheric pressure, or air pressure. It is the force exerted on a surface by the air above it as gravity pulls it to Earth.

Atmospheric pressure is commonly measured with a barometer. In a barometer, a column of mercury in a glass tube rises or falls as the weight of the atmosphere changes. Meteorologists describe the atmospheric pressure by how high the mercury rises.

An atmosphere (atm) is a unit of measurement equal to the average air pressure at sea level at a temperature of 15 degrees Celsius (59 degrees Fahrenheit). One atmosphere is 1,013 millibars, or 760 millimeters (29.92 inches) of mercury.

Atmospheric pressure drops as altitude increases. The atmospheric pressure on Denali, Alaska, is about half that of Honolulu, Hawai'i. Honolulu is a city at sea level. Denali, also known as Mount McKinley, is the highest peak in North America.

As the pressure decreases, the amount of oxygen available to breathe also decreases. At very high altitudes, atmospheric pressure and available oxygen get so low that people can become sick and even die.

Mountain climbers use bottled oxygen when they ascend very high peaks. They also take time to get used to the altitude because quickly moving from higher pressure to lower pressure can cause decompression sickness. Decompression sickness, also called "the bends", is also a problem for scuba divers who come to the surface too quickly.

Aircraft create artificial pressure in the cabin so passengers remain comfortable while flying. Atmospheric pressure is an indicator of weather. When a low-pressure system moves into an area, it usually leads to cloudiness, wind, and precipitation. High-pressure systems usually lead to fair, calm weather.

Atmospheric Pressure and Wind

Atmospheric pressure and wind are both significant controlling factors of Earth's weather and climate. Although these two physical variables may at first glance appear to be quite different, they are in fact closely related. Wind exists because of horizontal and vertical differences (gradients) in pressure, yielding a correspondence that often makes it possible to use the pressure distribution as an alternative representation of atmospheric motions. Pressure is the force exerted on a unit area, and atmospheric pressure is equivalent to the weight of air above a given area on Earth's surface or within its atmosphere. This pressure is usually expressed in millibars (mb; 1 mb equals 1,000 dynes per square cm) or in kilopascals (kPa; 1 kPa equals 10,000 dynes per square cm). Distributions of pressure on a map are depicted by a series of curved lines called isobars, each of which connects points of equal pressure.

At sea level the mean pressure is about 1,000 mb (100 kPa), varying by less than 5 percent from this value at any given location or time. Mean sea-level pressure values for the mid-winter months in the Northern Hemisphere are summarized in this first diagram, and mean sea-level pressure values for the mid-summer months are illustrated in the next diagram. Since charts of atmospheric pressure often represent average values over several days, pressure features that are relatively consistent day after day emerge, while more transient, short-lived features are removed. Those that remain are known as semipermanent pressure centres and are the source regions for major, relatively uniform bodies of air known as air masses. Warm, moist maritime tropical (mT) air forms over tropical and subtropical ocean waters in association with the high-pressure regions prominent there. Cool, moist maritime polar (mP) air, on the other hand, forms over the colder subpolar ocean waters just south and east of the large, winter oceanic low-pressure regions. Over the continents, cold dry continental polar (cP) air and extremely cold dry continental arctic (cA) air forms in the high-pressure regions that are especially pronounced in winter, while hot dry continental tropical (cT) air forms over hot desertlike continental domains in summer in association with low-pressure areas, which are sometimes called heat lows.

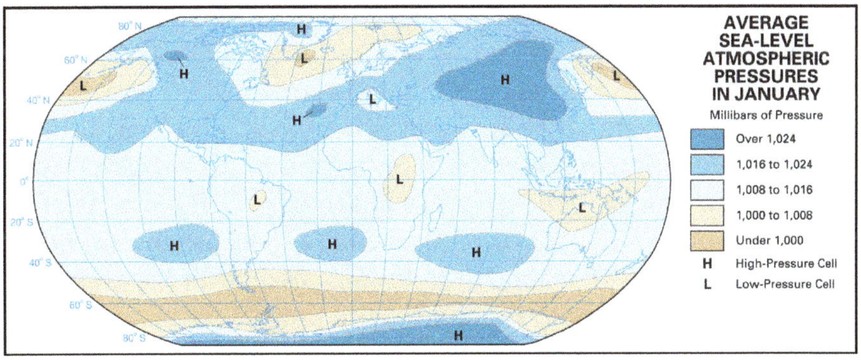

Average atmospheric pressure at sea level.

A closer examination of the diagrams above reveals some interesting features. First, it is clear that sea-level pressure is dominated by closed high- and low-pressure centres, which are largely caused by differential surface heating between low and high latitudes and between continental and oceanic regions. High pressure tends to be amplified over the colder surface features. Second, because of seasonal changes in surface heating, the pressure centres exhibit seasonal changes in their characteristics. For example, the Siberian High, Aleutian Low, and Icelandic Low that are so prominent in the winter virtually disappear in summer as the continental regions warm relative to surrounding bodies of water. At the same time, the Pacific and Atlantic highs amplify and migrate northward.

At altitudes well above Earth's surface, the monthly average pressure distributions show much less tendency to form in closed centres but rather appear as quasi-concentric circles around the poles. This more symmetrical appearance reflects the dominant role of meridional (north-south) differences in radiative heating and cooling. Excess heating in tropical latitudes, in contrast to polar areas, produces higher pressure at upper levels in the tropics as thunderstorms transfer air to higher levels. In addition, the greater heating/cooling contrast in winter yields stronger pressure differences during this season. Perfect symmetry between the tropics and the poles is interrupted by wavelike atmospheric disturbances associated with migratory and semipermanent high- and low-pressure surface weather systems. These weather systems are most pronounced over the Northern Hemisphere, with its more prominent land-ocean contrasts and orographic (high-elevation) features.

Wind

Relationship of Wind to Pressure and Governing Forces

The changing wind patterns are governed by Newton's second law of motion, which states that the sum of the forces acting on a body equals the product of the mass of that body and the acceleration caused by those forces. The basic relationship between atmospheric pressure and horizontal wind is revealed by disregarding friction and any changes in wind direction and speed to yield the mathematical relationship:

$$fu = -\frac{1}{\rho}\frac{\varphi p}{\varphi y} \text{ and } fv = \frac{1}{\rho}\frac{\varphi p}{\varphi x},$$

Where u is the zonal wind speed (+ eastward), v the meridional wind speed (+ northward), f = 2ω sin φ (Coriolis parameter), ω the angular velocity of Earth's rotation, φ the latitude, ρ the air density (mass per unit volume), p the pressure, and x and y the distances toward the east and north, respectively. This simple non-accelerating flow is known as geostrophic balance and yields a motion field known as the geostrophic wind. Equation expresses, for both the x and y directions, a balance between the force created by horizontal differences in pressure (the horizontal pressure-gradient force) and an apparent force that results from Earth's rotation (the Coriolis force). The pressure-gradient force expresses the tendency of pressure differences to effectuate air movement from higher to lower pressure. The Coriolis force arises because the air motions are observed on a rotating nearly spherical body. The total motion of a parcel of air has two parts: (1) the motion relative to Earth as if the planet were fixed, and (2) the motion given to the parcel of air by the planet's rotation. When the atmosphere is viewed from a fixed point in space, Earth's rotation is apparent. An observer in space would witness the total motion of the atmosphere. Conversely, an observer on the ground sees and measures only the relative motion of the atmosphere, because he is also rotating and cannot see directly the rotational motion applied by Earth. Instead, the observer on the ground sees the effect of the rotation as a deviation applied to the relative motion. The quantity that describes this deviation is the Coriolis force. Because the Coriolis force results from a ground-level frame of reference on a rotating planet, it is not a true force.

More specifically, the observer on the ground experiences the Coriolis force as a deflection of the relative motion to the right in the Northern Hemisphere and to the left in the Southern Hemisphere. Of particular significance in this simple model of wind-pressure relationships is the fact that the geostrophic wind blows in a direction parallel to the isobars, with the low pressure on the observer's left as he looks downwind in the Northern Hemisphere and on his right in the Southern Hemisphere.

Wind speed increases as the distance between isobars decreases (or pressure gradient increases). Curvature (i.e., changes in wind direction) can be added to this model with relative ease in a flow representation known as the gradient wind. The basic wind-pressure relationships, however, remain qualitatively the same. Of greatest importance is the fact that large-scale, observed winds tend to behave much as the geostrophic- or gradient-flow models predict in most of the atmosphere. The most notable exceptions occur in low latitudes, where the Coriolis parameter becomes very small—equation cannot be used to provide a reliable wind estimate—and in the lowest kilometre of the atmosphere, where friction becomes important. The friction induced by airflow over the underlying surface reduces the wind speed and alters the simple balance of forces such that the wind blows with a component toward lower pressure.

Cyclones and Anticyclones

Cyclones and anticyclones are regions of relatively low and high pressure, respectively. They occur over most of Earth's surface in a variety of sizes ranging from the very large semipermanent examples described above to smaller, highly mobile systems.

Common to both cyclones and anticyclones are the characteristic circulation patterns. The geostrophic-wind and gradient-wind models dictate that, in the Northern Hemisphere, flow around a cyclone—cyclonic circulation—is counterclockwise, and flow around an anticyclone—anticyclonic circulation—are clockwise. Circulation directions are reversed in the Southern Hemisphere. In the presence of friction, the superimposed component of motion toward lower pressure produces a "spiraling" effect toward the low-pressure centre and away from the high-pressure centre.

The cyclones that form outside the equatorial belt, known as extratropical cyclones, may be regarded as large eddies in the broad air currents that flow in the general direction from west to east around the middle and higher latitudes of both hemispheres. They are an essential part of the mechanism by which the excess heat received from the Sun in Earth's equatorial belt is conveyed toward higher latitudes. These higher latitudes radiate more heat to space than they receive from the Sun, and heat must reach them by winds from the lower latitudes if their temperature is to be continually cool rather than cold. If there were no cyclones and anticyclones, the north-south movements of the air would be much more limited, and there would be little opportunity for heat to be carried poleward by winds of subtropical origin. Under such circumstances the temperature of the lower latitudes would increase, and the polar regions would cool; the temperature gradient between them would intensify.

Strong horizontal gradients of temperature are particularly favourable for the formation and development of cyclones. The temperature difference between polar regions and the Equator builds up until it becomes sufficiently intense to generate new cyclones. As their associated cold fronts sweep equatorward and their warm fronts move poleward, the new cyclones reduce the temperature difference. Thus, the wind circulation on Earth represents a balance between the heating effects of solar radiation occurring in the polar regions and at the Equator. Wind circulation, through the effect of cyclones, anticyclones, and other wind systems, also periodically destroys this temperature contrast.

Cyclones of a somewhat different character occur closer to the Equator, generally forming in latitudes between 10° to 30° N and S over the oceans. They generally are known as tropical cyclones when their winds equal or exceed 74 miles (119 km) per hour. They are also known as hurricanes if they occur in the Atlantic Ocean and the Caribbean Sea, as typhoons in the western Pacific Ocean and the China Sea, and as cyclones off the coasts of Australia. These storms are of smaller diameter than the extratropical cyclones, ranging from 100 to 500 km (60 to 300 miles) in diameter, and are accompanied by winds that sometimes reach extreme violence.

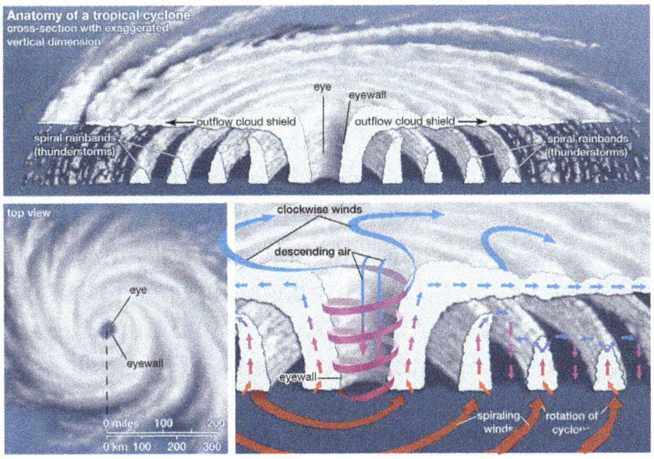

A top view and vertical cross section of a tropical cyclone.

Extratropical Cyclones

Of the two types of large-scale cyclones, extratropical cyclones are the most abundant and exert influence on the broadest scale; they affect the largest percentage of Earth's surface. Furthermore, this class of cyclones is the principal cause of day-to-day weather changes experienced in middle and high latitudes and thus is the focal point of much of modern weather forecasting. The seeds for many current ideas concerning extratropical cyclones were sown between 1912 and 1930 by a group of Scandinavian meteorologists working in Bergen, Nor. This so-called Bergen school, founded by Norwegian meteorologist and physicist Vilhelm Bjerknes, formulated a model for a cyclone that forms as a disturbance along a zone of strong temperature contrast known as a front, which in turn constitutes a boundary between two contrasting air masses. In this model the masses of polar and mid-latitude air around the globe are separated by the polar front (the transition region separating warmer tropical air from colder polar air). This region possesses a strong temperature gradient, and thus it is a reservoir of potential energy that can be readily tapped and converted into the kinetic energy associated with extratropical cyclones.

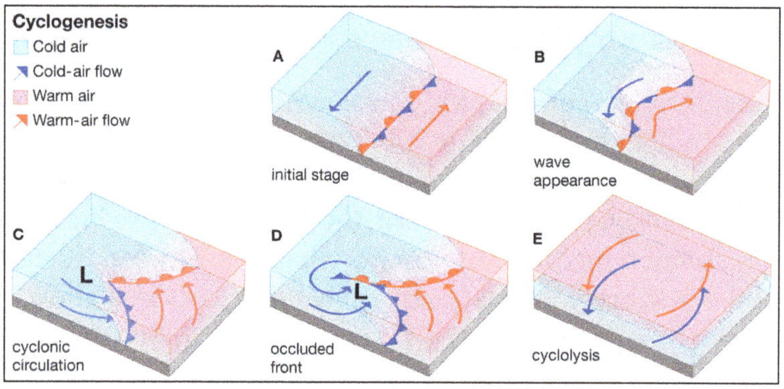

Evolution of a wave (frontal) cyclone.

For this reservoir to be tapped, a cyclone (called a wave, or frontal, cyclone) must develop much in the way shown in the diagram. The feature that is of primary importance prior to cyclone development (cyclogenesis) is a front, represented in the initial stage (A) as a heavy black line with alternating triangles or semicircles attached to it. This stationary or very slow-moving front forms a boundary between cold and warm air and thus is a zone of strong horizontal temperature gradient (sometimes referred to as a baroclinic zone). Cyclone development is initiated as a disturbance along the front, which distorts the front into the wavelike configuration (B; wave appearance). As the pressure within the disturbance continues to decrease, the disturbance assumes the appearance of a cyclone and forces poleward and equatorward movements of warm and cold air, respectively, which are represented by mobile frontal boundaries. As depicted in the cyclonic circulation stage (C), the front that signals the advancing cold air (cold front) is indicated by the triangles, while the front corresponding to the advancing warm air (warm front) is indicated by the semicircles. As the cyclone continues to intensify, the cold dense air streams rapidly equatorward, yielding a cold front with a typical slope of 1 to 50 and propagation speed that is often 8 to 15 metres per second (about 18 to 34 miles per hour) or more. At the same time, the warm less-dense air moving in a northerly direction flows up over the cold air east of the cyclone to produce a warm front with a typical slope of 1 to 200 and a typically much slower propagation speed of about 2.5 to 8 metres per second (6 to 18 miles per hour). This difference in propagation speeds between the two fronts allows the cold front to overtake the warm front and produce yet another, more complicated frontal structure, known as an occluded front. An occluded front (D) is represented by a line with alternating triangles and semicircles on the same side. This occlusion process may be followed by further storm intensification. The separation of the cyclone from the warm air toward the Equator, however, eventually leads to the storm's decay and dissipation (E) in a process called cyclolysis.

The life cycle of such an event is typically several days, during which the cyclone may travel from several hundred to a few thousand kilometres. In its path and wake occur dramatic weather changes. A typical sequence of weather possibly resulting from the approach and passage of a cyclone and its fronts through an area is depicted in the diagram. Shown in the occluded-front stage of the cyclogenesis diagram is a cross section of the clouds and precipitation that usually occur along line ab. Warm frontal weather is most frequently characterized by stratiform clouds, which ascend as the front approaches and potentially yield rain or snow. The passing of a warm front brings a rise in air temperature and clearing skies. The warmer air, however, may also harbour the ingredients for rain shower or thunderstorm formation, a condition that is enhanced as the cold front approaches.

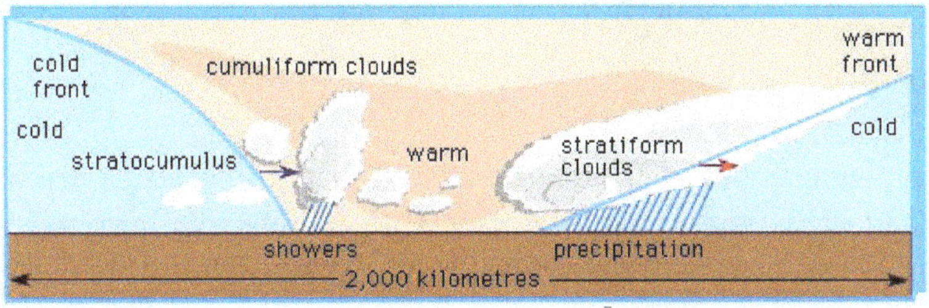

Cross section of clouds and precipitation often found along the cross-sectional line ab in panel D in the cyclogenesis diagram; the direction of frontal movement is indicated by the arrows.

The passage of the cold front is marked by the influx of colder air, the formation of stratocumulus clouds with some lingering rain or snow showers, and then eventual clearing. While this is an oft-repeated scenario, it is important to recognize that many other weather sequences can also occur. For example, the stratiform clouds of a warm front may have imbedded cumulus formations and thunderstorms; the warm sector might be quite dry and yield few or no clouds; the pre-cold-front weather may closely resemble that found ahead of the warm front; or the post-cold-front air may be completely cloud-free. Cloud patterns oriented along fronts and spiraling around the cyclone vortex are consistently revealed in satellite pictures of Earth.

The actual formation of any area of low pressure requires that mass in the column of air lying above Earth's surface be reduced. This loss of mass then reduces the surface pressure. In the late 1930s and early '40s, three members of the Bergen school—Norwegian American meteorologists Jacob Bjerknes and Jørgen Holmboe and Swedish American meteorologist Carl-Gustaf Rossby—recognized that transient surface disturbances were accompanied by complementary wave features in the flow in the middle and higher atmospheric layers associated with the jet stream. These wave features are accompanied by regions of mass divergence and convergence that support the growth of surface-pressure fields and direct their movement.

While extratropical cyclones form and intensify in association with fronts, there are small-scale cyclones that appear in the middle of a single air mass. A notable example is a class of cyclones, generally smaller than the frontal variety that forms in polar air streams in the wake of a frontal cyclone. These so-called polar lows are most prominent in subpolar marine environments and are thought to be caused by the transfer of heat and moisture from the warmer water surface into the overlying polar air and by supporting middle-tropospheric circulation features. Other small-scale cyclones form on the lee side of mountain barriers as the general westerly flow is disturbed by the mountain. These "lee cyclones" may produce major windstorms and dust storms downstream of a mountain barrier.

Satellite image of a large dust storm.

Anticyclones

While cyclones are typically regions of inclement weather, anticyclones are usually meteorologically quiet regions. Generally larger than cyclones, anticyclones exhibit persistent downward motions and yield dry stable air that may extend horizontally many hundreds of kilometres.

In most cases, an actively developing anticyclone forms over a ground location in the region of cold air behind a cyclone as it moves away. This anticyclone forms before the next cyclone advances into the area. Such an anticyclone is known as a cold anticyclone. A result of the downward air motion in an anticyclone, however, is compression of the descending air. As a consequence of this compression, the air is warmed. Thus, after a few days, the air composing the anticyclone at levels 2 to 5 km (1 to 3 miles) above the ground tends to increase in temperature, and the anticyclone is transformed into a warm anticyclone.

Warm anticyclones move slowly, and cyclones are diverted around their periphery. During their transformation from cold to warm status, anticyclones usually move out of the main belt followed by cyclones in middle latitudes and often amalgamate with the quasi-permanent bands of relatively high pressure found in both hemispheres around latitude 20° to 30°—the so-called subtropical anticyclones. On some occasions the warm anticyclones remain in the belt normally occupied by the mid-latitude westerly winds. The normal cyclone tracks are then considerably modified; atmospheric depressions (areas of low pressure) are either blocked in their eastward progress or diverted to the north or south of the anticyclone. Anticyclones that interrupt the normal circulation of the westerly wind in this way are called blocking anticyclones, or blocking highs. They frequently persist for a week or more, and the occurrence of a few such blocking anticyclones may dominate the character of a season. Blocking anticyclones are particularly common over Europe, the eastern Atlantic, and the Alaskan area.

The descent and warming of the air in an anticyclone might be expected to lead to the dissolution of clouds and the absence of rain. Near the centre of the anticyclone, the winds are light and the air can become stagnant. Air pollution can build up as a result. The city of Los Angeles, for example, often has poor air quality because it is frequently under a stationary anticyclone. In winter the ground cools, and the lower layers of the atmosphere also become cold. Fog may be formed as the air is cooled to its dew point in the stagnant air. Under other circumstances, the air trapped in the first kilometre above Earth's surface may pick up moisture from the sea or other moist surfaces, and layers of cloud may form in areas near the ground up to a height of about 1 km (0.6 mile). Such layers of cloud can be persistent in anticyclones (except over the continents in summer), but they rarely grow thick enough to produce rain. If precipitation occurs, it is usually drizzle or light snow.

Anticyclones are often regions of clear skies and sunny weather in summer; at other times of the year, cloudy and foggy weather—especially over wet ground, snow cover, and the ocean—may be more typical. Winter anticyclones produce colder than average temperatures at the surface, particularly if the skies remain clear. Anticyclones are responsible for periods of little or no rain, and such periods may be prolonged in association with blocking highs.

Cyclone and Anticyclone Climatology

Migrating cyclones and anticyclones tend to be distributed around certain preferred regions, known as tracks, that emanate from preferred cyclogenetic and anticyclogenetic regions. The contrast between the winter and summer mean sea-level pressure diagrams also indicates the typical cyclone tracks for both January and July. Favoured cyclogenetic regions in the Northern Hemisphere are found on the lee side of mountains and off the east coasts of continents. Cyclones then track east or southeast before eventually turning toward the northeast and decaying. The tracks are displaced farther northward in July, reflecting the more northward position of the polar front in summer. Continental cyclones usually intensify at a rate of 0.5 mb (0.05 kPa) per hour or less, although more dramatic examples can be found. Marine cyclones, on the other hand, often experience explosive development in excess of 1 mb (0.1 kPa) per hour, particularly in winter.

Anticyclones tend to migrate equatorward out of the cold air mass regions and then eastward before decaying or merging with a warm anticyclone. Like cyclones, warm anticyclones also slowly migrate poleward with the warm season.

In the Southern Hemisphere, where most of Earth's surface is covered by oceans, the cyclones are distributed fairly uniformly through the various longitudes. Typically, cyclones form initially in latitudes 30° to 40° S and move in a generally southeastward direction, reaching maturity in latitudes near 60° S. Thus, the Antarctic continent is usually ringed by a number of mature or decaying cyclones. The belt of ocean from 40° to 60° S is a region of persistent, strong westerly winds that form part of the circulation to the north of the main cyclone centres; These are the "roaring forties," where the westerly winds are interrupted only at intervals by the passage southeastward of developing cyclones.

Local Winds

Scale Classes

Organized wind systems occur in spatial dimensions ranging from tens of metres to thousands of kilometres and possess residence times that vary from seconds to weeks. The concept of scale considers the typical size and lifetime of a phenomenon. Since the atmosphere exhibits such a large variety of both spatial and temporal scales, efforts have been made to group various phenomena into scale classes. The class describing the largest and longest-lived of these phenomena is known as the planetary scale. Such phenomena are typically a few thousand kilometres in size and have lifetimes ranging from several days to several weeks. Examples of planetary-scale phenomena include the semipermanent pressure centres and certain globe-encircling upper-air waves.

A second class is known as the synoptic scale. Spanning smaller distances, a few hundred to a few thousand kilometres, and possessing shorter lifetimes, a few to several days, this class contains the migrating cyclones and anticyclones that control day-to-day weather changes. Sometimes the planetary and synoptic scales are combined into a single classification termed the large-scale, or macroscale. Large-scale wind systems are distinguished by the predominance of horizontal motions over vertical motions and by the preeminent importance of the Coriolis force in influencing wind characteristics. Examples of large-scale wind systems include the trade winds and the westerlies.

There is a third class of phenomena of even smaller size and shorter lifetime. In this class, vertical motions may be as significant as horizontal movement, and the Coriolis force often plays a less important role. Known as the mesoscale, this class is characterized by spatial dimensions of ten to a few hundred kilometres and lifetimes of a day or less. Because of the shorter time scale and because the other forces may be much larger, the effect of the Coriolis force in mesoscale phenomena is sometimes neglected. Two of the best-known examples of mesoscale phenomena are the thunderstorm and its devastating by-product, the tornado.

Local Wind Systems

The so-called sea and land breeze circulation is a local wind system typically encountered along coastlines adjacent to large bodies of water and is induced by differences that occur between the heating or cooling of the water surface and the adjacent land surface. Water has a higher heat capacity (i.e., more units of heat are required to produce a given temperature change in a volume of water) than do the materials in the land surface. Daytime solar radiation penetrates to several metres into the water, the water vertically mixes, and the volume is slowly heated. In contrast, daytime solar radiation heats the land surface more quickly because it does not penetrate more than a few centimetres below the land surface. The land surface, now at a higher temperature relative to the air adjacent to it, transfers more heat to its overlying air mass and creates an area of low pressure. Thus, a circulation cell much like that depicted in the diagram is induced. It should be noted that the surface flow is from the water toward the land and thus is called a sea breeze.

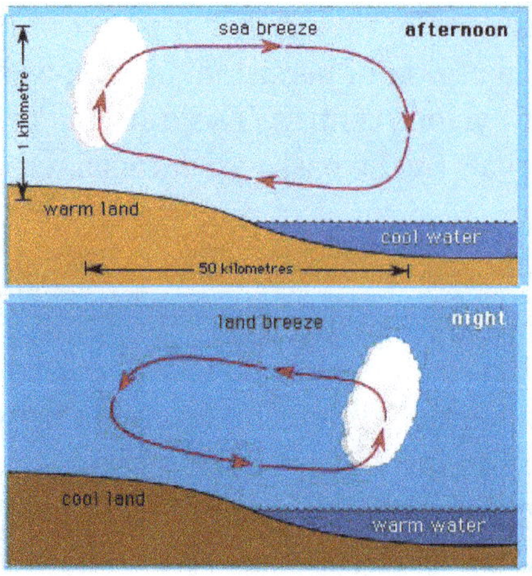

Typical sea-breeze (afternoon) and land-breeze (night) circulations with associated cloud formations.

Since the landmass possesses a lower heat capacity than water, the land cools more rapidly at night than does the water. Consequently, at night the cooler landmass yields a cooler overlying air mass and creates a zone of relatively higher pressure. This produces a circulation cell with air motions opposite to those found during the day. This flow from land to water is known as a land breeze. The land breeze is typically shallower than the sea breeze since the cooling of the atmosphere over land is confined to a shallower layer at night than the heating of the air during the day.

Sea and land breezes occur along the coastal regions of oceans or large lakes in the absence of a strong large-scale wind system during periods of strong daytime heating or night time cooling. Those who live within 10 to 20 km (6 to 12 miles) of the coastline often experience the cooler 19- to 37-km-per-hour (12- to 23-mile-per-hour) winds of the sea breeze on a sunny afternoon only to find it turn into a sultry land breeze late at night. One of the features of the sea and land breeze is a region of low-level air convergence in the termination region of the surface flow. Such convergence often induces local upward motions and cloud formations. Thus, in sea and land breeze regions, it is not uncommon to see clouds lying off the coast at night; these clouds are then dissipated by the daytime sea breeze, which forms new clouds, perhaps with showers occurring over land in the afternoon.

Another group of local winds is induced by the presence of mountain and valley features on Earth's surface. One subset of such winds, known as mountain winds or breezes, is induced by differential heating or cooling along mountain slopes. During the day, solar heating of the sunlit slopes causes the overlying air to move upslope. These winds are also called anabatic flow. At night, as the slopes cool, the direction of airflow is reversed, and cool downslope drainage motion occurs. Such winds may be relatively gentle or may occur in strong gusts, depending on the topographic configuration. These winds are one type of katabatic flow. In an enclosed valley, the cool air that drains into the valley may give rise to a thick fog condition. Fog persists until daytime heating reverses the circulation and creates clouds associated with the upslope motion at the mountain top.

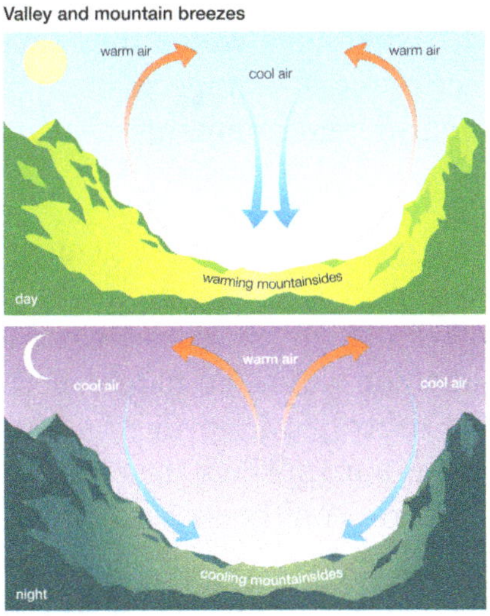

When the valley floor warms during the day, warm air rises up the slopes of surrounding mountains and hills to create a valley breeze. At night, denser cool air slides down the slopes to settle in the valley, producing a mountain breeze.

Another subset of katabatic flow, called foehn winds (also known as chinook winds east of the Rocky Mountains and as Santa Ana winds in southern California), is induced by adiabatic temperature changes occurring as air flows over a mountain. Adiabatic temperature changes are those that occur without the addition or subtraction of heat; they occur in the atmosphere when bundles of air are moved vertically. When air is lifted, it enters a region of lower pressure and expands. This expansion is accompanied by a reduction of temperature (adiabatic cooling). When air subsides, it contracts and experiences adiabatic warming. As air ascends on the windward side of the mountain, its cooling rate may be moderated by heat that is released during the formation of precipitation. However, having lost much of its moisture, the descending air on the leeward side of the mountain adiabatically warms faster than it was cooled on the windward ascent. Thus, the effect of this wind, if it reaches the surface, is to produce warm, dry conditions. Usually, such winds are gentle and produce a slow warming. On occasion, however, foehn winds may exceed 185 kilometres (115 miles) per hour and produce air-temperature increases of tens of degrees (sometimes more than 20 °C [36 °F]) within only a few hours.

Other types of katabatic wind can occur when the underlying geography is characterized by a cold plateau adjacent to a relatively warm region of lower elevation. Such conditions are satisfied in areas in which major ice sheets or cold elevated land surfaces border warmer large bodies of water. Air over the cold plateau cools and forms a large dome of cold dense air. Unless held back by background wind conditions, this cold air will spill over into the lower elevations with speeds that vary from gentle (a few kilometres per hour) to intense (93 to 185 km [58 to 115 miles] per hour), depending on the incline of the slope of the terrain and the distribution of the background pressure field. Two special varieties of katabatic wind are well known in Europe. One is the bora, which blows from the highlands of Croatia, Bosnia and Herzegovina, and Montenegro to the Adriatic Sea; the other is the mistral, which blows out of central and southern France to the Mediterranean Sea. Creating blizzard conditions, intense katabatic winds often blow northward off the Antarctic Ice Sheet.

Zonal Surface Winds

The diagrams of January and July mean sea-level pressure reveal that, on the average, certain geographic locations can expect to experience winds that emanate from one prevailing direction largely dictated by the presence of major semipermanent pressure systems. Such prevailing winds have long been known in marine environments because of their influence on the great sailing ships.

Tropical and subtropical regions are characterized by a general band of low pressure lying near the Equator. This band is bounded by centres of high pressure that may extend poleward into the middle latitudes. Between these low- and high-pressure regions is the region of the tropical winds. Of these the most extensive are the trade winds. So named because of their favourable influence on trade ships traveling across the subtropical North Atlantic, trade winds flow westward and somewhat in the direction of the Equator on the equatorward side of the subtropical high-pressure centres. The "root of the trades," occurring on the eastern side of a subtropical high-pressure centre, is characterized by subsiding air. This produces the very warm, dry conditions above a shallow layer of oceanic stratus clouds found in the eastern extremes of the subtropical Atlantic and Pacific ocean basins. As the trade winds progress westward, however, subsidence abates, the air mass becomes more humid, and scattered showers appear. These showers occur particularly on islands with elevated terrain features that interrupt the flow of the warm moist air. The equatorward flow of the trade winds of the Northern and Southern hemispheres often results in a convergence of the two air streams in a region known as the intertropical convergence zone (ITCZ). Deep convective clouds, showers, and thunderstorms occur along the ITCZ.

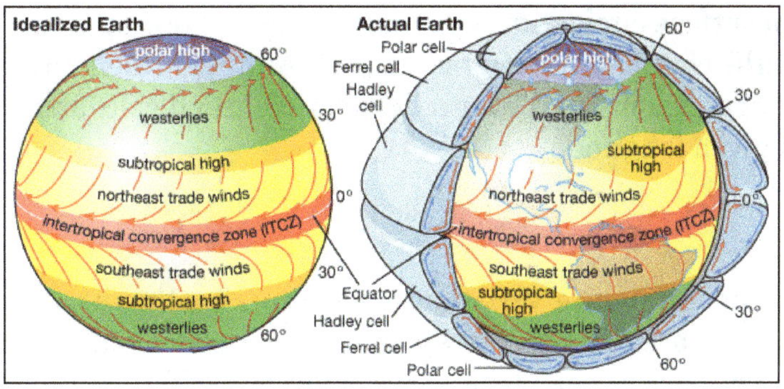

General patterns of atmospheric circulation over an idealized Earth with a uniform surface (left) and the actual Earth (right).

When the air reaches the western extreme of the high-pressure centre, it turns poleward and then eventually returns eastward in the middle latitudes. The poleward-moving air is now warm and laden with moist maritime tropical air (mT); it gives rise to the warm, humid, showery climate characteristic of the Caribbean region, eastern South America, and the western Pacific island chains. The westerlies are associated with the changeable weather common to the middle latitudes. Migrating extratropical cyclones and anticyclones associated with contrasting warm moist air moving poleward from the tropics and cold dry air moving equatorward from polar latitudes yield periods of rain (sometimes with violent thunderstorms), snow, sleet, or freezing rain interrupted by periods of dry, sunny, and sometimes bitterly cold conditions. Furthermore, these patterns are seasonally dependent, with more intense cyclones and colder air prevailing in winter but with a higher incidence of thunderstorms common in spring and summer. In addition, these migrations and the associated climate are complicated by the presence of landmasses and major mountain features, particularly in the Northern Hemisphere.

The westerlies lie on the equatorward side of the semipermanent subpolar centres of low pressure. Poleward of these centres, the surface winds turn westward again over significant portions of the subpolar latitudes. As in the middle latitudes, the presence of major landmasses, notably in the Northern Hemisphere, results in significant variations in these polar easterlies. In addition, the wind systems and the associated climate are seasonally dependent. During the short summer season, the wind systems of the polar latitudes are greatly weakened. During the long winter months, these systems strengthen, and periods of snow alternate with long intervals of dry cold air characteristic of continental polar or continental arctic air masses.

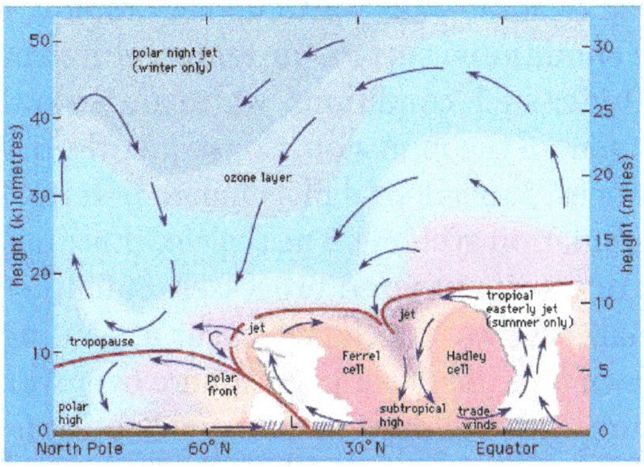

Positions of jet streams in the atmosphere: Arrows indicate directions of mean motions in a meridional plane.

These major regions of surface circulation and their associated pressure fields are related to mean meridional (north-south) circulation patterns as well. Although their presence is discernible in long-term mean statistics accumulated over a hemisphere, such cells are often difficult to detect on a daily basis at any given longitude.

Monsoons

Particularly strong seasonal pressure variations occur over continents, as shown in the January and July maps of sea-level atmospheric pressure. Such seasonal fluctuations, commonly called monsoons, are more pronounced over land surfaces because these surfaces are subject to more significant seasonal temperature variations than are water bodies. Since land surfaces both warm and cool faster than water bodies, they often quickly modify the temperature and density characteristics of air parcels passing over them.

Monsoon storm, Apache Junction, Ariz.

Monsoons blow for approximately six months from the northeast and six months from the southwest, principally in South Asia and parts of Africa; however, similar conditions also occur in Central America and the area between Southeast Asia and Australia. Summer monsoons have a dominant westerly component and a strong tendency to converge, rise, and produce rain. Winter monsoons have a dominant easterly component and a strong tendency to diverge, subside, and cause drought. Both are the result of differences in annual temperature trends over land and sea.

Diurnal Variability

Landmasses in regions affected by monsoons warm up very rapidly in the afternoon hours, especially on days with cloud-free conditions; surface air temperatures between 35 and 40 °C (95 and 104 °F) are not uncommon. Under such conditions, warm air is slowly and continually steeped in the moist and cloudy environment of the monsoon. Consequently, over the course of a 24-hour period, energy from this pronounced diurnal, or daily, change in terrestrial heating is transferred to the cloud, rain, and diurnal circulation systems. The scale of this diurnal change extends from that of coastal sea breezes to that of continent-sized processes. Satellite observations have confirmed that the effects of rapid diurnal temperature change occur at continental scales. For example, air from surrounding areas is drawn into the lower troposphere over warmer land areas of South Asia during summer afternoon hours. This buildup of afternoon heating is accompanied by the production of clouds and rain. In contrast, a reverse circulation, characterized by suppressed clouds and rain, is noted in the early morning hours.

Intra-annual Variability

Monsoon rainfall and dry spells alternate on several timescales. One such well-known timescale is found around periods of 40–50 or 30–60 days. This is called the Madden-Julian oscillation (MJO), named for American atmospheric scientists Roland Madden and Paul Julian in 1971. This phenomenon comes in the form of alternating cyclonic and anticyclonic regions that enhance and suppress rainfall, respectively, and flow eastward along the Equator in the Indian and Pacific oceans. The MJO has the ability to influence monsoonal circulation and rainfall by adding moisture during its cyclonic (wet) phase and reducing convection during its anticyclonic (dry) phase. At the surface in monsoon regions, both dry and wet spells result. These periods may alternate locally on the order of two or more weeks per phase.

Interannual Variability

The variability of monsoon-driven rainfall in the Indian Ocean and Australia appears to parallel El Niño episodes. During El Niño events, which occur about every two to seven years, ocean temperatures rise over the central equatorial Pacific Ocean by about 3 °C (5.4 °F). Atypical conditions characterized by increased rising air motion, convection, and rain are created in the western equatorial Pacific. At the same time, a compensating lobe of descending air, producing below-normal rainfall, appears in the vicinity of eastern Australia, Malaysia, and India. The graph illustrates a well-known El Niño–monsoon rainfall relationship. Here, precipitation figures from above- and below-normal monsoon rainfall periods over India are expressed as a function of years. Years characterized by El Niño events are marked by darkened histogram barbs. The graph shows that many of the years with below-normal monsoon rainfall coincide with El Niño years. This illustration provides only limited guidance to seasonal forecasters since monsoon rainfall is close to normal during many El Niño and La Niña years.

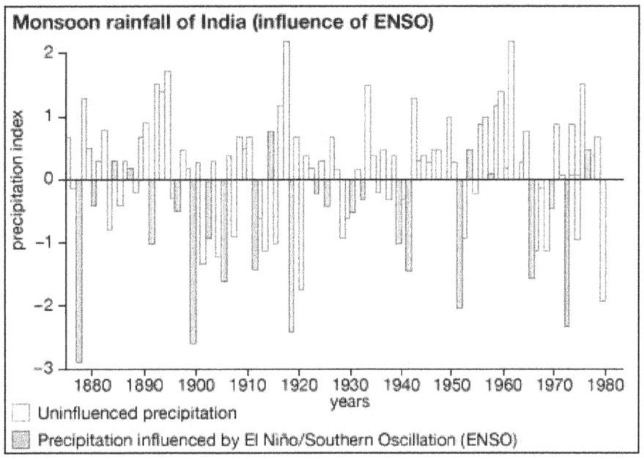

Graph depicting the influence of El Niño/Southern Oscillation (ENSO) on rainfall produced by the Indian summer monsoon. During years when ENSO is active, monsoon-driven precipitation over India often declines.

Many other factors, aside from equatorial Pacific Ocean surface temperatures, contribute to the interannual variability of monsoon rainfall. Excessive spring snow and ice cover on the Plateau of Tibet is related to the deficient monsoon rainfall that occurs during the following summer season in India. Furthermore, strong evidence exists that relates excessive snow and ice cover in western Siberia to deficient Asian summer rainfall. Warmer than normal sea surface temperatures over the Indian Ocean may also contribute somewhat to above-normal rainfall in South Asia. The interplay among these many factors makes forecasting monsoon strength a challenging problem for researchers.

A rather clear signature on the decadal variability of Indian rainfall has been documented by the Indian Weather Service. Decadal-scale variability appears in the graph as an annual running mean that combines average rainfall anomalies (totals as a departure from normal rainfall amounts) occurring at all Indian rain gauge sites. Periods of heavier-than-normal rainfall are followed by decades of somewhat less rainfall.

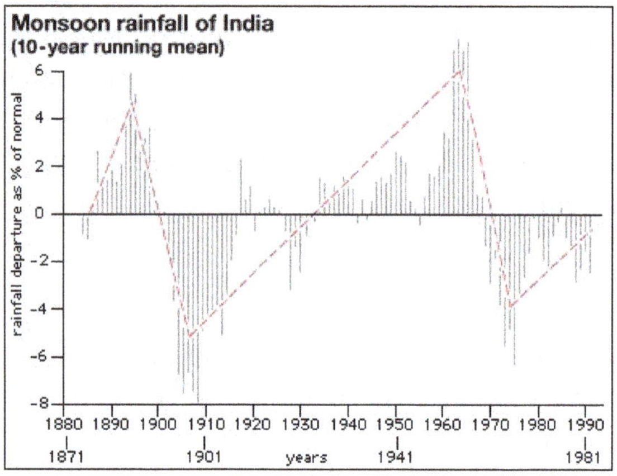

Graph of monsoon rainfall in India, 1871–1981. Annual rainfall amounts are depicted as percentages departing from the 110-year average. The red line superimposed on the graph suggests a recurring trend over this time period.

Upper-level Winds

Characteristics

The flow of air around the globe is greatest in the higher altitudes, or upper levels. Upper-level airflow occurs in wavelike currents that may exist for several days before dissipating. Upper-level wind speeds generally occur on the order of tens of metres per second and vary with height. The characteristics of upper-level wind systems vary according to season and latitude and to some extent hemisphere and year. Wind speeds are strongest in the midlatitudes near the tropopause and in the mesosphere.

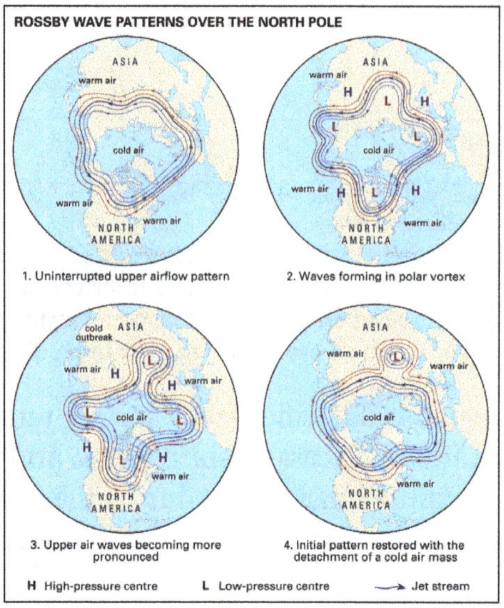

Rossby wave patterns over the North Pole depicting the formation of an outbreak of cold air over Asia.

Upper-level wind systems, like all wind systems, may be thought of as having parts consisting of uniform flow, rotational flow (with cyclonic or anticyclonic curvature), convergent or divergent flow (in which the horizontal area of masses of air shrinks or expands), and deformation (by which the horizontal area of air masses remains constant while experiencing a change in shape). Upper-level wind systems in the midlatitudes tend to have a strong component of uniform flow from west to east ("westerly" flow), though this flow may change during the summer. A series of cyclonic and anticyclonic vortices superimposed on the uniform west-to-east flow make up a wave train (a succession of waves occurring at periodic intervals). The waves are called Rossby waves after Swedish American meteorologist C.G. Rossby, who first explained fundamental aspects of their behaviour in the 1930s. Waves whose wavelengths are about 6,000 km (3,700 miles) or less are called short waves, while those with longer wavelengths are called long waves. In addition, short waves progress in the same direction as the mean airflow, which is from west to east in the midlatitudes; long waves retrogress (that is, move in the opposite direction of the mean flow). Although the undulating current of air is composed of a number of waves of varying wavelength, the dominant wavelength is usually around several thousand kilometres. Near and underneath the tropopause, regions of divergence are found over regions of gently rising air at the surface, while regions of convergence aloft are found over regions of sinking air below. These regions are usually much more difficult to detect than the regions of rotational and uniform flow. While the horizontal wind speed is typically in the range of 10–50 metres per second (about 20–110 miles per hour), the vertical wind speed associated with the waves is only on the order of centimetres per second.

The characteristics of upper-level wind systems are known mainly from an operational worldwide network of rawinsonde observations. (A rawinsonde is a type of radiosonde designed to track upper-level winds and whose position can be tracked by radar.) Winds measured from Doppler-radar wind profilers, aircraft navigational systems, and sequences of satellite-observed cloud imagery have also been used to augment data from the rawinsonde network; the latter two have been especially useful for defining the wind field over data-sparse regions, such as over the oceans.

A weather balloon being released at a weather station at the South Pole. Balloon-borne instrument packages designed to track upper-level winds and capable of being tracked by radar are called rawinsondes.

The winds at upper levels, where surface friction does not occur, tend to be approximately geostrophic. In other words, there is a near balance between the pressure gradient force, which directs air from areas of relatively high pressure to areas of relatively low pressure, and the Coriolis force, which deflects air from its straight-line path to the right in the Northern Hemisphere and to the left in the Southern Hemisphere. An important consequence of this geostrophic balance is that the winds blow parallel to isobars (cartographic lines indicating areas of equal pressure), and, according to Buys Ballott's law, lower pressures will be found to the left of the direction of the wind in the Northern Hemisphere and to the right of the wind in the Southern Hemisphere. Furthermore, wind speed increases as the spacing between isobars decreases. In a wave train of westerly flow, the regions of cyclonic flow are associated with troughs of low pressure, whereas anticyclonic flow are characterized by ridges of high pressure. Rising motions tend to be found downstream from the troughs and upstream from the ridges, while sinking motions tend to be found downstream from the ridges and upstream from the troughs. The areas of rising motion tend to be associated with clouds and precipitation (inclement weather), whereas the areas of sinking motion tend to be associated with clear skies (fair weather).

The vertical variation of the structure of the waves depends upon the temperature pattern. In general, because of the net difference in incoming shorter-wavelength solar radiation and outgoing longer-wavelength infrared radiation between the polar and the equatorial regions, there is a horizontal temperature gradient in the troposphere. At both the surface and upper levels, the troposphere is warmest at low latitudes and coldest at high latitudes. The atmosphere is mainly in hydrostatic balance, or equilibrium, between the upward-directed pressure gradient force and the downward-directed force of gravity. This circumstance is expressed in the following relationship:

$$\partial p/\partial z = -\rho g$$

where $\partial p/\partial z$ is the partial derivative of p with respect to z, p is the pressure, z is the height, ρ is the density of the air, and g is the acceleration of gravity. A consequence of this hydrostatic relationship is that the pressure at any level is equal to the weight of the column of air above. According to the ideal gas law,

$$p = \rho RT$$

where R is the gas constant and T is the temperature. At any given pressure, the density varies inversely with temperature. Therefore, relatively cold air is heavier than relatively warm air at the same pressure. It follows from $\partial p/\partial z = -\rho g$ and $p = \rho RT$ that pressure decreases more rapidly with height at high latitudes in the colder air than it does at lower latitudes in the warmer air. If there is a westerly geostrophic wind at midlevels in the troposphere, then pressure decreases with increasing latitude. Consequently, the horizontal spacing between isobars decreases with height. Thus, the geostrophic wind speed, which approximates the actual wind speed, increases with height. Above the tropopause the pole-to-Equator temperature gradient is reversed as air temperature increases with height, so that the westerlies decrease in intensity in the stratosphere. Thus, the strongest westerly current of winds is located near the tropopause.

The aforementioned relationship can be analyzed quantitatively by considering the vertical variation in the geostrophic wind, which is found from the hydrostatic equation $\partial p/\partial z = -\rho g$ the ideal gas law $p = \rho RT$ and the geostrophic wind formula, approximately as follows:

$$\partial u_g/\partial z = -\,g/fT\,\partial T/\partial y \text{ and } \partial v_g/\partial z = -\,g/fT\,\partial T/\partial z$$

where $\partial u_g/\partial z$ is the partial derivative of u_g with respect to z, u_g and v_g are the components of the geostrophic wind in the zonal (straight from west to east) and meridional (north to south) directions, respectively, and f is the Coriolis parameter. The equations given in above equation are known as the thermal wind relations. The difference between the geostrophic wind at some higher level and the geostrophic wind below is called the thermal wind. It follows that the thermal wind vector is oriented so that in the colder air it lies to the left in the Northern Hemisphere and to the right in the Southern Hemisphere.

In addition to the general pole-to-Equator temperature gradient found in the troposphere, there are zonally oriented temperature variations that are wavelike. In fact, to a first approximation, the isotherms (cartographic lines indicating areas of equal temperature) are nearly parallel to the isobars in the upper levels of the troposphere. Most frequently, relatively cold air lies just upstream from upper-level troughs and just downstream from upper-level ridges, while relatively warm air lies just upstream from upper-level ridges and just downstream from upper-level troughs. The thermal wind relation above equation indicates that the wave train of troughs and ridges tilts with height to the west. In the midlatitudes during the summer and in some locations within the midlatitudes during the winter, the meridional temperature gradient weakens so much that the westerlies become weak or nonexistent. As a result, the wavelike wind field disappears and the flow pattern is that of cyclones and anticyclones "cut off" from the flow. When cold air is colocated with the upper-level cyclones and warm air is colocated with the upper-level anticyclones, according to above equation both circulation patterns increase in intensity with height and are called cold-core and warm-core systems, respectively. Tropical cyclones, on the other hand, are warm-core systems that are most intense at the surface and that decrease in intensity with height.

The vertical structure of upper-level waves has an important effect on smaller-scale features that may be embedded within them. The susceptibility of the atmosphere to vertical overturning (a mixing of lower-level warmer air with higher-level colder air) through deep cumulus convection (e.g., thunderstorms) depends on the rate at which temperature decreases with height. When regions of relatively cold air aloft associated with upper-level troughs or cyclones become superimposed during the winter over relatively warm ocean surfaces or during the summer over hot and humid landmasses, then convective storms can form. The type of mesoscale convective system (MCS) that can form depends in large part on the vertical wind shear. When the vertical shear is very strong, supercells and tornadoes may be spawned, especially during the warmer months. During the winter, bands of precipitation sometimes line up along the vertical shear vector through a process known as slantwise convection.

Propagation and Development of Waves

Upper-level waves in the westerlies in midlatitudes usually move from west to east, in part as a result of advection (a process in which the airflow transports a property of the atmosphere [warmth, cold, etc.] downstream) and in part as a result of propagation, which acts in the opposite direction, toward the west. Rossby showed that to a good approximation,

$$c = U - \beta / (2\pi/L)^2,$$

Where c is the phase speed of the waves, U is the speed from west to east of the component of upper-level wind due to uniform flow, β is the meridional, or north-south, gradient of the Coriolis parameter (f), and L is the zonal wavelength (the distance between successive troughs or ridges). According to equation, since the magnitude of f increases toward the poles, β is positive, and hence waves whose wavelengths are short have a relatively small component due to propagation. In this situation, advection overwhelms the effect of propagation, and the waves move on downstream.

On the other hand, if in midlatitudes the wavelength is very long, then the effects of propagation may exactly cancel the effects of advection, and the waves may become stationary; or if the wavelength becomes even longer, then the waves may become retrograde. From just above equation it can be seen that Rossby waves owe their propagation characteristics to the north-south variation of f. In nature, temperature effects and heating and cooling over warm and cold surfaces can modify above equation somewhat.

The physical basis for above equation and for the development of upper-level systems and how they relate to surface systems is described by an elegant theory developed in the late 1940s called quasigeostrophic theory. A measure of the tendency for a fluid to rotate is known as vorticity and is given by the following equation:

$$\zeta = \partial v/\partial x - \partial u/\partial y,$$

where ζ is the relative vorticity with respect to Earth's surface. The variables x and y are the coordinate axes for space and correspond to the measurements to the east and north, respectively. The variables u and v are zonal and meridional components (the components of motion in the easterly and northerly directions), respectively, of the wind. On the rotating Earth, the vorticity is the sum of the relative vorticity with respect to Earth's surface, given by the aforementioned expression, and Earth's vorticity, given by f, the Coriolis parameter. Troughs are associated with cyclonic vorticity, and ridges are associated with anticyclonic vorticity. In a wave train, the pressure falls downstream from troughs, where the wind is directed from the region of maximum vorticity along the trough to the region of minimum vorticity, which is along the ridge, and the pressure rises downstream from ridges. On the other hand, pressure can rise east of troughs (and west of ridges) where there is a component of motion from the Equator to the pole. For example, pressure rises from regions of low magnitude of to higher magnitude of f (from low values of Earth's vorticity to higher values of Earth's vorticity). Likewise, pressure falls west of troughs (and east of ridges) where there is a component of motion from one of the poles to the Equator—from relatively high magnitude of to lower magnitude of f. The effect of pressure increases and decreases is greatest when the wavelength is relatively short, such as when the effects of the advection of Earth's vorticity are overwhelmed by the effects of advection of relative vorticity.

The development and amplification of Rossby waves is typically a result of the advection of warmer or colder air at low levels. When warm air is advected underneath a layer of air not experiencing much, if any, advection, the pressure at the top of the layer rises. Conversely, pressure falls when cold air advects under a similar layer of air. If the wave train tilts to the west with height so that cold air lines to the west of troughs and thus east of ridges, the pressure aloft in the troughs decreases. Similarly, when warm air lines to the east of troughs and thus west of ridges, the pressure aloft in the ridges increases. As a result, the amplitude of the waves in the wave train increases, thereby enhancing the temperature advection process, so that there is a positive feedback mechanism that makes the waves continue to amplify. In this process, called baroclinic instability, potential energy is converted into kinetic energy—which occurs as wind—as warm, light air rises and cold, heavy air sinks. Since baroclinic instability is associated with horizontal temperature gradients, according to the thermal wind relation, there must be vertical wind shear.

It is also possible for Rossby waves to amplify through a process called barotropic instability. Barotropic instability, however, requires horizontal shear, not vertical shear; kinetic energy for the waves comes from the mean kinetic energy associated with the westerly wind current. The waves grow in amplitude at the expense of the mean flow. Barotropic instability can occur when the horizontal shear varies with latitude such that the sum of Earth's vorticity and the relative vorticity associated with the horizontal shear is small with respect to latitude.

Relationships to Surface Features

Rossby waves propagating through the upper and middle troposphere cause disturbances to form at the surface. According to quasigeostrophic theory, when there is a wave train embedded within a zone of pole-to-Equator temperature gradient, air rises east of upper-level troughs (and west of upper-level ridges) and sinks west of upper-level troughs (and east of upper-level ridges). These vertical air motions are required to maintain the approximate geostrophic and hydrostatic balance, which are necessary for quasigeostrophic equilibrium. Air converges at the surface underneath the rising current of air to compensate for the upward loss of mass and diverges at the surface underneath a sinking current of air to compensate for the downward gain of mass. As a consequence of the lateral deviation of the air by the Coriolis force, Earth's vorticity is converted into cyclonic relative vorticity where air converges and anticyclonic relative vorticity where air diverges. According to the geostrophic wind relation, cyclonic gyres are associated with low-pressure centres, whereas anticyclonic gyres are connected with areas of high pressure. Thus, low-pressure areas form at the surface downstream from upper-level troughs and upstream from upper-level ridges, whereas the reverse is true for high-pressure areas. These surface low- and high-pressure areas thereby create a westward tilt with height of the waves in pressure. Since there tends to be a pole-to-Equator-directed geostrophic wind west of surface lows and east of surface highs, and an Equator-to-pole-directed geostrophic wind east of surface lows and west of surface highs, there is cold advection underneath upper-level troughs and warm advection underneath upper-level ridges; the baroclinic instability process is thus facilitated.

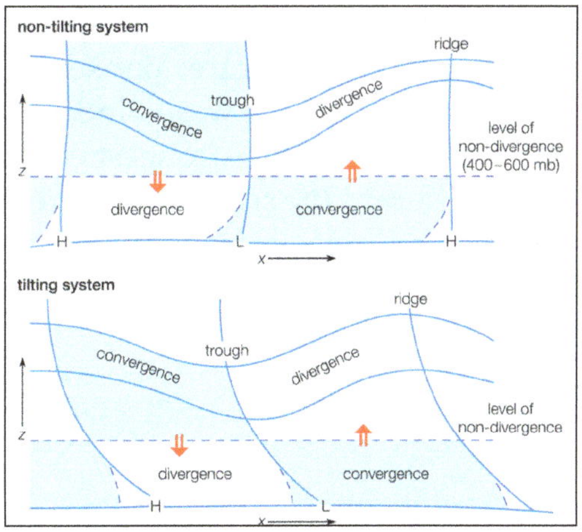

Vertical cross sections through a wave system depicting typical divergence and convergence distributions for non-tilting and tilting systems.

Jet Streams

The upper-level wind flow described above is frequently concentrated into relatively narrow bands called jet streams, or jets. The jets, whose wind speeds are usually in excess of 30 metres per second (about 70 miles per hour) but can be as high as 107 metres per second (about 240 miles per hour), act to steer upper-level waves. Jet streams are of great importance to air travel because they affect the ground speed, the velocity relative to the ground, of aircraft. Since strong upper-level flow is usually associated with strong vertical wind shear, jet streams in midlatitudes are accompanied by strong horizontal temperature gradients, as required by the thermal wind relation. Some regions of high vertical wind shear are marked by clear-air turbulence (CAT). Jet streams whose extents are relatively isolated are called jet streaks. Well-defined circulation patterns of rising and sinking air are usually found just upstream and downstream, respectively, from jet streaks (that are not too curved). Rising motion is found to the left and right just downstream and upstream, respectively, and sinking motion is found to the right and left just downstream and upstream, respectively. Jets tend to be strongest near the tropopause where the horizontal temperature gradient reverses.

The polar front jet moves in a generally westerly direction in midlatitudes, and its vertical wind shear which extends below its core is associated with horizontal temperature gradients that extend to the surface. As a consequence, this jet manifests itself as a front that marks the division between colder air over a deep layer and warmer air over a deep layer. The polar front jet can be baroclinically unstable and break up into waves. The subtropical jet is found at lower latitudes and at slightly higher elevation, because of the increase in height of the tropopause at lower latitudes. The associated horizontal temperature gradients of the subtropical jet do not extend to the surface, so that a surface front is not evident. In the tropics an easterly jet is sometimes found at upper levels, especially when a landmass is located poleward of an ocean, so the temperature increases with latitude. The polar front jet and the subtropical jet play a role in maintaining Earth's general circulation. They are slightly different in each hemisphere because of differences in the distribution of landmasses and oceans.

Winds in the Stratosphere and Mesosphere

The winds in the stratosphere and mesosphere are usually estimated from temperature data collected by satellites. The winds at these high levels are assumed to be geostrophic. Overall, in the midlatitudes, they have a westerly component in the winter and an easterly component in the summer. The highest zonal winds are around 60–70 metres per second (135–155 miles per hour) at 65–70 km (40–43 miles) above Earth's surface. The west-wind component is stronger during the winter in the Southern Hemisphere. The axes of the strongest easterly and westerly wind components in the Southern Hemisphere tilt toward the south with increased altitude during the Northern Hemisphere winter and the Southern Hemisphere summer. The zonal component of the thermal wind shear is in accord with the zonal distribution of temperature.

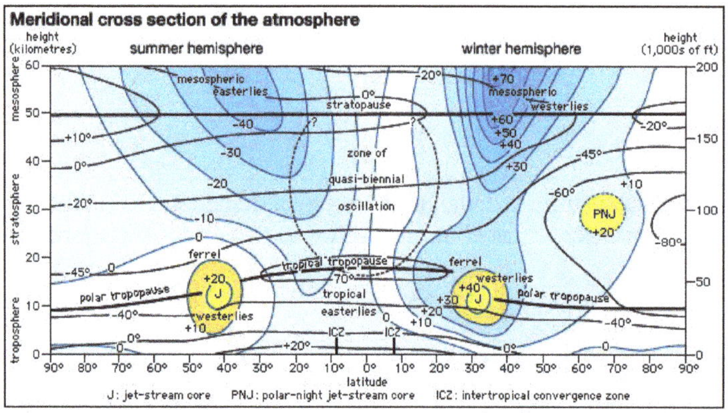

Meridional cross section of the atmosphere to a height of 60 km (37 miles) in Earth's summer and winter hemispheres, showing seasonal changes.

Numerical values for wind are in units of metres per second and are typical of the Northern Hemisphere, but the structure is much the same in the Southern Hemisphere. Positive and negative signs indicate winds of opposite direction.

During the winter there is, in the mean, an intense cyclonic vortex about the poles in the lower stratosphere. Over the North Pole this vortex has an embedded mean trough over northeastern North America and over northeastern Asia, whereas over the Pacific there is a weak anticyclonic vortex. The winter cyclonic vortex over the South Pole is much more symmetrical than the one over the North Pole. During the summer there is an anticyclone above each pole that is much weaker than the wintertime cyclone.

In the stratosphere, deviations from the mean behaviour of the winds occur during events called sudden warmings, when the meridional temperature gradient reverses on timescales as short as several days. This also has the effect of reversing the zonal wind direction. Sudden warmings tend to occur during the early and middle parts of the winter and the transition period from winter to spring. The latter marks the changeover from the cold winter polar cyclone to the warm summer polar anticyclone. It is noteworthy that long waves from the troposphere can propagate into the stratosphere during the winter when westerlies and sudden warmings occur, but this is not the case during the summer when easterly winds prevail.

The zonal component of the winds in the stratosphere above equatorial and tropical regions is, in the mean, relatively weak. This is not necessarily the case at any given time, because they reverse direction on the average every 13–14 months. This phenomenon, which is known as the quasi-biennial oscillation (QBO), is caused by the interaction of vertically propagating waves with the mean flow. Its effect is greatest about 27 km (17 miles) above Earth's surface in the equatorial region. The strongest easterlies are stronger than the strongest westerlies.

References

- Atmospheric-physics, modelling-and-prediction: ecmwf.int, Retrieved 14 June, 2019
- Atmospheric-physics: thefreedictionary.com, Retrieved 5 January, 2019
- Solar-radiation-basics: energy.gov, Retrieved 25 July, 2019
- Solar-radiation-and-temperature, climate-meteorology, science: britannica.com, Retrieved 9 February, 2019
- Atmospheric-temperature: encyclopedia.com, Retrieved 19 August, 2019
- Temperature, climate-meteorology, science: britannica.com, Retrieved 29 March, 2019
- Atmospheric-humidity-and-precipitation, climate-meteorology, science: britannica.com, Retrieved 3 March, 2019
- Precipitation: nationalgeographic.org, Retrieved 11 April, 2019
- Precipitation, climate-meteorology, science: britannica.com, Retrieved 14 June, 2019
- Atmospheric-pressure: nationalgeographic.org, Retrieved 8 May, 2019
- Atmospheric-pressure-and-wind, science: britannica.com, Retrieved 5 July, 2019

Chapter 5

Meteorology

The branch of atmospheric science which focuses on weather forecasting by using elements of atmospheric chemistry and atmospheric physics is called meteorology. The chapter closely examines the key concepts of meteorology as well as weather forecasting to provide an extensive understanding of the subject.

Meteorology is the study of the atmosphere, atmospheric phenomena, and atmospheric effects on our weather. The atmosphere is the gaseous layer of the physical environment that surrounds a planet. Earth's atmosphere is roughly 100 to 125 kilometers (65-75 miles) thick. Gravity keeps the atmosphere from expanding much farther.

Meteorology is a subdiscipline of the atmospheric sciences, a term that covers all studies of the atmosphere. A subdiscipline is a specialized field of study within a broader subject or discipline. Climatology and aeronomy are also subdisciplines of the atmospheric sciences. Climatology focuses on how atmospheric changes define and alter the world's climates. Aeronomy is the study of the upper parts of the atmosphere, where unique chemical and physical processes occur. Meteorology focuses on the lower parts of the atmosphere, primarily the troposphere, where most weather takes place.

Meteorologists use scientific principles to observe, explain, and forecast our weather. They often focus on atmospheric research or operational weather forecasting. Research meteorologists cover several subdisciplines of meteorology to include: climate modeling, remote sensing, air quality, atmospheric physics, and climate change. They also research the relationship between the atmosphere and Earth's climates, oceans, and biological life.

Forecasters use that research, along with atmospheric data, to scientifically assess the current state of the atmosphere and make predictions of its future state. Atmospheric conditions both at the Earth's surface and above are measured from a variety of sources: weather stations, ships, buoys, aircraft, radar, weather balloons, and satellites. This data is transmitted to centers throughout the world that produce computer analyses of global weather. The analyses are passed on to national and regional weather centers, which feed this data into computers that model the future state of the atmosphere. This transfer of information demonstrates how weather and the study of it take place in multiple, interconnected ways.

Scales of Meteorology

Weather occurs at different scales of space and time. The four meteorological scales are: microscale, mesoscale, synoptic scale, and global scale. Meteorologists often focus on a specific scale in their work.

Microscale Meteorology

Microscale meteorology focuses on phenomena that range in size from a few centimeters to a few kilometers, and that have short life spans (less than a day). These phenomena affect very small geographic areas, and the temperatures and terrains of those areas.

Microscale meteorologists often study the processes that occur between soil, vegetation, and surface water near ground level. They measure the transfer of heat, gas, and liquid between these surfaces. Microscale meteorology often involves the study of chemistry.

Tracking air pollutants is an example of microscale meteorology. MIRAGE-Mexico is collaboration between meteorologists in the United States and Mexico. The program studies the chemical and physical transformations of gases and aerosols in the pollution surrounding Mexico City. MIRAGE-Mexico uses observations from ground stations, aircraft, and satellites to track pollutants.

Mesoscale Meteorology

Mesoscale phenomena range in size from a few kilometers to roughly 1,000 kilometers (620 miles). Two important phenomena are mesoscale convective complexes (MCC) and mesoscale convective systems (MCS). Both are caused by convection, an important meteorological principle. Convection is a process of circulation. Warmer, less-dense fluid rises, and colder, denser fluid sinks. The fluid that most meteorologists study is air. (Any substance that flows is considered a fluid.) Convection results in a transfer of energy, heat, and moisture—the basic building blocks of weather.

In both an MCC and MCS, a large area of air and moisture is warmed during the middle of the day—when the sun angle is at its highest. As this warm air mass rises into the colder atmosphere, it condenses into clouds, turning water vapor into precipitation. An MCC is a single system of clouds that can reach the size of the state of Ohio and produce heavy rainfall and flooding. An MCS is a smaller cluster of thunderstorms that lasts for several hours. Both react to unique transfers of energy, heat, and moisture caused by convection.

Synoptic Scale Meteorology

Synoptic-scale phenomena cover an area of several hundred or even thousands of kilometers. High- and low-pressure systems seen on local weather forecasts are synoptic in scale. Pressure, much like convection, is an important meteorological principle that is at the root of large-scale weather systems as diverse as hurricanes and bitter cold outbreaks.

Low-pressure systems occur where the atmospheric pressure at the surface of the Earth is less than its surrounding environment. Wind and moisture from areas with higher pressure seek low-pressure systems. This movement, in conjunction with the Coriolis force and friction, causes the system to rotate counter-clockwise in the Northern Hemisphere and clockwise in the Southern Hemisphere, creating a cyclone. Cyclones have a tendency for upward vertical motion. This allows moist air from the surrounding area to rise, expand and condense into water vapor, forming clouds. This movement of moisture and air causes the majority of our weather events.

Hurricanes are a result of low-pressure systems (cyclones) developing over tropical waters in the Western Hemisphere. The system sucks up massive amounts of warm moisture from the sea, causing convection to take place, which in turn causes wind speeds to increase and pressure to fall. When these winds reach speeds over 119 kilometers per hour (74 miles per hour), the cyclone is classified as a hurricane.

Hurricanes can be one of the most devastating natural disasters in the Western Hemisphere. The National Hurricane Center, in Miami, Florida, regularly issues forecasts and reports on all tropical weather systems. During hurricane season, hurricane specialists issue forecasts and warnings for every tropical storm in the western tropical Atlantic and eastern tropical Pacific. Businesses and government officials from the United States, the Caribbean, Central America, and South America rely on forecasts from the National Hurricane Center.

High-pressure systems occur where the atmospheric pressure at the surface of the Earth is greater than its surrounding environment. This pressure has a tendency for downward vertical motion, allowing for dry air and clear skies.

Extremely cold temperatures are a result of high-pressure systems that develop over the Arctic and move over the Northern Hemisphere. Arctic air is very cold because it develops over ice and snow-covered ground. This cold air is so dense that it pushes against Earth's surface with extreme pressure, preventing any moisture or heat from staying within the system.

Meteorologists have identified many semi-permanent areas of high-pressure. The Azores high, for instance, is a relatively stable region of high pressure around the Azores, an archipelago in the mid-Atlantic Ocean. The Azores high is responsible for arid temperatures of the Mediterranean basin, as well as summer heat waves in Western Europe.

Global Scale Meteorology

Global scale phenomena are weather patterns related to the transport of heat, wind, and moisture from the tropics to the poles. An important pattern is global atmospheric circulation the large-scale movement of air that helps distribute thermal energy (heat) across the surface of the Earth.

Global atmospheric circulation is the fairly constant movement of winds across the globe. Winds develop as air masses move from areas of high pressure to areas of low pressure. Global atmospheric circulation is largely driven by Hadley cells. Hadley cells are tropical and equatorial convection patterns. Convection drives warm air high in the atmosphere, while cool, dense air pushes lower in a constant loop. Each loop is a Hadley cell.

Hadley cells determine the flow of trade winds, which meteorologists forecast. Businesses, especially those exporting products across oceans, pay close attention to the strength of trade winds because they help ships travel faster. Westerly's are winds that blow from the west in the midlatitudes. Closer to the Equator, trade winds blow from the northeast (north of the Equator) and the southeast (south of the Equator).

Meteorologists study long-term climate patterns that disrupt global atmospheric circulation. Meteorologists discovered the pattern of El Nino, for instance. El Niño involves ocean currents and trade winds across the Pacific Ocean. El Niño occurs roughly every five years, disrupting global atmospheric circulation and affecting local weather and economies from Australia to Peru.

El Niño is linked with changes in air pressure in the Pacific Ocean known as the Southern Oscillation. Air pressure drops over the eastern Pacific, near the coast of the Americas, while air pressure rises over the western Pacific, near the coasts of Australia and Indonesia. Trade winds weaken. Eastern Pacific nations experience extreme rainfall. Warm ocean currents reduce fish stocks, which depend on nutrient-rich upwelling of cold water to thrive. Western Pacific nations experience drought, devastating agricultural production. Understanding the meteorological processes of El Niño helps farmers, fishers, and coastal residents prepare for the climate pattern.

Weather

Weather is what is going on in the atmosphere at a particular place at a particular time. Weather can change rapidly. A location's weather depends on air temperature; air pressure; fog; humidity; cloud cover; precipitation; wind speed and direction. All of these are directly related to the amount of energy that is in the system and where that energy is. The ultimate source of this energy is the sun.

Elements of Weather

Humidity

Humidity is the amount of water vapor in the air in a particular spot. We usually use the term to mean relative humidity, the percentage of water vapor a certain volume of air is holding relative to the maximum amount it can contain. If the humidity today is 80%, it means that the air contains 80% of the total amount of water it can hold at that temperature. What will happen if the humidity increases to more than 100%? The excess water condenses and forms precipitation.

Since warm air can hold more water vapor than cool air, raising or lowering temperature can change air's relative humidity. The temperature at which air becomes saturated with water is called the air's dew point. This term makes sense, because water condenses from the air as dew, if the air cools down overnight and reaches 100% humidity.

Clouds

Clouds have a big influence on weather by preventing solar radiation from reaching the ground; absorbing warmth that is re-emitted from the ground; and as the source of precipitation.

When there are no clouds, there is less insulation. As a result, cloudless days can be extremely hot, and cloudless nights can be very cold. For this reason, cloudy days tend to have a lower range of temperatures than clear days.

There are a variety of conditions needed for clouds to form. First, clouds form when air reaches its dew point. This can happen in two ways: (1) Air temperature stays the same but humidity increases.

This is common in locations that are warm and humid. (2) Humidity can remain the same, but temperature decreases. When the air cools enough to reach 100% humidity, water droplets form. Air cools when it comes into contact with a cold surface or when it rises.

Rising air creates clouds when it has been warmed at or near the ground level and then is pushed up over a mountain or mountain range or is thrust over a mass of cold, dense air. Water vapor is not visible unless it condenses to become a cloud. Water vapor condenses around a nucleus, such as dust, smoke, or a salt crystal. This forms a tiny liquid droplet. Billions of these water droplets together make a cloud.

Clouds are classified in several ways. The most common classification used today divides clouds into four separate cloud groups, which are determined by their altitude and if precipitation is occurring or not. High-level clouds form from ice crystals where the air is extremely cold and can hold little water vapor. Cirrus, cirrostratus, and cirrocumulusare all names of high clouds. Cirrocumulus clouds are small, white puffs that ripple across the sky, often in rows. Cirrus clouds may indicate that a storm is coming.

Middle-level clouds, including altocumulus and altostratus clouds, may be made of water droplets, ice crystals or both, depending on the air temperatures. Thick and broad altostratus clouds are gray or blue-gray. They often cover the entire sky and usually mean a large storm, bearing a lot of precipitation, is coming.

Low-level clouds are nearly all water droplets. Stratus, stratocumulus and nimbostratus clouds are common low clouds. Nimbostratus clouds are thick and dark that produce precipitation. Clouds with the prefix 'cumulo' grow vertically instead of horizontally and have their bases at low altitude and their tops at high or middle altitude. Clouds grow vertically when strong unstable air currents are rising upward. Common clouds include cumulus humilis, cumulus mediocris, cumulus congestus, and cumulonimbus.

Fog

Fog is a cloud located at or near the ground. When humid air near the ground cools below its dew point, fog is formed. The several types of fog that each form in a different way.

Radiation fog forms at night when skies are clear and the relative humidity is high. As the ground cools, the bottom layer of air cools below its dew point. Tule fog is an extreme form of radiation fog found in some regions. San Francisco, California, is famous for its summertime advection fog. Warm, moist Pacific Ocean air blows over the cold California current and cools below its dew point. Sea breezes bring the fog onshore. Steam fogappears in autumn when cool air moves over a warm lake. Water evaporates from the lake surface and condenses as it cools, appearing like steam. Warm humid air travels up a hillside and cools below its dew point to create upslope fog.

Precipitation

Precipitation is an extremely important part of weather. Some precipitation forms in place. The most common precipitation comes from clouds. Rain or snow droplets grow as they ride air currents in a cloud and collect other droplets. They fall when they become heavy enough to escape from the rising air currents that hold them up in the cloud. One million cloud droplets will combine to make only one rain drop. If temperatures are cold, the droplet will hit the ground as a snowflake.

Air Masses

An air mass is a large mass of air that has similar characteristics of temperature and humidity within it. An air mass acquires these characteristics above an area of land or water known as its source region. When the air mass sits over a region for several days, or longer, it picks up the distinct temperature and humidity characteristics of that region.

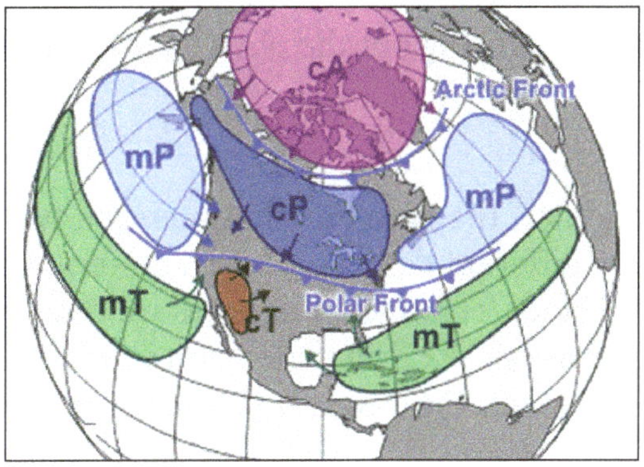

Air Mass Formation

Where an air mass receives it's characteristics of temperature and humidity is called the source region. Air masses are slowly pushed along by high-level winds, when an air mass moves over a new region it shares its temperature and humidity with that region. So the temperature and humidity of a particular location depends partly on the characteristics of the air mass that sits over it.

Storms arise if the air mass and the region it moves over have different characteristics. For example, when a colder air mass moves over warmer ground, the bottom layer of air is heated. That air rises, forming clouds, rain, and sometimes thunderstorms. How would a moving air mass form an inversion? When a warmer air mass travels over colder ground, the bottom layer of air cools and, because of its high density, is trapped near the ground.

In general, cold air masses tend to flow toward the equator and warm air masses tend to flow toward the poles. This brings heat to cold areas and cools down areas that are warm. It is one of the many processes that act towards balancing out the planet's temperatures. Air masses are slowly pushed along by high-level winds. When an air mass moves over a new region, it shares its temperature and humidity with that region. So the temperature and humidity of a particular location depends partly on the characteristics of the air mass that sits over it. Air masses are classified based on their temperature and humidity characteristics. Below are examples of how air masses are classified over North America.

- Maritime tropical (mT) – moist, warm air mass.
- Continental tropical (cT) – dry, warm air mass.
- Maritime polar (mP) – moist, cold air mass.
- Continental polar (cP) – dry, cold air mass.

Storms arise if the air mass and the region it moves over have different characteristics. For example, when a colder air mass moves over warmer ground, the bottom layer of air is heated. That air rises, forming clouds, rain, and sometimes thunderstorms. How would a moving air mass form an inversion? When a warmer air mass travels over colder ground, the bottom layer of air cools and, because of its high density, is trapped near the ground.

In general, cold air masses tend to flow toward the equator and warm air masses tend to flow toward the poles. This brings heat to cold areas and cools down areas that are warm. It is one of the many processes that act towards balancing out the planet's temperatures.

Weather Front

When two air masses meet together, the boundary between the two is called a weather front. At a front, the two air masses have different densities, based on temperature, and do not easily mix. One air mass is lifted above the other, creating a low pressure zone. If the lifted air is moist, there will be condensation and precipitation. Winds are common at a front. The greater the temperature difference between the two air masses, the stronger the winds will be. Fronts are the main cause of stormy weather.

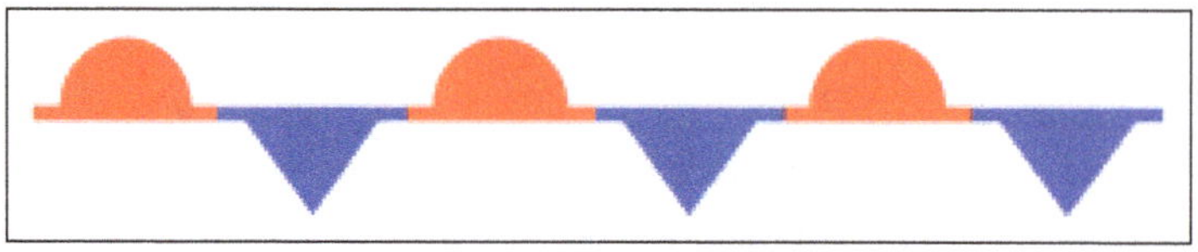

Stationary Fronts

At a stationary front the air masses do not move. A front may become stationary if an air mass is stopped by a barrier, such as a mountain range. A stationary front may bring days of rain, drizzle, and fog. Winds usually blow parallel to the front, but in opposite directions. After several days, the front will likely break apart. When a cold air mass takes the place of a warm air mass, there is a cold front.

Cold Fronts

Cold Front

Imagine that you are standing in one spot as a cold front approaches. Along the cold front, the denser, cold air pushes up the warm air, causing the air pressure to decrease. If the humidity is high enough, some types of cumulus clouds will grow. High in the atmosphere, winds blow ice crystals from the tops of these clouds to create cirrostratus and cirrus clouds. At the front, there will be a line of rain showers, snow showers, or thunderstorms with blustery winds. A squall line is a line of severe thunderstorms that forms along a cold front. Behind the front is the cold air mass. This mass is drier so precipitation stops. The weather may be cold and clear or only partly cloudy. Winds may continue to blow into the low pressure zone at the front. The weather at a cold front varies with the season.

- Spring and summer: The air is unstable so thunderstorms or tornadoes may form.
- Spring: If the temperature gradient is high, strong winds blow.
- Autumn: Strong rains fall over a large area.
- Winter: The cold air mass is likely to have formed in the frigid arctic so there are frigid temperatures and heavy snows.

Warm Fronts

Warm Front

Along a warm front, a warm air mass slides over a cold air mass. When warm, less dense air moves over the colder, denser air, the atmosphere is relatively stable.

Imagine that you are on the ground in the wintertime under a cold winter air mass with a warm front approaching. The transition from cold air to warm air takes place over a long distance so the first signs of changing weather appear long before the front is actually over you. Initially, the air is cold: the cold air mass is above you and the warm air mass is above it. High cirrus clouds mark the transition from one air mass to the other.

Over time, cirrus clouds become thicker and cirrostratus clouds form. As the front approaches, altocumulus and altostratus clouds appear and the sky turns gray. Since it is winter, snowflakes fall. The clouds thicken and nimbostratus clouds form. Snowfall increases. Winds grow stronger as the low pressure approaches. As the front gets closer, the cold air mass is just above you but the warm air mass is not too far above that. The weather worsens. As the warm air mass approaches, temperatures rise and snow turns to sleet and freezing rain. Warm and cold air mix at the front, leading to the formation of stratus clouds and fog.

Occluded Fronts

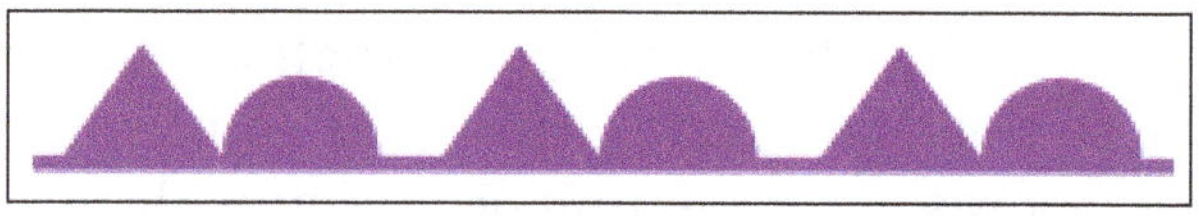

An occluded front usually forms around a low pressure system. The occlusion starts when a cold front catches up to a warm front. The air masses, in order from front to back, are cold, warm, and then cold again.

Coriolis Effect curves the boundary where the two fronts meet towards the pole. If the air mass that arrives third is colder than either of the first two air masses, that air mass slips beneath them both. This is called a cold occlusion. If the air mass that arrives third is warm, that air mass rides over the other air mass. This is called a warm occlusion.

The weather at an occluded front is especially fierce right at the occlusion. Precipitation and shifting winds are typical. The Pacific Coast has frequent occluded fronts.

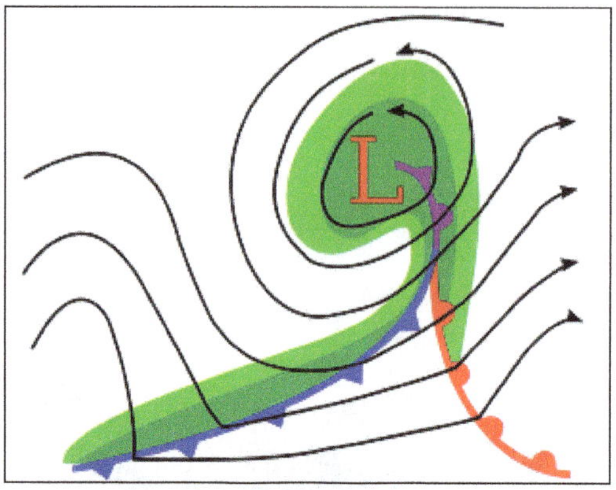

Remember, a weather front is basically the boundary between two air masses of different densities. At the center of each air mass is typically a high pressure. This means that weather is typically sunny within air masses, but their temperatures could vary with the season and humidity could vary based on the source region of the air mass.

Weather Forecasting

Weather forecasting refers to the prediction of the weather through application of the principles of physics, supplemented by a variety of statistical and empirical techniques. In addition to predictions of atmospheric phenomena themselves, weather forecasting includes predictions of changes on Earth's surface caused by atmospheric conditions—e.g., snow and ice cover, storm tides, and floods.

Measurements and Ideas as the basis for Weather Prediction

The observations of few other scientific enterprises are as vital or affect as many people as those related to weather forecasting. From the days when early humans ventured from caves and other natural shelters, perceptive individuals in all likelihood became leaders by being able to detect nature's signs of impending snow, rain, or wind, indeed of any change in weather. With such information they must have enjoyed greater success in the search for food and safety, the major objectives of that time.

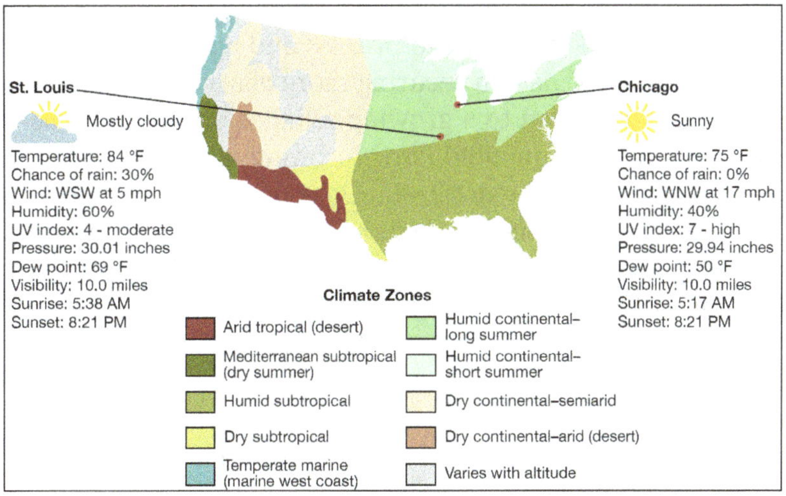

Weather conditions: Comparing weather conditions for St. Louis and Chicago on a given day.

In a sense, weather forecasting is still carried out in basically the same way as it was by the earliest humans—namely, by making observations and predicting changes. The modern tools used to measure temperature, pressure, wind, and humidity in the 21st century would certainly amaze them, and the results obviously are better. Yet, even the most sophisticated numerically calculated forecast made on a supercomputer requires a set of measurements of the condition of the atmosphere—an initial picture of temperature, wind, and other basic elements, somewhat comparable to that formed by our forebears when they looked out of their cave dwellings. The primeval approach entailed insights based on the accumulated experience of the perceptive observer, while the modern technique consists of solving equations. Although seemingly quite different, there are underlying similarities between both practices. In each case the forecaster asks "What is?" in the sense of "What kind of weather prevails today?" and then seeks to determine how it will change in order to extrapolate what it will be.

Because observations are so critical to weather prediction, an account of meteorological measurements and weather forecasting is a story in which ideas and technology are closely intertwined, with creative thinkers drawing new insights from available observations and pointing to the need for new or better measurements, and technology providing the means for making new observations and for processing the data derived from measurements. The basis for weather prediction started with the theories of the ancient Greek philosophers and continued with Renaissance scientists, the scientific revolution of the 17th and 18th centuries, and the theoretical models of 20th- and 21st-century atmospheric scientists and meteorologists. Likewise, it tells of the development of the "synoptic" idea—that of characterizing the weather over a large region at exactly the same time in order to organize information about prevailing conditions. In synoptic meteorology, simultaneous observations for a specific time are plotted on a map for a broad area whereby a general view of the weather in that region is gained. The so-called synoptic weather map came to be the principal tool of 19th-century meteorologists and continues to be used today in weather stations and on television weather reports around the world.

Since the mid-20th century, digital computers have made it possible to calculate changes in atmospheric conditions mathematically and objectively—i.e., in such a way that anyone can obtain the same result from the same initial conditions. The widespread adoption of numerical weather prediction models brought a whole new group of players—computer specialists and experts in numerical processing and statistics—to the scene to work with atmospheric scientists and meteorologists. Moreover, the enhanced capability to process and analyze weather data stimulated the long-standing interest of meteorologists in securing more observations of greater accuracy. Technological advances since the 1960s led to a growing reliance on remote sensing, particularly the gathering of data with specially instrumented Earth-orbiting satellites. By the late 1980s, forecasts of the weather were largely based on the determinations of numerical models integrated by high-speed supercomputers—except for some shorter-range predictions, particularly those related to local thunderstorm activity, which were made by specialist's directly interpreting radar and satellite measurements. By the early 1990s a network of next-generation Doppler weather radar (NEXRAD) was largely in place in the United States, which allowed meteorologists to predict severe weather events with additional lead time before their occurrence. During the late 1990s and early 21st century, computer processing power increased, which allowed weather bureaus to produce more-sophisticated ensemble forecasts—that is, sets of multiple model runs whose results limit the range of uncertainty with respect to a forecast.

Practical Applications

Systematic weather records were kept after instruments for measuring atmospheric conditions became available during the 17th century. Undoubtedly these early records were employed mainly by those engaged in agriculture. Planting and harvesting obviously can be planned better and carried out more efficiently if long-term weather patterns can be estimated. In the United States, the foundations of the national weather services were laid down by American physicist Joseph Henry, the first head of the Smithsonian Institution. In 1849 Henry created a network of volunteer weather observers to help improve storm prediction in the U.S. The first national weather services were provided by the U.S. Army Signal Corps beginning on February 9, 1870, which also incorporated Henry's volunteer weather observers by 1874. These operations were taken over by the Department of Agriculture in 1891. By the early 1900s free mail service and telephone were providing forecasts daily to millions of American farmers. The U.S. Weather Bureau established a Fruit-Frost (forecasting) Service during World War I, and by the 1920s radio broadcasts to agricultural interests were being made in most states.

Weather forecasting became an important tool for aviation during the 1920s and '30s. Its application in this area gained in importance after Francis W. Reichelderfer was appointed chief of the U.S. Weather Bureau (USWB) in 1939. Reichelderfer had previously modernized the U.S. Navy's meteorological service and made it a model of support for naval aviation. During World War II the discovery of very strong wind currents at high altitudes (the jet streams, which can affect aircraft speed) and the general susceptibility of military operations in Europe to weather led to a special interest in weather forecasting.

One of the most famous wartime forecasting problems was for Operation Overlord, the invasion of the European mainland at Normandy by Allied forces. An unusually intense June storm brought high seas and gales to the French coast, but a moderation of the weather that was successfully predicted by Col. J.M. Stagg of the British forces (after consultation with both British and American forecasters) enabled Gen. Dwight D. Eisenhower, supreme commander of the Allied Expeditionary Forces, to make his critical decision to invade on June 6, 1944.

The second half of the 20th century saw a reorganization of the country's weather bureau. The USWB was part of the Department of Agriculture until 1940, when it was added to the Department of Commerce. On October 9, 1970, the USWB became the National Weather Service.

In addition, the later part of the 20th century was a time of unprecedented growth of commercial weather-forecasting firms in the United States and elsewhere. Marketing organizations and stores hire weather-forecasting consultants to help with the timing of sales and promotions of products ranging from snow tires and roofing materials to summer clothes and resort vacations. Many oceangoing shipping vessels as well as military ships use optimum ship routing forecasts to plan their routes in order to minimize lost time, potential damage, and fuel consumption in heavy seas. Similarly, airlines carefully consider atmospheric conditions when planning long-distance flights so as to avoid the strongest head winds and to ride with the strongest tail winds.

International trading of foodstuffs such as wheat, corn (maize), beans, sugar, cocoa, and coffee can be severely affected by weather news. For example, in 1975 a severe freeze in Brazil caused the price of coffee to increase substantially within just a few weeks, and in 2017 Georgia peach growers blamed the combination of warm winter temperatures and a spring freeze on the loss of nearly 80 percent of the state's peach crop. In addition, extreme heat and drought can affect production; one study estimated that 9–10 percent of cereal crops between 1964 and 2007 were lost to these phenomena. Weather-forecasting organizations are thus frequently called upon by banks, commodity traders, and food companies to give them advance knowledge of the possibility of such sudden changes. The cost of all sorts of commodities and services, whether they are tents for outdoor events or plastic covers for the daily newspapers, can be reduced or eliminated if reliable information about possible precipitation can be obtained in advance.

Forecasts must be quite precise for applications that are tailored to specific industries. Gas and electric utilities, for example, may require forecasts of temperature within one or two degrees a day ahead of time, or ski-resort operators may need predictions of night time relative humidity on the slopes within 5 to 10 percent in order to schedule snow making.

Principles and Methodology of Weather Forecasting

Short-range Forecasting

- Objective Predictions

When people wait under a shelter for a downpour to end, they are making a very-short-range weather forecast. They are assuming, based on past experience that such hard rain usually does not last very long. In short-term predictions the challenge for the forecaster is to improve on what the layperson can do. For years the type of situation represented in the above example proved particularly vexing for forecasters, but since the mid-1980s they have been developing a method called nowcasting to meet precisely this sort of challenge. In this method, radar and satellite observations of local atmospheric conditions are processed and displayed rapidly by computers to project weather several hours in advance. The U.S. National Oceanic and Atmospheric Administration operates a facility known as PROFS (Program for Regional Observing and Forecasting Services) in Boulder, Colo., specially equipped for nowcasting.

Meteorologists can make somewhat longer-term forecasts (those for 6, 12, 24, or even 48 hours) with considerable skill because they are able to measure and predict atmospheric conditions for large areas by computer. Using models that apply their accumulated expert knowledge quickly, accurately, and in a statistically valid form, meteorologists are now capable of making forecasts objectively. As a consequence, the same results are produced time after time from the same data inputs, with all analysis accomplished mathematically. Unlike the prognostications of the past made with subjective methods, objective forecasts are consistent and can be studied, reevaluated, and improved.

Another technique for objective short-range forecasting is called MOS (for Model Output Statistics). Conceived by Harry R. Glahn and D.A. Lowry of the U.S. National Weather Service, this method involves the use of data relating to past weather phenomena and developments to extrapolate the values of certain weather elements, usually for a specific location and time period. It overcomes the weaknesses of numerical models by developing statistical relations between model forecasts and observed weather. These relations are then used to translate the model forecasts directly to specific weather forecasts. For example, a numerical model might not predict the occurrence of surface winds at all, and whatever winds it did predict might always be too strong. MOS relations can automatically correct for errors in wind speed and produce quite accurate forecasts of wind occurrence at a specific point, such as Heathrow Airport near London. As long as numerical weather prediction models are imperfect, there may be many uses for the MOS technique.

- Predictive skills and procedures

Short-range weather forecasts generally tend to lose accuracy as forecasters attempt to look farther ahead in time. Predictive skill is greatest for periods of about 12 hours and is still quite substantial for 48-hour predictions. An increasingly important group of short-range forecasts are economically motivated. Their reliability is determined in the marketplace by the economic gains they produce (or the losses they avert).

Weather warnings are a special kind of short-range forecast; the protection of human life is the forecaster's greatest challenge and source of pride. The first national weather forecasting service in the United States (the predecessor of the Weather Bureau) was in fact formed, in 1870, in response to the need for storm warnings on the Great Lakes. Increase Lapham of Milwaukee urged Congress to take action to reduce the loss of hundreds of lives incurred each year by Great Lakes shipping during the 1860s. The effectiveness of the warnings and other forecasts assured the future of the American public weather service.

Weather warnings are issued by government and military organizations throughout the world for all kinds of threatening weather events: tropical storms variously called hurricanes, typhoons, or tropical cyclones, depending on location; great oceanic gales outside the tropics spanning hundreds of kilometres and at times packing winds comparable to those of tropical storms; and, on land, flash floods, high winds, fog, blizzards, ice, and snowstorms.

A particular effort is made to warn of hail, lightning, and wind gusts associated with severe thunderstorms, sometimes called severe local storms (SELS) or simply severe weather. Forecasts and warnings also are made for tornadoes, those intense, rotating windstorms that represent the most violent end of the weather scale. Destruction of property and the risk of injury and death are extremely high in the path of a tornado, especially in the case of the largest systems (sometimes called maxi-tornadoes).

Because tornadoes are so uniquely life-threatening and because they are so common in various regions of the United States, the National Weather Service operates a National Severe Storms Forecasting Center (NSSFC) in Kansas City, Mo., where SELS forecasters survey the atmosphere for the conditions that can spawn tornadoes or severe thunderstorms. This group of SELS forecasters, assembled in 1952, monitors temperature and water vapour in an effort to identify the warm, moist regions where thunderstorms may form and studies maps of pressure and winds to find regions where the storms may organize into mesoscale structures. The group also monitors jet streams and dry air aloft that can combine to distort ordinary thunderstorms into rare rotating ones with tilted chimneys of upward rushing air that, because of the tilt, are unimpeded by heavy falling rain. These high-speed updrafts can quickly transport vast quantities of moisture to the cold upper regions of the storms, thereby promoting the formation of large hailstones. The hail and rain drag down air from aloft to complete a circuit of violent, cooperating updrafts and downdrafts.

By correctly anticipating such conditions, SELS forecasters are able to provide time for the mobilization of special observing networks and personnel. If the storms actually develop, specific warnings are issued based on direct observations. This two-step process consists of the tornado or severe thunderstorm watch, which is the forecast prepared by the SELS forecaster, and the warning, which is usually released by a local observing facility. The watch may be issued when the skies are clear, and it usually covers a number of counties. It alerts the affected area to the threat but does not attempt to pinpoint which communities will be affected.

By contrast, the warning is very specific to a locality and calls for immediate action. Radar of various types can be used to detect the large hailstones, the heavy load of raindrops, the relatively clear region of rapid updraft, and even the rotation in a tornado. These indicators, or an actual sighting, often trigger the tornado warning. In effect, a warning is a specific statement that danger is imminent, whereas a watch is a forecast that warnings may be necessary later in a given region.

Long-range Forecasting

- Techniques

Extended-range or long-range, weather forecasting has had a different history and a different approach from short- or medium-range forecasting. In most cases, it has not applied the synoptic method of going forward in time from a specific initial map. Instead, long-range forecasters have tended to use the climatological approach, often concerning themselves with the broad weather picture over a period of time rather than attempting to forecast day-to-day details.

There is good reason to believe that the limit of day-to-day forecasts based on the "initial map" approach is about two weeks. Most long-range forecasts thus attempt to predict the departures from normal conditions for a given month or season. Such departures are called anomalies. A forecast might state that "spring temperatures in Minneapolis have a 65 percent probability of being above normal." It would likely be based on a forecast anomaly map, which shows temperature anomaly patterns. The maps do not attempt to predict the weather for a particular day, but rather forecast trends (i.e., warmer than normal) for an extended amount of time, such as a season (i.e., spring).

The U.S. Weather Bureau began making experimental long-range forecasts just before the beginning of World War II, and its successor, the National Weather Service, continues to express such predictions in probabilistic terms, making it clear that they are subject to uncertainty. Verification shows that forecasts of temperature anomalies are more reliable than those of precipitation that monthly forecasts are better than seasonal ones, and that winter months are predicted somewhat more accurately than other seasons.

Prior to the 1980s the technique commonly used in long-range forecasting relied heavily on the analog method, in which groups of weather situations (maps) from previous years were compared to those of the current year to determine similarities with the atmosphere's present patterns (or "habits"). An association was then made between what had happened subsequently in those "similar" years and what was going to happen in the current year. Most of the techniques were quite subjective, and there were often disagreements of interpretation and consequently uneven quality and marginal reliability.

Persistence (warm summers follow warm springs) or anti-persistence (cold springs follow warm winters) also were used, even though, strictly speaking, most forecasters consider persistence forecasts "no-skill" forecasts. Yet, they too have had limited success.

- Prospects for new procedures

In the last quarter of the 20th century the approach of and prospects for long-range weather forecasting changed significantly. Stimulated by the work of Jerome Namias, who headed the U.S. Weather Bureau's Long-Range Forecast Division for 30 years, scientists began to look at ocean-surface temperature anomalies as a potential cause for the temperature anomalies of the atmosphere in succeeding seasons and at distant locations. At the same time, other American meteorologists, most notably John M. Wallace, showed how certain repetitive patterns of atmospheric flow were related to each other in different parts of the world. With satellite-based observations available, investigators began to study the El Niño phenomenon. Atmospheric scientists also revived the work of Gilbert Walker, an early 20th-century British climatologist who had studied the Southern Oscillation, the aforementioned up-and-down fluctuation of atmospheric pressure in the Southern Hemisphere. Walker had investigated related air circulations (later called the Walker Circulation) that resulted from abnormally high pressures in Australia and low pressures in Argentina or vice versa.

All of this led to new knowledge about how the occurrence of abnormally warm or cold ocean waters and of abnormally high or low atmospheric pressures could be interrelated in vast global connections. Knowledge about these links—El Niño/Southern Oscillation (ENSO)—and about the behaviour of parts of these vast systems enables forecasters to make better long-range predictions, at least in part, because the ENSO features change slowly and somewhat regularly. This approach of studying interconnections between the atmosphere and the ocean may represent the beginning of a revolutionary stage in long-range forecasting.

Since the mid-1980s, interest has grown in applying numerical weather prediction models to long-range forecasting. In this case, the concern is not with the details of weather predicted 20 or 30 days in advance but rather with objectively predicted anomalies. The reliability of long-range forecasts, like that of short- and medium-range projections, has improved substantially in recent years. Yet, many significant problems remain unsolved, posing interesting challenges for all those engaged in the field.

Importance of Weather Forecasting

Weather forecasting is used in many situations like severe weather alerts and advisories, predicting the behavior of the cloud for air transport, prediction of waterways in a sea, agricultural development and avoiding forest fire.

Severe Weather Alerts and Advisories

A major part of modern weather forecasting is the severe weather alerts and advisories which are the national weather service's issue in anticipation of severe or hazardous weather are expected. This is done to protect life and property. Some of the most commonly known of severe weather advisories are the severe thunderstorm and tornado warning, as well as the severe thunderstorm and tornado watch. Other forms of these advisories include winter weather, high wind, flood, tropical cyclone, and fog. Severe weather advisories and alerts are broadcast through the media, including radio, using emergency systems as the Emergency Alert System which breaks into regular programming.

Predicting the Behavior of the Cloud for Air Transport

The aviation industry is especially sensitive to the weather and accurate weather forecasting is essential. Fog or exceptionally low ceilings can prevent many aircraft from landing and taking off. Turbulence and icing are also significant in-flight hazards. Thunderstorms are a problem for all aircrafts because of severe turbulence due to their updrafts and outflow boundaries, icing due to the heavy precipitation, as well as large hail, strong winds, and lightning, all of which can cause severe damage to an aircraft in flight. Volcanic ash is also a significant problem for aviation, as aircraft can lose engine power within ash clouds.

Prediction of Waterways in a Sea

Commercial and recreational use of waterways can be limited significantly by wind direction, speed, wave periodicity, high tides and precipitation. These factors can each influence the safety of marine transit. Consequently, a variety of codes have been established to efficiently transmit detailed marine weather forecasts to vessel pilots via radio, for example marine forecast. Typical weather forecasts can be received at sea through the use of Radio fax.

Agricultural Development

Weather plays an important role in agricultural production. It has a profound influence on the growth, development and yields of a crop, incidence of pests and diseases, water needs and fertilizer requirements in terms of differences in nutrient mobilization due to water stresses and timeliness and effectiveness of prophylactic and cultural operations on crops. Weather aberrations may cause (i) physical damage to crops and (ii) soil erosion. The quality of the crop produced during movement from field to storage and transport to market depends on weather. Bad weather may affect the quality of the produce during transport and viability and vigor of seeds and planting material during storage.

Avoiding Forest Fire

Weather forecasting of wind, precipitations and humidity is essential for preventing and controlling wildfires. Different indices, like the Forest fire weather index and the Haines Index, have been developed to predict the areas more at risk to experience fire from natural or human causes. Conditions for the development of harmful insects can also be predicted by weather forecasting.

Military applications

Military weather forecasters present weather conditions to the war fighter community. Military weather forecasters provide pre-flight and in-flight weather briefs to pilots and provide real time resource protection services for military installations. Naval forecasters cover the waters and ship weather forecasts. The Navy provides a special service to both themselves and the rest of the federal government by issuing forecasts for tropical cyclone across the Pacific and Indian Oceans through their Joint Typhoon Warning Center.

Air Force

Air Force Weather provides weather forecasting for the Air Force and the Army. Air Force forecasters cover air operations in both wartime and peacetime operations and provide Army support. Military and civilian forecasters actively cooperate in analyzing and creating weather forecast products.

Types of Weather Forecasting

A daily weather forecast involves the work of thousands of observers and meteorologists all over the world. Modern computers make forecasts more accurate than ever, and weather satellites orbiting the earth take photograph of cloud from space.

Forecasters use the observations from ground and space, along with formulas and rules based on experience of what has happened in the past, and then make their forecast.

Meteorologists actually use a combination of several different methods to come up with their daily weather forecasts. They are:

a. Persistence Forecasting
b. Synoptic Forecasting
c. Statistical Forecasting
d. Computer forecasting.

Persistence Forecasting

The simplest method of forecasting the weather is persistence forecasting. It relies upon today's conditions to forecast the conditions tomorrow. This can be a valid way of forecasting the weather when it is in a steady state, such as during the summer season in the tropics. This method of forecasting strongly depends upon the presence of a stagnant weather pattern. It can be useful in both short range forecasts and long range forecasts. This assumes that what the weather is doing now is what it will continue to do. To find out what the weather is doing, meteorologists make weather observations.

Synoptic Forecasting

This method uses the basic rules for forecasting. Meteorologists take their observations, and apply those rules to make a short-term forecast.

Statistical Forecasting

Records of average temperatures, average rainfall and average snowfall over the years give forecasters an idea of what the weather is "supposed to be like" at a certain time of the year.

Computer Forecasting

Forecasters take their observations and plug the numbers into complicated equations. Several ultra-high-speed computers run these various equations to make computer "models" which give a forecast for the next several days. Often, different equations produce different results, so meteorologists must always use the other forecasting methods along with this one.

Using all the above methods, forecasters come up with their "best guess" as to what weather conditions will be over the next few days.

Weather forecasting now has a wide range of operational products that traditionally are classified under the following groups:

a. Very short-range forecast
b. Short-range forecast

c. Medium-range forecast

d. Long-range forecast.

Each weather forecast can be defined on the basis of the following criteria:

a. Dominant technology

b. Temporal range of validity after emission

c. Characteristics of input and output time and space resolution

d. Broadcasting needs

e. Accuracy.

Chapter 6

Climatology

Climate refers to the weather conditions averaged over a long period of time. The branch of science which is involved in the study of climate is called climatology. There are a number of important areas of study within this field such as climate change and global warming. The topics elaborated in this chapter will help in gaining a better perspective about these areas of study within climatology.

Climate

Climate refers to the conditions of the atmosphere at a particular location over a long period of time; it is the long-term summation of the atmospheric elements (and their variations) that, over short time periods, constitutes weather. These elements are solar radiation, temperature, humidity, precipitation (type, frequency, and amount), atmospheric pressure, and wind (speed and direction).

The best modern definitions of climate regard it as constituting the total experience of weather and atmospheric behaviour over a number of years in a given region. Climate is not just the "average weather" (an obsolete, and always inadequate, definition). It should include not only the average values of the climatic elements that prevail at different times but also their extreme ranges and variability and the frequency of various occurrences. Just as one year differs from another, decades and centuries are found to differ from one another by a smaller, but sometimes significant, amount. Climate is therefore time-dependent, and climatic values or indexes should not be quoted without specifying what years they refer to.

Climate and the Oceans

Hurricane Isabel: Hurricane Isabel over the Atlantic Ocean.

The atmosphere and the oceans are intimately related. They affect one another primarily through the transfer of heat and moisture. Heat energy moves from the oceans to the atmosphere through the processes of direct heat transfer and evaporation, and energy from the atmosphere flows to the oceans in the form of precipitation. Many ocean currents are driven by surface-level winds; they move warm water from the tropics to the poles and cold water from the poles toward the tropics. Warm water plays a substantial role in the development of tropical cyclones and extratropical cyclones, and warm and cold ocean currents alike strongly influence the dominant climate patterns of coastal areas. In addition, the complex interactions between the oceans and the atmosphere periodically alter certain large-scale climate patterns, such as the El Niño/Southern Oscillation (ENSO).

The notion of a connection between the temperature of the surface layers of the oceans and the circulation of the lowest layer of the atmosphere, the troposphere, is a familiar one. The surface mixed layer of the ocean is a huge reservoir of heat when compared with the overlying atmosphere. The heat capacity of an atmospheric column of unit area cross section extending from the ocean surface to the outermost layers of the atmosphere is equivalent to the heat capacity of a column of seawater of 2.6-metre (8.5-foot) depth. The surface layer of the oceans is continuously being stirred by the overlying winds and waves, and thus a surface mixed layer is formed that has vertically uniform properties in temperature and salinity. This mixed layer, which is in direct contact with the atmosphere, has a minimum depth of 20 metres (about 66 feet) in summer and a maximum depth exceeding 100 metres (about 330 feet) in late winter in the midlatitudes. At lower latitudes the seasonal variation in the mixed layer is less marked than at higher latitudes, except in regions such as the Arabian Sea where the onset of the southwestern Indian monsoon may produce large changes in the depth of the mixed layer. Temperature anomalies (i.e., deviations from the normal seasonal temperature) in the surface mixed layer have a long residence time compared with those of the overlying turbulent atmosphere. Hence, they may persist for a number of consecutive seasons and even for years.

The Ocean Surface and Climate Anomalies

Observational studies to investigate the relationship between anomalies in ocean surface temperature and the tropospheric circulation have been undertaken primarily in the Pacific and Atlantic. They have identified large-scale ocean surface temperature anomalies that have similar spatial scales to monthly and seasonal anomalies in atmospheric circulation. The longevity of the ocean surface temperature anomalies, as compared with the shorter dynamical and thermodynamical "memory" of the atmosphere, has suggested that they may be an important predictor for seasonal and interannual climate anomalies.

First, it is useful to consider some examples of the association between anomalies in ocean surface temperature and irregular changes in climate. The Sahel, a region that borders the southern fringe of the Sahara in Africa, experienced a number of devastating droughts during the 1970s and '80s, which can be compared with a much wetter period during the 1950s. Data was obtained that showed the difference in ocean surface temperature during the period from July to September between the "driest" and "wettest" rainfall seasons in the Sahel after 1950. Of particular note were the higher-than-normal surface temperatures in the tropical South Atlantic, Indian, and Southeast Pacific oceans and the lower-than-normal temperatures in the North Atlantic and Pacific oceans. This example illustrates that climate anomalies in one region of the world may be linked to ocean surface temperature changes on a global scale.

Shorter-lived climate anomalies, on timescales of months to one or two years, also have been related to ocean surface temperature anomalies. The equatorial oceans have the largest influence on these climate anomalies because of the evaporation of water. A relatively small change in ocean surface temperature—of, perhaps, 1 °C (1.8 °F)—may result in a large change in the evaporation of water into the atmosphere. The increased water vapour in the lower atmosphere is condensed in regions of upward motion known as convergence zones. This process liberates latent heat of condensation, which in turn provides a major fraction of the energy to drive tropical circulation and is one of the mechanisms responsible for the El Niño/Southern Oscillation (ENSO) phenomenon.

Given the sensitivity of the tropical atmosphere to variations in tropical sea surface temperature, there also has been considerable interest in their influence on extratropical circulation. The sensitivity of the tropospheric circulation to surface temperature in both the tropical Pacific and Atlantic oceans has been shown in both theoretical and observational studies. Figures were prepared to demonstrate the correlation between the equatorial ocean surface temperature in the eastern Pacific (the location of El Niño) and the atmospheric circulation in the middle troposphere during winter. The atmospheric pattern was a characteristic circulation type known as the Pacific–North American (PNA) mode. Such patterns are intrinsic modes of the atmosphere, which may be forced by thermal anomalies in the tropical atmosphere and which in their turn is forced by tropical ocean surface temperature anomalies. As noted earlier, enhanced tropical sea surface temperatures increase evaporation into the atmosphere. In the 1982–83 and 1997–98 El Niño events, a pattern of circulation anomalies occurred throughout the Northern Hemisphere during winter. These modes of the atmosphere, however, account for much less than 50 percent of the variability of the circulation in midlatitudes, though in certain regions (northern Japan, southern Canada, and the southern United States) they may have sufficient amplitude for them to be used for predicting seasonal surface temperature perhaps up to two seasons in advance.

The response of the atmosphere to midlatitude ocean surface anomalies has been difficult to detect unambiguously because of the complexity of the turbulent westerly flow between 20° and 60° latitude in both hemispheres. This flow has many properties of nonlinear chaotic systems and thus exhibits behaviour that is difficult to predict beyond a couple of weeks. The atmosphere alone can exhibit large fluctuations on seasonal and longer timescales without any change in external forcing conditions, such as ocean surface temperature. Notwithstanding this inherent problem, some effects of ocean surface temperature anomalies on the atmosphere have been observed and modeled.

The influence of the oceans on the atmosphere in the midlatitudes is greatest during autumn and early winter when the ocean mixed layer releases to the atmosphere the large quantities of heat that it has stored up over the previous summer. Anomalies in ocean surface temperature are indicative of either a surplus or a deficiency of heat available to the atmosphere. The response of the atmosphere to ocean surface temperature, however, is not geographically random. The circulation over the North Atlantic and northern Europe during early winter has been found to be sensitive to large ocean surface temperature anomalies south of Newfoundland. When a warm positive anomaly exists in this region, an anomalous surface anticyclone occurs in the atmosphere above the central Atlantic at similar latitude to the temperature anomaly, and an anomalous cyclonic circulation is located over the North Sea, Scandinavia, and central Europe. With colder-than-normal water south of Newfoundland, the circulation patterns are reversed, producing cyclonic circulation over the central Atlantic and anticyclonic circulation over Europe. The sensitivity of the atmosphere to ocean surface temperature anomalies in this particular region south of Newfoundland is thought to be related to the position of the overlying storm tracks and jet stream. This region is the most active in the Northern Hemisphere for the growth of storms associated with very large heat fluxes from the surface layer of the ocean.

Another example of a similar type of air-sea interaction event has been documented over the North Pacific Ocean. A statistical seasonal relationship exists between the summer ocean temperature anomaly in the Gulf of Alaska and the atmospheric circulation over the Pacific and North America during the following autumn and winter. The presence of warmer-than-normal ocean surface water in the Gulf of Alaska results in increased cyclone development during the subsequent autumn and winter. The relationship has been established by means of monthly sea surface temperature and atmospheric pressure data collected over 30 years in the North Pacific Ocean.

The air-sea interaction events in both the North Pacific and North Atlantic oceans discussed above raise questions as to how the anomalies in ocean surface temperature in these areas are initiated, how they are maintained, and whether they yield useful information for atmospheric prediction beyond the normal timescales of weather forecasting (i.e., one to two weeks). Statistical analysis of previous case studies has shown that ocean surface temperature anomalies initially develop in response to anomalous atmospheric forcing. Once developed, however, the temperature anomaly of the ocean surface tends to reinforce and thereby maintain the anomalous atmospheric circulation. The mechanisms thought to be responsible for this behaviour in the ocean are the surface wind drift, wind mixing, and the interchange of heat between the ocean and atmosphere. The question of prediction is therefore difficult to answer, as these events depend on a synchronous and interconnected behaviour between the atmosphere and the surface layer of the ocean, which allows for positive feedback between the two systems.

Formation of Tropical Cyclones

Tropical cyclones represent still another example of air-sea interactions. These storm systems are known as hurricanes in the North Atlantic and eastern North Pacific and as typhoons in the western North Pacific. The winds of such systems revolve around a centre of low pressure in an counter clockwise direction in the Northern Hemisphere and in a clockwise direction in the Southern Hemisphere. The winds attain velocities in excess of 115 km (71 miles) per hour, or 65 knots, in most cases. Tropical cyclones may last from a few hours to as long as two weeks, the average lifetime being six days.

The oceans provide the source of energy for tropical cyclones both by direct heat transfer from their surface (known as sensible heat) and by the evaporation of water. This water is subsequently condensed within a storm system, thereby releasing latent heat energy. When a tropical cyclone moves over land, this energy is severely depleted and the circulation of the winds is consequently weakened.

Such storms are truly phenomena of the tropical oceans. They originate in two distinct latitude zones, between 4° and 22° S and between 4° and 35° N. They are absent in the equatorial zone between 4° S and 4° N. Most tropical cyclones are spawned on the poleward side of the region known as the intertropical convergence zone (ITCZ).

More than two-thirds of observed tropical cyclones originate in the Northern Hemisphere. The North Pacific has more than one-third of all such storms, while the southeast Pacific and South Atlantic are normally devoid of them. Most Northern Hemispheric tropical cyclones occur between May and November, with peak periods in August and September. The majority of Southern Hemispheric cyclones occur between December and April, with peaks in January and February.

Conditions associated with Cyclone Formation

The formation of tropical cyclones is strongly influenced by the temperature of the underlying ocean or, more specifically, by the thermal energy available in the upper 60 metres (about 200 feet) of ocean waters. Typically, the underlying ocean should have a temperature in excess of 26 °C (about 79 °F) in this layer. This temperature requirement, however, is only one of five that need to be met for a tropical cyclone to form and develop. The other preconditions relate to the state of the tropical atmosphere between the sea surface and a height of 16 km (about 10 miles), the boundary of the tropical troposphere. They can be summarized as follows:

- A deep convergence of air must occur in the troposphere between the surface and a height of 7 km (about 4 miles) that produces a cyclonic circulation in the lower troposphere overlain by an anticyclonic circulation in the upper troposphere. The stronger the inflow, or convergence, of the air, the more favourable are the conditions for tropical cyclone formation.
- The vertical shear of the horizontal wind velocity between the lower troposphere and the upper troposphere should be at minimum. Under this condition the heat and moisture are retained rather than being exchanged and diluted with the surrounding air. Monsoonal and trade wind flows are characterized by a large vertical shear of the horizontal wind and so are not generally conducive to tropical cyclone development.
- A strong vertical coupling of the flow patterns between the upper and lower troposphere is required. This is achieved by large-scale deep convection associated with cumulonimbus clouds.
- A high humidity level in the middle troposphere from 3 to 6 km (1.8 to 3.7 miles) in height is more conducive to the production of deep cumulonimbus convection and therefore to stronger vertical coupling in the troposphere.

All these conditions may be met but still not lead to cyclone formation. It is thought that the most important factor is the presence of a large-scale cyclonic circulation in the lower troposphere. The above conditions occur for a period of 5 to 15 days and are followed by less-favourable conditions for duration of 10 to 20 days.

Once a tropical cyclone has formed, it usually follows certain distinct stages during its lifetime. In its formative stage the winds are below hurricane force, and the central pressure is about 1,000 millibars, or 750 mm (29.53 inches) of mercury. The formative period is extremely variable in length, ranging from 12 hours to a few days. This stage is followed by a period of intensification, when the central pressure drops rapidly below 1,000 millibars. The winds increase rapidly, and they may achieve hurricane force within a radius of 30 to 50 km (19 to 31 miles) of the storm centre. At this stage the cloud and rainfall patterns become well organized into narrow bands that spiral inward toward the centre. In the mature phase the central pressure stops falling and, as a consequence, the winds no longer increase. The region of hurricane-force winds, however, expands to occupy a radius of 300 km (186 miles) or more. This expansion is not symmetrical around the storm centre; the strongest winds occur toward the right-hand side of the centre in the direction of the cyclone's path. The period of maturity may last one to three days. The terminal stage of a tropical cyclone is usually reached when the storm strikes land, which causes an increase in energy dissipation by surface friction and a reduction in its energy supply of moisture. A reduction in moisture input into the storm system may also take place when it moves over a colder segment of the ocean. Similarly, the storm can regain its strength over warmer water. This process was observed in the case of Hurricane Katrina, a catastrophic tropical cyclone that passed through the Gulf of Mexico in 2005. A tropical cyclone may regenerate in higher latitudes as an extratropical depression, but it loses its identity as a tropical storm in the process. The typical lifetime of a tropical cyclone from its birth to death is about six days.

The paths of tropical cyclones show a wide variation. In both the North Atlantic and the North Pacific, the paths tend to be initially northwestward and then recurve toward the northeast at higher latitudes. It is now known that the tracks of tropical cyclones are largely determined by the large-scale tropospheric flow. This fact, coupled with the aid of high-resolution numerical models, makes more-accurate predictions of their tracks feasible. Polar-orbiting and geostationary satellites make it possible to accurately track cyclones over the remotest areas of the tropical oceans.

Effects of Tropical Cyclones on Ocean Waters

A tropical cyclone can affect the thermal structure and currents in the surface layer of the ocean waters in its path. Cooling of the surface layer occurs in the wake of such a storm. Maximum cooling occurs on the right of a hurricane's path in the Northern Hemisphere. In the wake of Hurricane Hilda's passage through the Gulf of Mexico in 1964 at a translational speed of only five knots, the surface waters were cooled by as much as 6 °C (10.8 °F). Tropical cyclones that have higher translational velocities cause less cooling of the surface. The surface cooling is caused primarily by wind-induced upwelling of cooler water from below the surface layer. The warm surface water is simultaneously transported toward the periphery of the cyclone, where it downwells into the deeper ocean layers. Heat loss across the air-sea interface and the wind-induced mixing of the surface water with those of the cooler subsurface layers make a significant but smaller contribution to surface cooling.

In addition to surface cooling, tropical cyclones may induce large horizontal surge currents (storm surges) and vertical displacements of the thermocline. The surge currents have their largest amplitude at the surface, where they may reach velocities approaching 1 metre (about 3 feet) per second. The horizontal currents and the vertical displacement of the thermocline observed in the wake of a tropical cyclone oscillate close to the inertial period. These oscillations remain for a few days after the passage of the storm and spread outward from the rear of the system as an internal wake on the thermocline. The vertical motion may transport nutrients from the deeper layers into the sunlit surface waters, which in turn promotes phytoplankton blooms (i.e., the rapid growth of diatoms and other minute one-celled organisms). The ocean surface temperature normally recovers to its pre-cyclone value within 10 days of a storm's passage.

Influence on Atmospheric Circulation and Rainfall

Tropical cyclones play an important role in the general circulation of the atmosphere, accounting for 2 percent of the global annual rainfall and between 4 and 5 percent of the global rainfall in August and September at the height of the Northern Hemispheric cyclone season. For a local area the occurrence of a single tropical cyclone can have a major impact on the annual rainfall. Furthermore, tropical cyclones contribute approximately 2 percent of the kinetic energy of the general circulation of the atmosphere, some of which is exported from the tropics to higher latitudes.

Poleward Transfer of Heat

Thermohaline Circulation

A significant characteristic of the large-scale North Atlantic circulation is the poleward transport of heat. Heat is transferred in a northward direction throughout the North Atlantic. This heat is absorbed by the tropical waters of the Pacific and Indian oceans as well as of the Atlantic and is then transferred to the high latitudes, where it is finally given up to the atmosphere.

The mechanism for the heat transfer is principally by thermohaline circulation rather than by wind-driven circulation. Circulation of the thermohaline type involves a large-scale overturning of the ocean, with warm and saline water in the upper 1,000 metres (3,300 feet) moving northward and being cooled in the Labrador, Greenland, and Norwegian seas. The density of the water in contact with the atmosphere is increased by surface cooling, and the water subsequently sinks below the surface layer to the lowest depths of the ocean. This water is mixed with the surrounding water masses by a variety of processes to form the North Atlantic Deep Water. The water moves slowly southward as the lower limb of the thermohaline circulation. It is this overturning circulation that is responsible for the warm winter climate of northwestern Europe (notably the British Isles and Norway) rather than the horizontal wind-driven circulation discussed above. The North Atlantic Drift, which is an extension of the Gulf Stream system to the south, provides this northward flow of warm and saline waters into the polar seas. This feature makes the circulation of the North Atlantic Ocean uniquely different from that of the Pacific Ocean, which has a less effective thermohaline circulation. Although there is a northward transfer of heat in the North Pacific, the subtropical wind-driven gyre in the upper ocean is mainly responsible for it. The Kuroshio on the western boundary of the North Pacific gyre is principally driven by the surface wind circulation of the North Pacific.

Studies of the sediment cores obtained from the ocean floor have indicated that the ocean surface temperature was as much as 10 °C (18 °F) cooler than today in the northernmost region of the North Atlantic Ocean during the last glacial maximum some 18,000 years ago. This difference in surface temperature would indicate that the warm North Atlantic Drift was much reduced compared with what it is at present, and hence the thermohaline circulation was considerably weaker. In contrast, the Gulf Stream was probably more intense than it is today and exhibited a large shift from its present path to an eastward flow at 40° N.

The Gulf Stream

This major current system is a western boundary current that flow poleward along a boundary separating the warm and more saline waters of the Sargasso Sea to the east from the colder, slightly fresher continental slope waters to the north and west. The warm, saline Sargasso Sea, composed of a water mass known as North Atlantic Central Water, has a temperature that ranges from 8 to 19 °C (46.4 to 66.2 °F) and a salinity between 35.10 and 36.70 parts per thousand (ppt). This is one of the two dominant water masses of the North Atlantic Ocean; the other is the North Atlantic Deep Water, which has a temperature of 2.2 to 3.5 °C (4 to 6.3 °F) and a salinity between 34.90 and 34.97 ppt and which occupies the deepest layers of the ocean (generally below 1,000 metres [about 3,300 feet]). The North Atlantic Central Water occupies the upper layer of the North Atlantic Ocean between roughly 20° and 40° N. The "lens" of this water is at its lowest depth of 1,000 metres in the northwest Atlantic and becomes progressively shallower to the east and south. To the north it shallows abruptly and outcrops at the surface in winter, and it is at this point that the Gulf Stream is most intense.

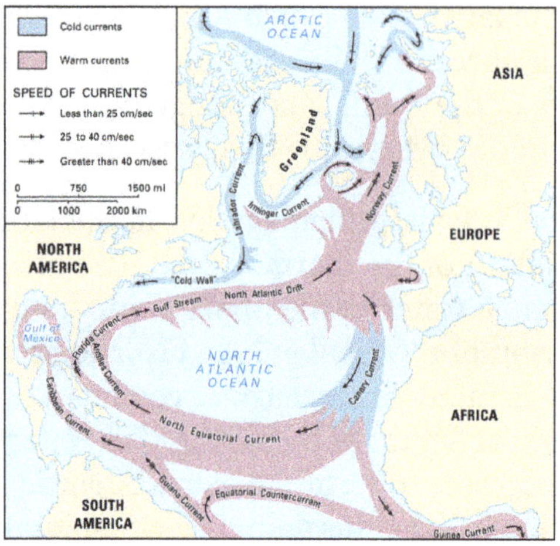

Major warm and cold currents of the North Atlantic Ocean.

The Gulf Stream flows northward along the rim of the warm North Atlantic Central Water, from the Florida Straits along the continental slope of North America to Cape Hatteras. There it leaves the continental slope and turns northeastward as an intense meandering current that extends toward the Grand Banks of Newfoundland. Its maximum velocity is typically between 1 and 2 metres (about 3 to 7 feet) per second. At this stage a part of the current loops back onto itself, flowing south and east. Another part flows eastward toward Spain and Portugal, while the remaining water flows northeastward as the North Atlantic Drift (also called the North Atlantic Current) into the northernmost regions of the North Atlantic Ocean between Scotland and Iceland.

The southward-flowing currents are generally weaker than the Gulf Stream and occur in the eastern part of the North Atlantic Central Water lens or the subtropical gyre. The circulation to the south on the southern rim of the subtropical gyre is completed by the westward-flowing North Equatorial Current, part of which flows into the Gulf of Mexico; the remaining part flows northward as the Antilles Current. This subtropical gyre of warm North Atlantic Central Water is the hub of the energy that drives the North Atlantic circulation. It is principally forced by the overlying atmospheric circulation, which at these latitudes is dominated by the clockwise circulation of a subtropical anticyclone. This circulation is not steady and fluctuates in particular on its poleward side where extratropical cyclones in the westerlies periodically make incursions into the region. On the western side, hurricanes (during the period from May to November) occasionally disturb the atmospheric circulation. Because of the energy of the subtropical gyre and its associated currents, these short-term fluctuations have little influence on it, however. The gyre obtains most of its energy from the climatological wind distribution over periods of one or two decades. This wind distribution drives a system of surface currents in the uppermost 100 metres of the ocean. Nonetheless, these currents are not simply a reflection of the surface wind circulation, as they are influenced by the Coriolis force. The wind-driven current decays with depth, becoming negligible below 100 metres. The water in this surface layer is transported to the right and perpendicular to the surface wind stress because of the Coriolis force. Hence an eastward-directed wind on the poleward side of the subtropical anticyclone would transport the surface layer of the ocean to the south. On the equatorward side of the anticyclone the trade winds would cause a contrary drift of the surface layer to the north and west. Thus, surface waters under the subtropical anticyclone are driven toward the midlatitudes at about 30° N. These surface waters, which are warmed by solar heating and have a high salinity by virtue of the predominance of evaporation over precipitation at these latitudes, then converge and are forced downward into the deeper ocean.

Over many decades this process forms a deep lens of warm, saline North Atlantic Central Water. The shape of the lens of water is distorted by other dynamic effects, the principal one being the change in the vertical component of the Coriolis force with latitude known as the beta effect. This effect involves the displacement of the warm water lens toward the west, so that the deepest part of the lens is situated to the north of the island of Bermuda rather than in the central Atlantic Ocean. This warm lens of water plays an important role, establishing a horizontal pressure gradient force in and below the wind-driven current. The sea level over the deepest part of the lens is about one metre higher than outside the lens. The Coriolis force in balance with this horizontal pressure gradient force gives rise to a dynamically induced geostrophic current, which occurs throughout the upper layer of warm water. The strength of this geostrophic current is determined by the horizontal pressure gradient through the slope in sea level. The slope in sea level across the Gulf Stream has been measured by satellite radar altimeter to be one metre over a horizontal distance of 100 km (62 miles), which is sufficient to cause a surface geostrophic current of one metre per second at 43° N.

The large-scale circulation of the Gulf Stream system is, however, only one aspect of a far more complex and richer structure of circulation. Embedded within the mean flow is a variety of eddy structures that not only put kinetic energy into circulation but also carry heat and other important properties, such as nutrients for biological systems. The best known of these eddies are the Gulf Stream rings, which develop in meanders of the current east of Cape Hatteras. Though the eddies were mentioned as early as 1793 by Jonathan Williams, a grandnephew of American scientist and statesman Benjamin Franklin, they were not systematically studied until the early 1930s by the oceanographer Phil E. Church. Intensive research programs were finally undertaken during the 1970s. Gulf Stream rings have either warm or cold cores. The warm-core rings are typically 100 to 300 km (62 to 186 miles) in diameter and have a clockwise rotation. They consist of waters from the Gulf Stream and Sargasso Sea and form when the meanders in the Gulf Stream pinch off on its continental slope side. They move generally westward and are reabsorbed into the Gulf Stream at Cape Hatteras after a typical lifetime of about six months. The cold-core rings, composed of a mixture of Gulf Stream and continental slope waters, are formed when the meanders pinch off to the south of the Gulf Stream. They are a little larger than their warm-core counterparts, characteristically having diameters of 200 to 300 km (124 to 186 miles) and a counterclockwise rotation. They move generally southwestward into the Sargasso Sea and have lifetimes of one to two years. The cold-core rings are usually more numerous than warm-core rings, typically 10 each year as compared with five warm-core rings annually.

The Kuroshio

This western boundary current is similar to the Gulf Stream in that it produces both warm and cold rings. The warm rings are generally 150 km (93 miles) in diameter and have a lifetime similar to their Gulf Stream counterparts. The cold rings form at preferential sites and in most cases drift southwestward into the Western Pacific Ocean. Occasionally a cold ring has been observed to move northwestward and eventually be reabsorbed into the Kuroshio.

El Niño/Southern Oscillation and Climatic Change

As was explained earlier, the oceans can moderate the climate of certain regions. Not only do they affect such geographic variations, but they also influence temporal changes in climate. The timescales of climate variability range from a few years to millions of years and include the so-called ice age cycles that repeat every 20,000 to 40,000 years, interrupted by interglacial periods of "optimum" climate, such as the present. The climatic modulations that occur at shorter scales include such periods as the Little Ice Age from the early 14th to the mid-19th centuries, when the average temperature of the Northern Hemisphere was approximately 0.6 °C (1.1 °F) lower than it is today. Several climate fluctuations on the scale of decades occurred in the 20th century, such as warming from 1910 to 1940, cooling from 1940 to 1970, and the warming trend since 1970.

Although many of the mechanisms of climate change are understood, it is usually difficult to pinpoint the specific causes. Scientists acknowledge that climate can be affected by factors external to the land-ocean-atmosphere climate system, such as variations in solar brightness, the shading effect of aerosols injected into the atmosphere by volcanic activity, or the increased atmospheric concentration of greenhouse gases (e.g., carbon dioxide, nitrous oxide, methane, and chlorofluorocarbons) produced by human activities. However, none of these factors completely explains the periodic variations observed during the 20th century, which may simply be manifestations of the natural variability of climate. The existence of natural variability at many timescales makes the identification of causative factors such as human-induced warming more difficult. Whether change is natural or caused, the oceans play a key role and have a moderating effect on influencing factors.

The El Niño Phenomenon

The shortest, or interannual, timescale relates to natural variations that are perceived as years of unusual weather—e.g., excessive heat, drought, or storminess. Such changes are so common in many regions that any given year is about as likely to be considered exceptional as typical. The best example of the influence of the oceans on interannual climate anomalies is the occurrence of El Niño and La Niña conditions in the eastern Pacific Ocean at irregular intervals of about 3–8 years. The stronger El Niño episodes of enhanced ocean temperatures (2–8 °C [3.6–14.4 °F] above normal) are typically accompanied by altered weather patterns around the globe, such as droughts in Australia, northeastern Brazil, and the highlands of southern Peru, excessive summer rainfall along the coast of Ecuador and northern Peru, severe winter storminess along the coast of central Chile, and unusual winter weather along the west coast of North America.

The effects of El Niño have been documented in Peru since the Spanish conquest in 1525. The Spanish term "la corriente de El Niño" was introduced by fishermen of the Peruvian port of Paita in the 19th century, referring to a warm, southward ocean current that temporarily displaces the normally cool, northward-flowing Humboldt, or Peru, Current. The name is a pious reference to the Christ Child, chosen because of the typical appearance of the countercurrent during the Christmas season. By the end of the 19th century, Peruvian geographers recognized that every few years this countercurrent is more intense than normal, extends farther south, and is associated with torrential rainfall over the otherwise dry northern desert. The abnormal countercurrent also was observed to bring tropical debris, as well as such flora and fauna as bananas and aquatic reptiles, from the coastal region of Ecuador farther north. Increasingly during the 20th century, El Niño came to connote an exceptional year rather than the original annual event.

As Peruvians began to exploit the guano of marine birds for fertilizer in the early 20th century, they noticed El Niño-related deteriorations in the normally high marine productivity of the coast of Peru as manifested by large reductions in the bird populations that depend on anchovies and sardines for sustenance. The preoccupation with El Niño increased after mid-century, as the Peruvian fishing industry rapidly expanded to exploit the anchovies directly. Fish meal produced from the anchovies was exported to industrialized countries as a feed supplement for livestock. By 1971 the Peruvian fishing fleet had become the largest in its history; it had extracted very nearly 13 million metric tons of anchovies in that year alone. Peru was catapulted into first place among fishing nations, and scientists expressed serious concern that fish stocks were being depleted beyond self-sustaining levels, even for the extremely productive marine ecosystem of Peru. The strong El Niño of 1972–73 captured world attention because of the drastic reduction in anchovy catches to a small fraction of prior levels. The anchovy catch did not return to previous levels, and the effects of plummeting fish meal exports reverberated throughout the world commodity markets.

El Niño was only a curiosity to the scientific community in the first half of the 20th century, thought to be geographically limited to the west coast of South America. There was little data, mainly gathered coincidentally from foreign oceanographic cruises, and it was generally believed that El Niño occurred when the normally northward coastal winds off Peru, which cause the upwelling of cool, nutrient-rich water along the coast, decreased, ceased, or reversed in direction. When systematic and extensive oceanographic measurements were made in the Pacific in 1957–58 as part of the International Geophysical Year, it was found that El Niño had occurred during the same period and was also associated with extensive warming over most of the Pacific equatorial zone. Eventually tide-gauge and other measurements made throughout the tropical Pacific showed that the coastal El Niño was but one manifestation of basinwide ocean circulation changes that occur in response to a massive weakening of the westward-blowing trade winds in the western and central equatorial Pacific and not to localized wind anomalies along the Peru coast.

The Southern Oscillation

The wind anomalies are a manifestation of an atmospheric counterpart to the oceanic El Niño. At the turn of the century, the British climatologist Gilbert Walker set out to determine the connections between the Malaysian-Australian monsoon and other climatic fluctuations around the globe in an effort to predict unusual monsoon years that bring drought and famine to the Asian sector. Unaware of any connection to El Niño, he discovered a coherent interannual fluctuation of atmospheric pressure over the tropical Indo-Pacific region, which he termed the Southern Oscillation (SO). During years of reduced rainfall over northern Australia and Indonesia, the pressure in that region (e.g., at what are now Darwin and Jakarta) was anomalously high and wind patterns were altered. Simultaneously, in the eastern South Pacific pressures were unusually low, negatively correlated with those at Darwin and Jakarta. A Southern Oscillation Index (SOI), based on pressure differences between the two regions (east minus west), showed low, negative values at such times, which were termed the "low phase" of the SO. During more normal "high-phase" years, the pressures were low over Indonesia and high in the eastern Pacific, with high, positive values of the SOI. In papers published during the 1920s and '30s, Walker gave statistical evidence for widespread climatic anomalies around the globe being associated with the SO pressure "seesaw."

In the 1950s, years after Walker's investigations, it was noted that the low-phase years of the SOI corresponded with periods of high ocean temperatures along the Peruvian coast, but no physical connection between the SO and El Niño was recognized until Norwegian American meteorologist Jacob Bjerknes, in the early 1960s, tried to understand the large geographic scale of the anomalies observed during the 1957–58 El Niño event. Bjerknes formulated the first conceptual model of the large-scale ocean-atmosphere interactions that occur during El Niño episodes. His model has been refined through intensive research since the early 1970s.

During a year or two prior to an El Niño event (high-phase years of the SO), the westward trade winds typically blow more intensely along the Equator in the equatorial Pacific, causing warm upper-ocean water to accumulate in a thickened surface layer in the western Pacific where sea level rises. Meanwhile, the stronger, upwelling-favourable winds in the eastern Pacific induce colder surface water and lowered sea levels off South America. Toward the end of the year preceding an El Niño, the area of intense tropical storm activity over Indonesia migrates eastward toward the equatorial Pacific west of the International Date Line (which corresponds in general to the 180th meridian of longitude), bringing episodes of eastward wind reversals to that region of the ocean. These wind bursts excite extremely long ocean waves, known as Kelvin waves (imperceptible to an observer), that propagate eastward toward the coast of South America, where they cause the upper ocean layer of relatively warm water to thicken and sea level to rise.

The tropical storms of the western Pacific also occur in other years, though less frequently, and produce similar Kelvin waves, but an El Niño event does not result, and the waves continue poleward along the coast toward Chile and California, detectable only in tide-gauge measurements. Something else occurs prior to an El Niño that is not fully understood: as the Kelvin waves travel eastward along the Equator, an anomalous eastward current carries warm western Pacific water farther east, and the warm surface layer deepens in the central equatorial Pacific (east of the International Date Line). Additional surface warming takes place as the upwelling-favourable winds bring warmer subsurface water to the surface. (The subsurface water is warmer now, rather than cooler, because the overlying layer of warmer water is now significantly deeper than before.) The anomalous warming creates conditions favourable for the further migration of the tropical storm centre toward the east, giving renewed vigour to eastward winds, more Kelvin waves, and additional warming. Each increment of anomalies in one medium (e.g., the ocean) induces further anomalies in the other (the atmosphere) and vice versa, giving rise to an unstable growth of anomalies through a process of positive feedbacks. During this time the SO is found in its low phase.

After several months of these unstable ocean-atmosphere interactions, the entire equatorial zone becomes considerably warmer (2–5 °C [3.6–9 °F]) than normal, and a sizable volume of warm upper ocean water is transported from the western to the eastern Pacific. As a result, sea levels fall by 10–20 cm (about 4–8 inches) in the west and rise by larger amounts off the coast of South America, where sea surface temperature anomalies may vary from 2–8 °C above normal. Anomalous conditions typically persist for 10–14 months before returning to normal. The warming off South America occurs even though the upwelling-favourable winds there continue unabated: the upwelled water is warmer now, rather than cooler as before, and its associated nutrients are less plentiful, thereby failing to sustain the marine ecosystem at its prior productive levels.

The current focus of oceanographic research is on understanding the circumstances leading to the demise of the El Niño event and the onset of another such event several years later. The most widely held hypothesis is that a second class of long equatorial ocean waves—Rossby waves with a shallow surface layer—is generated by El Niño and that they propagate westward to the landmasses of Asia. There the Rossby waves reflect off the Asian coast eastward along the Equator in the form of upwelling Kelvin waves, resulting in a thinning of the upper ocean warm layer and a cooling of the upper ocean as the winds mix deeper, cooler water to the surface. This process is thought to initiate one to two years of colder-than-average conditions until westward-propagating Rossby waves are again generated, functioning as a switching mechanism, this time to start another El Niño sequence.

Another goal of scientists is to understand climate change on the scale of centuries or longer and to make projections about the changes that will occur within the next few generations. Yet, determinations of current climatic trends from recent data are made difficult by natural variability at shorter timescales, such as the El Niño phenomenon. Many scientists are attempting to understand the mechanisms of change during an El Niño event from improved global measurements so as to determine how the ocean-atmosphere engine operates at longer timescales. Others are studying prehistoric records preserved in trees, sediments, and fossil corals in an effort to reconstruct past variations, including those like El Niño. Their aim is to remove such short-term variations so as to be able to make more accurate estimates of long-term trends.

Climate and Life

The connection between climate and life arises from a two-way exchange of mass and energy between the atmosphere and the biosphere. In Earth's early history, before life evolved, only geochemical and geophysical processes determined the composition, structure, and dynamics of the atmosphere. Since life evolved on Earth, biochemical and biophysical processes have played a role in the determination of the composition, structure, and dynamics of the atmosphere. Humans, Homo sapiens, are increasingly shouldering this role by mediating interactions between the biosphere and the atmosphere.

The planet Earth.

The living organisms of the biosphere use gases from, and return "waste" gases to, the atmosphere, and the composition of the atmosphere is a product of this gas exchange. It is very likely that, prior to the evolution of life on Earth, 95 percent of the atmosphere was made up of carbon dioxide, and water vapour was the second most abundant gas. Other gases were present in trace amounts. This atmosphere was the product of geochemical and geophysical processes in the interior of Earth and was mediated by volcanic outgassing. It is estimated that the great mass of carbon dioxide in this early atmosphere gave rise to an atmospheric pressure 60 times that of modern times. Today only about 0.035 percent of Earth's atmosphere is carbon dioxide. Much of the carbon dioxide present in Earth's first atmosphere has been removed by photosynthesis, chemosynthesis, and weathering. Currently, most of the carbon dioxide now resides in Earth's limestone sedimentary rocks, in coral reefs, in fossil fuels, and in the living components of the present-day biosphere. In this transformation, the atmosphere and the biosphere coevolved through continuous exchanges of mass and energy.

Biogenic gases are gases critical for, and produced by, living organisms. In the contemporary atmosphere, they include oxygen, nitrogen, water vapour, carbon dioxide, carbon monoxide, methane, ozone, nitrogen dioxide, nitric acid, ammonia and ammonium ions, nitrous oxide, sulfur dioxide, hydrogen sulfide, carbonyl sulfide, dimethyl sulfide, and a complex array of non-methane hydrocarbons. Of these gases, only nitrogen and oxygen are not "greenhouse gases." Added to this roster of biogenic gases is a much longer list of human-generated gases from industrial, commercial, and cultural activities that reflect the diversity of the human enterprise on Earth.

The Gaia Hypothesis

The notion that the biosphere exerts important controls on the atmosphere and other parts of the Earth system has increasingly gained acceptance among earth and ecosystem scientists. While this concept has its origins in the work of American oceanographer Alfred C. Redfield in the mid-1950s, it was English scientist and inventor James Lovelock that gave it its modern currency in the late 1970s. Lovelock initially proposed that the biospheric transformations of the atmosphere support the biosphere in an adaptive way through a sort of "genetic group selection." This idea generated extensive criticism and spawned a steady stream of new research that has enriched the debate and advanced both ecology and environmental science. Lovelock called his idea the "Gaia Hypothesis" and defined Gaia as a complex entity involving Earth's biosphere, atmosphere, oceans, and soil; the totality constituting a feedback of cybernetic systems which seeks an optimal physical and chemical environment for life on this planet.

The Greek word Gaia, or Gaea, meaning "Mother Earth," is Lovelock's name for Earth, which is envisioned as a "superorganism" engaged in planetary biogeophysiology. The goal of this superorganism is to produce a homeostatic, or balanced, Earth system. The scientific process of research and debate will eventually resolve the issue of the reality of the "Gaian homeostatic superorganism," and Lovelock has since revised his hypothesis to exclude goal-driven genetic group selection. Nevertheless, it is now an operative norm in contemporary science that the biosphere and the atmosphere interact in such a way that an understanding of one requires an understanding of the other. Furthermore, the reality of two-way interactions between climate and life is well recognized.

The Evolution of Life and the Atmosphere

Life on Earth began at least as early as 3.5 billion years ago during the middle of the Archean Eon (about 4 billion to 2.5 billion years ago). It was during this interval that life first began to exercise certain controls on the atmosphere. The atmosphere's prebiological state is often characterized as being rich in water vapour and carbon dioxide. Though some nitrogen was also present, little if any oxygen was available. Chemical reactions with hydrogen sulfide, hydrogen, and reduced compounds of nitrogen and sulfur precluded any but the shortest lifetime for free oxygen in the atmosphere. As a result, life evolved in an atmosphere that was reducing (high hydrogen content) rather than oxidizing (high oxygen content). In addition to their chemically reducing character, the predominant gases of this prebiotic atmosphere, with the exception of nitrogen, were largely transparent to incoming sunlight but opaque to outgoing terrestrial infrared radiation. As a result, these gases are called, perhaps improperly, greenhouse gases because they are able to slow the release of outgoing radiation back into space.

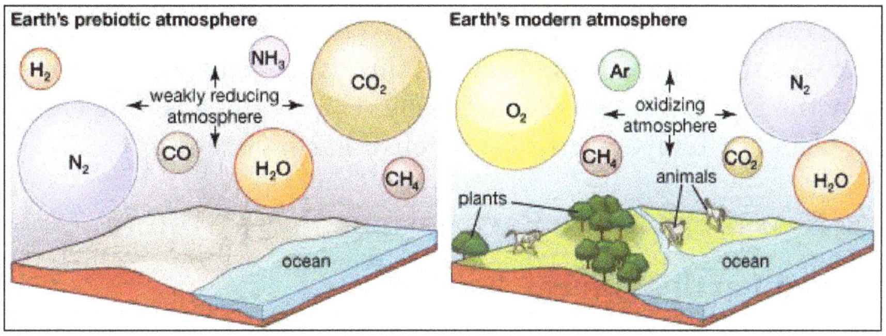

Earth's early and modern atmospheres.

Comparison of Earth's prebiotic and modern atmospheres: Before life began on the planet, Earth's atmosphere was largely made up of nitrogen and carbon dioxide gases. After photosynthesizing organisms multiplied on Earth's surface and in the oceans, much of the carbon dioxide was replaced with oxygen.

In the Archean Eon, the Sun produced as much as 25 percent less light than it does today; however, Earth's temperature was much like that of today. This is possible because the greenhouse gas-rich Archean atmosphere was effective in retarding the loss of terrestrial radiation to space. The resulting long residence time of energy within the Earth-atmosphere system resulted in a warmer atmosphere than would have been possible otherwise. The average temperature of Earth's surface in the early Archean Eon was warmer than the modern global average. It was, according to some sources, probably similar to temperatures found in today's tropics. Depending on the amount of nitrogen present during the Archean Eon, it has been suggested that the atmosphere may have held more than 1,000 times as much carbon dioxide than it does today.

Archean organisms included photosynthetic and chemosynthetic bacteria, methane-producing bacteria, and a more primitive group of organisms now called the "Archaea" (a group of prokaryotes more related to eukaryotes than to bacteria and found in extreme environments). Through their metabolic processes, organisms of the Archean Eon slowly changed the atmosphere. Hydrogen rose from trace amounts to about 1 part per million (ppm) of dry air. Methane concentrations increased from near zero to about 100 ppm. Oxygen increased from near zero to 1 ppm, whereas nitrogen concentrations rose to encompass 99 percent of all atmospheric molecules excluding water vapour. Carbon dioxide concentrations decreased to only 0.3 percent of the total; however, this was nearly 10 times the current concentration. The composition of the atmosphere, its radiation budget, its thermodynamics, and its fluid dynamics were transformed by life from the Archean Eon.

American geochemist Robert Garrels calculated that, in the absence of life and given the burial rate of carbon in rocks, oxygen would be unavailable to form water, and free hydrogen would be lost to space. Without the presence of life and compounded by this loss of hydrogen, there would be no oceans, and Earth would have become merely a dusty planet by the middle of the Archean Eon. By the end of the Archean Eon 2.5 billion years ago, both the pigment chlorophyll and photosynthetic organisms had evolved such that the production of oxygen increased rapidly. The atmosphere became transformed from a reducing atmosphere with carbon dioxide, limited oxygen, and anaerobic organisms (that is, life-forms that do not require oxygen for respiration) in control to one with an oxidizing atmosphere that was rich in oxygen, poor in carbon dioxide, and dominated by aerobic organisms (that is, life-forms requiring oxygen for respiration).

With the decline in carbon dioxide and a rise in oxygen, the greenhouse warming capacity of Earth's atmosphere was sharply reduced; however, this happened over a period of time when the energy produced by the Sun increased systematically. These compensating changes resulted in a relatively constant planetary temperature over much of Earth's history.

The Role of the Biosphere in the Earth-atmosphere System

The Biosphere and Earth's Energy Budget

Biogenic gases in the atmosphere play a role in the dynamics of Earth's planetary radiation budget, the thermodynamics of the planet's moist atmosphere, and, indirectly, the mechanics of the fluid flows that are Earth's planetary wind systems. In addition, human cultural and economic activities add a new dimension to the relationship between the biosphere and the atmosphere. While humans are biologically trivial compared with bacteria in the exchange of gases with the atmosphere, chemical compounds produced from human industrial activities and other economic enterprises are changing the gaseous composition of the atmosphere in climatically significant ways. The largest changes involve the harvesting of ancient carbon stores. This organic material has been transformed into fossil fuels (coal, petroleum, natural gas, and others) by geologic processes acting upon the remains of plants and animals over many millions of years. Different forms of carbon may be burned and thus used as energy sources. In so doing, organic carbon is converted into carbon dioxide. Additionally, humans are also burning trees, grasses, and other biomass for cooking purposes and clearing the land for agriculture and other activities. The combination of burning both fossil fuels and biomass is enriching the atmosphere with carbon dioxide and adding to the essential reservoir of greenhouse gases.

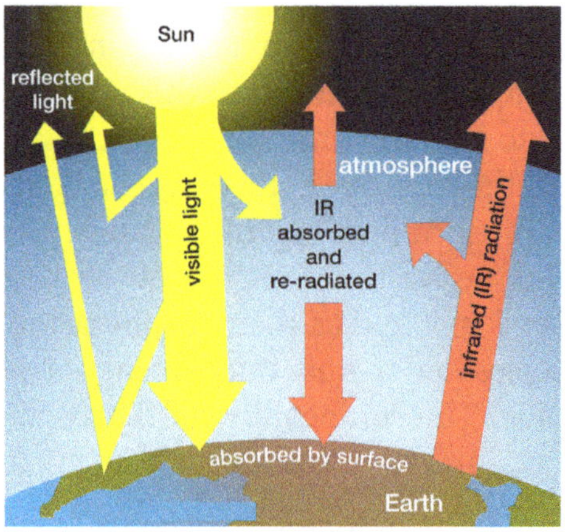

Greenhouse effect on Earth.

The greenhouse effect on Earth- Some incoming sunlight is reflected by Earth's atmosphere and surface, but most is absorbed by the surface, which is warmed. Infrared (IR) radiation is then emitted from the surface. Some IR radiation escapes to space, but some is absorbed by the atmosphere's greenhouse gases (especially water vapour, carbon dioxide, and methane) and reradiated in all directions, some to space and some back toward the surface, where it further warms the surface and the lower atmosphere.

Earth's atmosphere is largely transparent to sunlight. Of the sunlight absorbed by the entire Earth-atmosphere system, about one-third is absorbed by the atmosphere and two-thirds by Earth's surface. Sunlight is absorbed by the molecules of the atmosphere, by cloud droplets, and by dust and debris. Though oxygen and nitrogen make up nearly 99 percent of the atmosphere, these diatomic molecules do not vibrate in a way that permits them to absorb terrestrial radiation. They are largely transparent to outgoing terrestrial radiation as well as to incoming solar radiation.

Over the continents, the surface cover of vegetation is the principal absorbing medium of Earth's surface, although other surfaces such as bare rock, sand, and water also absorb solar radiation. At night, absorption at the surface (that is, below 1.2 metres [4 feet]) is reradiated, in the form of long-wave infrared radiation, away from Earth's surface back toward space. Most of this infrared radiation is absorbed by the principal biogenic trace gases of the atmosphere—the so-called greenhouse gases: water vapour, carbon dioxide, and methane. Without these biogenic greenhouse gases, Earth would be 33 °C (59 °F) colder on average than it is. A moderate-emission scenario from the 2007 Intergovernmental Panel on Climate Change (IPCC) report predicts that the continued addition of greenhouse gases from fossil fuels will increase the average global temperature by between 2.3 and 4.3 °C (4.1 and 7.7 °F) over the next century. Other scenarios, predicting greater greenhouse gas emissions, forecast even greater global warming.

The Cycling of Biogenic Atmospheric Gases

The cycling of oxygen, nitrogen, water vapour, and carbon dioxide, as well as the trace gases—methane, ammonia, various oxides of nitrogen and sulfur, and non-methane hydrocarbons—between the atmosphere and the biosphere results in relatively constant proportions of these compounds in the atmosphere over time. Without the continuous generation of these gases by the biosphere, they would quickly disappear from the atmosphere.

Average composition of the atmosphere			
gas	composition by volume (ppm)*	composition by weight (ppm)*	total mass (10^{20} g)
nitrogen	780,900	755,100	38.648
oxygen	209,500	231,500	11.841
argon	9,300	12,800	0.655
carbon dioxide	386	591	0.0299
neon	18	12.5	0.000636
helium	5.2	0.72	0.000037
methane	1.5	0.94	0.000043
krypton	1.0	2.9	0.000146
nitrous oxide	0.5	0.8	0.000040
hydrogen	0.5	0.035	0.000002
ozone**	0.4	0.7	0.000035
xenon	0.08	0.36	0.000018
*ppm = parts per million. **Variable, increases with height.			

The carbon cycle, as it relates to the biosphere, is simple in its essence. Inorganic carbon (carbon dioxide) is converted to organic carbon (the molecules of life). To complete the cycle, organic carbon is then converted back to inorganic carbon. Ultimately, the carbon cycle is powered by sunlight as green plants and cyanobacteria (blue-green algae) use sunlight to split water into oxygen and hydrogen and to fix carbon dioxide into organic carbon. Carbon dioxide is removed from the atmosphere, and oxygen is added. Animals engage in aerobic respiration, in which oxygen is consumed and organic carbon is oxidized to manufacture inorganic carbon dioxide. It should be noted that chemosynthetic bacteria, which are found in deep-ocean and cave ecosystems, also fix carbon dioxide and produce organic carbon. Instead of using sunlight as an energy source, these bacteria rely on the oxidation of either ammonia or sulfur.

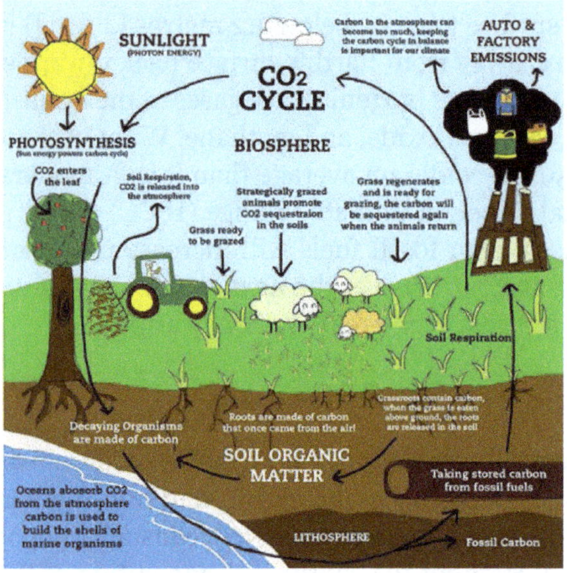

The carbon cycle is the complex path that carbon follows through the atmosphere, oceans, soil, and plants and animals.

The carbon cycle is fully coupled to the oxygen cycle. Each year, photosynthesis fixes carbon dioxide and releases 100,000 megatons of oxygen to the atmosphere. Respiration by animals and living organisms consumes about the same amount of oxygen and produces carbon dioxide in return. Oxygen and carbon dioxide are thus coupled in two linked cycles. On a seasonal basis, an enrichment of atmospheric carbon dioxide occurs in the winter half of the year, whereas a drawdown of atmospheric carbon dioxide takes place during the summer.

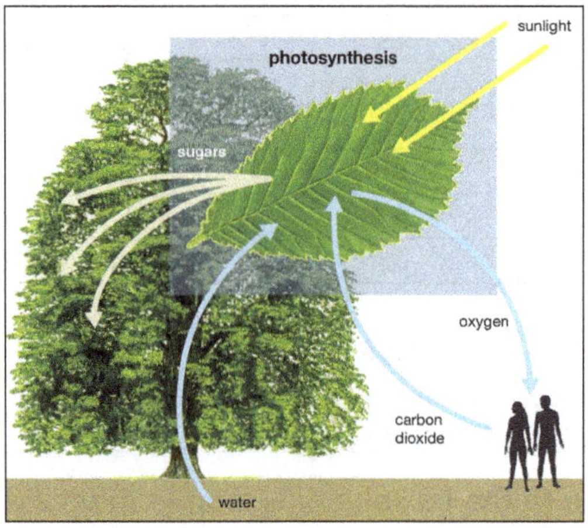

Green plants such as trees use carbon dioxide, sunlight, and water to create sugars.

The nitrogen cycle begins with the fixing of inorganic atmospheric nitrogen (N_2) into organic compounds. These nitrogen-containing compounds are used by organisms and, through the process of denitrification, are converted back to inorganic atmospheric nitrogen. Ammonia and ammonium ions are the products of nitrogen fixation and may be incorporated into some organisms as organic nitrogen-containing molecules. In addition, ammonium ions may be oxidized to form nitrites, which can be further oxidized by nitrifying bacteria into nitrates. Though both nitrites and nitrates may be used to make organic nitrogen-containing molecules, nitrates are especially useful for plant growth and are key compounds that support both terrestrial and aquatic food chains. Nitrates may also be denitrified by bacteria to produce nitrogen gas. This process completes the nitrogen cycle.

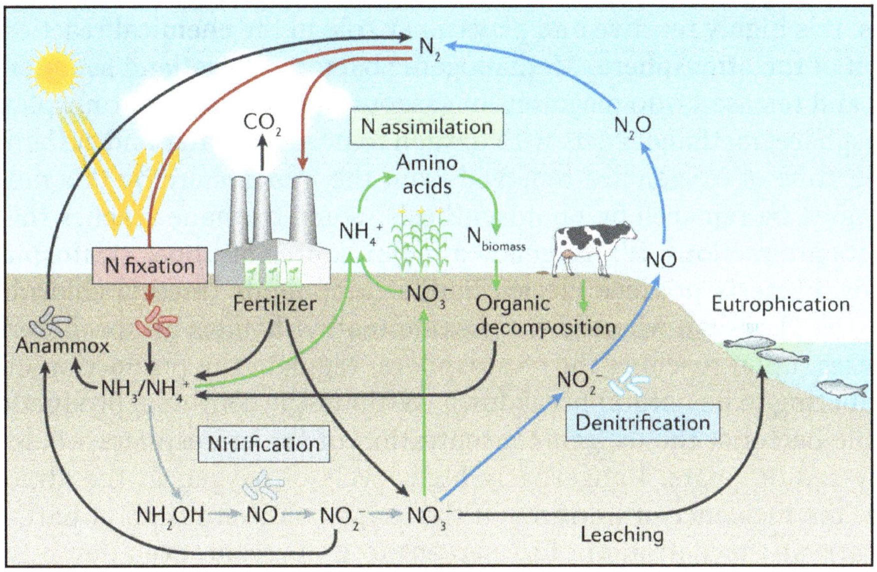

The nitrogen cycle.

The nitrogen cycle is coupled to both the carbon and oxygen cycles. Seventy-eight percent of the gases of the atmosphere by volume is diatomic nitrogen. Diatomic nitrogen is the most stable of the nitrogen-containing gases of the atmosphere. Only 300 megatons of nitrogen must be produced each year by denitrifying bacteria to account for losses. These losses mostly occur during lightning discharges and during nitrogen-fixation activities by blue-green algae and nitrogen-fixing bacteria. (The latter are found in the root nodules of certain plants called legumes.) Nitrogen-fixing bacteria use atmospheric nitrogen to produce oxides of nitrogen. Denitrifying bacteria convert nitrates in soils and wetlands to nitrogen gas, which is then returned to the atmosphere.

Nitrous oxide occurs in trace amounts (0.3 ppm) in the atmosphere. Between 100 and 300 megatons of nitrous oxide are produced by soil and marine bacteria each year to maintain this concentration. In the atmosphere, nitrous oxide is short-lived because it is quickly broken down by ultraviolet light. Nitric oxide (NO), a minor contributor in the breakdown of stratospheric ozone, is the by-product of this reaction.

Like nitrous oxide, ammonia is also produced in soils and marine waters by bacteria and escapes into the atmosphere. Nearly 1,000 megatons are added to the atmosphere each year. Ammonia decreases the acidity of precipitation and serves as a nutrient for plants when returned to the land via precipitation.

There are six major sulfur-containing biogenic atmospheric gases that are part of the sulfur cycle. They include hydrogen sulfide, carbon disulfide (CS_2), carbonyl sulfide (COS), dimethyl sulfide (DMS; C_2H_6S), dimethyl disulfide ($[CH_3S]_2$), and methyl mercaptan (CH_3SH). Sulfur dioxide (SO_2) is an oxidation product of these sulfur gases, and it is also added to the atmosphere by volcanoes, burning biomass, and anthropogenic sources (i.e., smelting metals and coal ignition). SO_2 is removed from the atmosphere and returned to the biosphere in rainfall. The increased acidity of rain and snow from anthropogenic additions of SO_2 and oxides of nitrogen is often referred to as "acid precipitation." The acidity of this precipitation and other phenomena, such as "acid fog," is partly cancelled by the release of ammonia in the atmosphere.

The concentration of methane at any one time in the atmosphere is only about 1.7 ppm. Though only a trace gas, it is highly reactive and plays a key role in the chemical reactions that control the composition of the atmosphere. Methanogenic bacteria in wetland sediments decompose organic matter and release 1,000 megatons of gaseous methane to the atmosphere per year. In the lower atmosphere, methane reacts with oxygen to produce water and carbon dioxide. Each year 2,000 megatons of oxygen are removed from the atmosphere by this mechanism. This loss of oxygen must be replaced by photosynthesis. Some methane reaches the upper stratosphere, where its interaction with oxygen is a major source of upper stratospheric moisture. Within wetlands, bacteria produce methyl halide compounds (methyl chloride [CH_3Cl] and methyl iodide [CH_3I] gases), whereas these same methyl halides are produced in forests by fungi. These gases, upon reaching the stratosphere, regulate the production of stratospheric ozone by contributing to its natural breakdown. Without the continual production of methane by methanogenic bacteria, the oxygen concentration of the atmosphere would increase by 1 percent in only 12,000 years. Dangerously high levels of oxygen in the atmosphere would greatly increase the incidence of wildfires. If the oxygen concentration of Earth's atmosphere rose from its current concentration of 21 percent to 25 percent, even damp twigs and grass would easily ignite. Non-methane hydrocarbons of terrestrial origin are generally well mixed in the free atmosphere above the planetary boundary layer. These organic particles weaken incoming solar radiation as it passes through the atmosphere, and reductions of 1 percent have been recorded.

Biosphere Controls on the Structure of the Atmosphere

Because the biosphere plays a key role in the flux of energy from the surface to the atmosphere, it also contributes to the structure of the atmosphere. Three major fluxes are important: the direct transfer of heat from the surface to the atmosphere by conduction and convection (sensible heating), the energy flux to the atmosphere carried by water vapour via evaporation and transpiration from the surface (latent heat energy), and the flux of radiant energy from the surface to the atmosphere (infrared terrestrial radiation). These fluxes differ in the altitude at which the heating of the air takes place and thus contribute to the thermal structuring of the atmosphere. Sensible heating primarily warms the planetary boundary layer (PBL) of the atmosphere. In marine areas, the PBL occurs in the lowest 1 km (3,300 feet); in heavily vegetated areas, the PBL occurs in the lowest 1 to 2 km (3,300 to 6,600 feet); and in arid regions, it occurs in the lowest 4 or 5 km (13,100 to 16,400 feet) of the atmosphere. In contrast, the latent heat of the atmosphere is released when the water vapour is converted into cloud droplets by condensation. Heating by latent energy release generally occurs above the PBL.

On the other hand, heating of the atmosphere by radiation from the surface depends on the density of the atmosphere and its water vapour content. Radiative heating from the surface declines with increasing altitude. The availability of water to evaporate from the surface limits the sensible heating of the air near the surface and so limits the maximum daytime surface air temperature.

Biosphere Controls on the Planetary Boundary Layer

The top of the planetary boundary layer (PBL) can be visually marked by the elevation of the base of the clouds. In addition, the PBL can also be denoted by a thin layer of haze often seen by passengers aboard airplanes during take-off from airports. During the day, the air within the PBL is thoroughly mixed by convection induced by the heating of Earth's surface. The thickness of the PBL depends on the intensity of this surface heating and the amount of water evaporated into the air from the biosphere. In general, the greater the heating of the surface, the deeper the PBL. Over deserts, the PBL may extend up to 4 or 5 km (13,100 or 16,400 feet) in altitude. In contrast, the PBL is less than 1 km (0.6 mile) thick over ocean areas, since little surface heating takes place there because of the vertical mixing of water. The wetter the air adverted into the region and the greater the additional water added by evaporation and transpiration, the lower the height of the top of the PBL. For every 1 °C (1.8 °F) increase in daily maximum surface temperature for a well-mixed PBL, the top of the PBL is elevated 100 metres (about 325 feet). In New England forests during the days following the spring leafing, it has been shown that the top of the PBL is lowered to between 200 and 400 metres (650 and 1,300 feet). By contrast, during the months before the leafing out, the PBL thickens from solar heating as the sun rises higher in the sky and day length increases.

If convective mixing of the air in the PBL is vigorous, convection currents may penetrate through the temperature inversion at the top of the PBL. The cooling of the lifting air initiates the condensation of water vapour and the development of miniscule particles of liquid water called cloud droplets. The small clouds just above the PBL are known as planetary boundary layer clouds. These clouds scatter direct sunlight. As the ratio of diffuse sunlight to direct beam sunlight increases, greater levels of photosynthetic productivity are favoured in the biosphere below. The result is a dynamic synergy between the atmosphere and biosphere.

The landscapes of most human-dominated ecosystems are decidedly "patchy" in their geography. Cities, suburbs, fields, forests, lakes, and shopping centres both heat and evaporate water into the air of the PBL according to the nature of the surfaces involved. Convection and the prospect of breaking through the top of the PBL vary markedly across such heterogeneous landscapes. These upward and downward currents or vertical eddies within the PBL transfer mass and energy upward from the surface. The frequency, timing, and strength of convective weather elements, including thunderstorms, vary according to the patchiness of the land use and land cover pattern of the area. In general, the greater the patchiness of the landscape and the earlier the hour in the day, the more frequent and more intense these rain-producing systems become.

In the absence of an organized storm in the region, the air above the PBL sinks gently and the air below lifts. At the top of the PBL, a small inversion, where temperatures increase with height, develops. This inversion essentially becomes a stable layer in the atmosphere. Emissions from the biosphere below are thus contained within the PBL and may build up below this layer over time. Consequently, the PBL may become quite turbid, hazy, or filled with smog.

When the sinking from above is vigorous, the PBL inversion grows in thickness. This situation has the effect of hindering the development of thunderstorms, which depend on rapidly rising air. This often occurs over southern California, and thus the chance of thunderstorms forming there is small. Emissions from both the biosphere and from anthropogenic activities accumulate in this part of the atmosphere, and pollution may build up to such an extent that health warnings may be required. In locations free of temperature inversions, convection processes are strong enough, particularly during the summer months, that emissions are scavenged and quickly lifted by thunderstorms to regions high above the PBL. Often, acidic compounds from these emissions are returned to the surface in the precipitation that falls.

Biosphere Controls on Maximum Temperatures by Evaporation and Transpiration

Solar radiation is converted to sensible and latent heat at Earth's surface. A change in sensible heat results in a change in the temperature of a medium, whereas energy stored as latent heat is used to drive a process, such as a phase change in a substance from its liquid to its gaseous state, and does not produce a change in temperature. Thus, the daily maximum surface temperature at a given location is dependent on the amount of radiant energy converted to sensible heat. Water available for evaporation increases latent heating by adding water vapour to the atmosphere. As a result, relatively little energy remains to heat the air, and thus the sensible heating of the air near the ground is minimized. In addition, daily maximum temperatures are not as high in locations with strong latent heating.

As day length increases from winter to summer, sensible heating and maximum surface temperatures rise. In the U.S. Midwest, prior to the leafing out of vegetation in the springtime and the resulting rise in evaporation and transpiration, sensible heating causes an average increase in maximum surface temperatures of only about 0.3 °C (0.5 °F) per day. The process of leaf production creates a surge in evaporation and transpiration and results in increased latent heating and reduced sensible heating. After leafing, since most of the available thermal energy is used to convert liquid water to water vapour rather than to heat the air, the average day-to-day rise in daily maximum temperatures is reduced to about 0.1 °C (0.2 °F) per day.

This effect extends upward through the atmosphere. Prior to leafing out, the one-kilometre-thick layer occurring between the 850-to-750-millibar pressure level (which typically occurs between 1,650 and 2,750 metres [5,400 and 9,000 feet]) in the Midwest warmed at the rate of 0.1 °C (0.2 °F) per day. Following leafing out, the warming rate fell to 0.02 °C (0.04 °F) per day. Scientists have used computer models of the atmosphere to study the effect of transpiration from vegetation on maximum surface air temperatures. In these models, the variable controlling transpiration by vegetation was "turned off," and the character of the resulting modeled climate was studied. By subtracting the effect of transpiration, temperatures in central North America and on the other continents were predicted to equilibrate at a very hot 45 °C (113 °F). Such warming is nearly realized in desert areas where moisture is unavailable for transpiration.

Biosphere Controls on Minimum Temperatures

During the late 1860s, British experimental physicist John Tyndall, based on his studies of the infrared radiation absorption by atmospheric gases, concluded that nighttime minimum temperatures were dependent on the concentration of trace gases in the atmosphere. Of these gases, water vapour had the greatest impact. To emphasize the significance of water vapour on decreases in air temperature during the night, he wrote that if all the water vapour in the air over England was removed even for a single night, it would be "attended by the destruction of every plant which a freezing temperature could kill." As a result, it follows that the greater the water content of the atmosphere, the lower the radiative loss of energy to the sky and the less the surface atmosphere is cooled. Thus, locations with substantial amounts of water vapour experience reduced nocturnal cooling.

Water vapour in the atmosphere also limits the extent to which temperatures fall at night. This limiting temperature is known as the dew point, which is defined as the temperature at which condensation begins. Over North America east of the 100th meridian (a line of longitude traditionally dividing the moist eastern part of North America from drier western areas), average nighttime minimum temperatures are within a degree or two of the dew point temperature. Upon nocturnal cooling, the dew point is reached, condensation begins, and latent energy is converted to heat. Additional temperature falls are retarded by this release of heat to the atmosphere. A significant fraction of the water in the atmosphere over the continents comes from the evaporation of water from soils and the transpiration from vegetation. Transpired water directly moderates temperature by increasing humidity and thus raising the dew point. As a consequence, the amount of outgoing terrestrial radiation released to space is reduced. These results in the elevation of the minimum temperature of the air above what it would otherwise be.

Dew often forms on grass during cool nights.

The effect of spring leafing on the buildup of humidity in the lower atmosphere has received the attention of researchers in recent years. In the late 1980s, American climatologists M.D. Schwartz and T.R. Karl used the superimposed epoch method to study the climate before and after the leafing out of lilac plants in the spring in the U.S. Midwest. (This method uses time series data from multiple locations, which can be compared to one another by adjusting each data set around the respective onset date of lilac blooming.) In the illustration, the x-axis marks days before and after leafing, whereas the y-axis shows the related changes in the vapour pressure of the atmosphere. A second y-axis follows the day-by-day changes in minimum temperatures. Prior to the average date of leafing, the atmospheric humidity (vapour pressure) is relatively constant and minimum temperatures hover near freezing. At leafing, there is an abrupt increase in atmospheric humidity. Following leafing, daily minimum temperatures also increase abruptly. Although frosts are possible until June 10 in many parts of the Midwest, the chances of frost decline as the atmosphere is humidified.

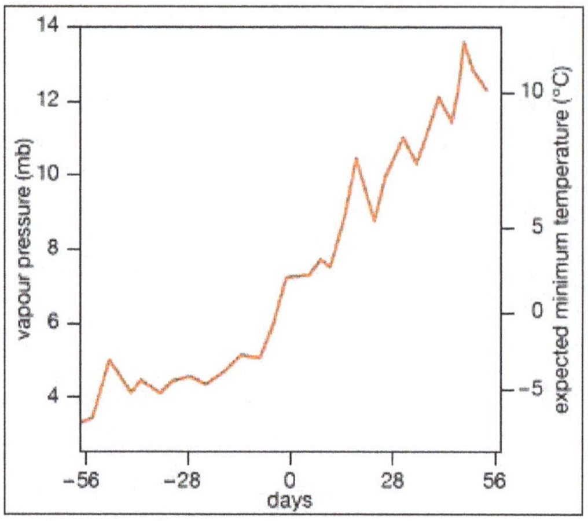

Lilac bud break, humidity, and temperature: Graph of atmospheric vapour pressure and expected minimum temperature for 56 days prior to and following the average day of the leafing out of lilac plants, based on data compiled by M.D. Schwartz and T.R. Karl, 1990.

Climate and Changes in the Albedo of the Surface

The amount of solar energy available at the surface for sensible and latent heating of the atmosphere depends on the albedo, or the reflectivity, of the surface. Surface albedos vary by location, season, and land cover type. The albedo of unvegetated ground devoid of snow ranges from 0.1 to 0.6 (10 to 60 percent), while the albedo of fully forested lands ranges from 0.08 to 0.15. An increase of 0.1 in regional albedo has been associated with a 20 percent decline in rainfall events connected with thunderstorms. Equivalent reductions in both evaporation and transpiration have also been reported in areas with sudden increases in albedo.

The greatest changes in albedo occur in regions undergoing desertification and deforestation. Depending on the albedo of the underlying soil, reductions in vegetative land cover may give rise to albedo increases of as much as 0.2. Model studies of the vegetative zone known as the Sahel in Africa reveal that albedo increased from 0.14 to 0.35 due to desertification occurring during the 20th century. This coincided with a 40 percent decrease in rainfall. In addition, it is likely that the clearing of forests and prairies for agricultural crops over the past several hundred years has altered the albedo of extensive regions of the middle latitudes.

Deforestation: Smoldering remains of a plot of deforested land in the Amazon Rainforest of Brazil. Annually, it is estimated that net global deforestation accounts for about two gigatons of carbon emissions to the atmosphere.

Sahelian landscape near Zinder, Niger.

Contemporary agricultural practices give rise to large variations in albedo from season to season as the land passes through the cycle of tilling, planting, crop growth, and harvest. At larger scales, an agricultural mosaic often emerges as each different plot of ground is covered by plantings of a single species. Viewed from the air, landscapes in the middle latitudes appear as a heterogeneous mix of forests, grasslands, meadows, water bodies, farmlands, wetlands, and urban types. The resultant patchiness in the landscape produces a patchiness in surface albedo. The mosaic of land use types creates a mix in the fluxes of sensible and latent heat to the atmosphere. Such changes to the heat flux have been shown to cause changes in the timing, intensity, and frequency of summer thunderstorms.

The Effect of Vegetation Patchiness on Mesoscale Climates

The establishment of vegetation bands or patches 50 to 100 km (30 to 60 miles) in width in semi-arid regions could increase atmospheric convection and precipitation beyond that expected over areas of uniform vegetation. This convection creates spatial differences in the upward and downward wind velocities and contributes to the development of mesoscale (20 to 200 km [12 to 120 miles]) circulation in the atmosphere. For example, when creating models for forecasting atmospheric conditions on the Great Plains and along the Front Range of the Rocky Mountains, the mix of land cover and vegetation types must be specified to properly relate the fluxes of momentum and sensible and latent heat to the larger-scale circulation of the atmosphere. Proper calculations are also necessary to estimate rainfall. In addition, the specific location and hour of the day that thunderstorms occur depend on the heterogeneity of the vegetation cover of this region. Field observations have shown that the heterogeneity of surface roughness (small-scale irregularities in topography), soil moisture, forest coverage, and transpiration affect the location and pace of the formation of convective clouds and rainfall. Both convection and thunderstorm development tend to occur earlier in the day in heterogeneous landscapes.

Biosphere Controls on Surface Friction and Localized Winds

Averaged annually over Earth's entire surface, the Sun provides about 345 watts per square metre of energy. About 30 percent of this energy is reflected away to space and is never used in the Earth-atmosphere system. Of that which remains, a little less than 1 percent (3.1 watts per square metre) accelerates the air by generating winds. An equal amount of energy must eventually be lost, or else wind speeds would perpetually increase.

Earth as a thermodynamic system is dissipative—the mechanical energy of the winds is eventually converted to heat through friction. Over the continents, it is the combination of terrain and the veneer of vegetation that offers the frictional roughness to dissipate the surface winds and convert this kinetic energy into heat. Marine winds approaching the British Isles average about 12 metres per second (27 miles per hour), but they are decelerated to 6 metres per second (13 miles per hour) because of the friction of the landscape's surface shortly after the winds make landfall. Without vegetation cover, the continents would offer much less friction to the wind, and wind speeds in unvegetated landscapes would be nearly twice as fast as those in vegetated landscapes.

The correct specification of Earth's surface roughness due to vegetation, for use in computer models of the atmosphere, is critical to proper model performance. If the height of the terrain and vegetation are not specified correctly, the patterns of Earth's winds, global geography, and rainfall will be poorly modeled. When modeling newly desertified areas, such as the Sahel, it is important to understand that desertification creates vegetation of lower stature and thus lower surface roughness values. As a result, both wind velocities and wind direction could change from previous patterns over landscapes with taller vegetation.

The extent of this impact of the biosphere on the atmosphere is revealed in climate model studies. One such study modeled the influence of reduced vegetation on surface roughness over the Indian subcontinent and provided evidence for a weaker monsoon and reduced rainfall. Given that much of the northwest third of India underwent a severe desertification and cultural collapse near the beginning of historical times, the role cultures play in vegetation reduction and climate change should not be ignored.

The vegetation cover of the continents is not passive in response to the winds. Greenhouse-grown trees subjected to mechanical forces designed to mimic the winds lay down new woody tissue called "reaction wood," which results in a stiffer tree over time. This material helps trees become more resilient and offer more frictional resistance to wind. This negative feedback, where increased winds result in stiffer vegetation and thereby subsequently reduced wind speeds, might well apply at the global scale by balancing the energy used to heat and accelerate the air (3.1 watts per square metre) with the surface friction needed to dissipate it.

Biosphere Impacts on Precipitation Processes

Cloud Condensation Nuclei

The formation and subsequent freezing of cloud droplets depend on the presence of cloud condensation nuclei and ice nuclei, respectively. Significantly, the biosphere is a major source of both of these kinds of nuclei. Over the continents, condensation nuclei are readily available and are of biogenic as well as anthropogenic origin. Examples of condensation nuclei include sea salt, small soil particles, and dust.

As atmospheric convection increases with the heating of the day, cloud condensation nuclei are mixed into and above the planetary boundary layer and into the troposphere. In the bottom 0.5 km (the lowest 1,600 feet or so) of the atmosphere, nuclei typically number 2.2 × 1010 per cubic metre. In the next 0.5 km (between 1,600 and 3,300 feet) above, half as many nuclei are found. The number of condensation nuclei continues to decline with increased altitude. Furthermore, in general, the number of nuclei in the air over land is 10 times higher than over the oceans.

Cloud condensation nuclei are generally abundant. They do not limit cloud formation over the continents; however, low numbers of condensation nuclei over the oceans may limit cloud formation there. In addition to natural sources, particulates from fuel combustion and sulfur dioxide gas resulting from high sulfur fuels also contribute to the load of condensation nuclei over the continents. Both the number and kind of condensation nuclei present in the atmosphere affect the cloudiness and the brightness of clouds in a given region. In this way, condensation nuclei play a significant role in determining both regional and global albedo.

There is a type of condensation nuclei that forms in the marine air over the margins of continents. Though these nuclei are often few in number they play a large role in cloud formation near the coasts of continents and may contribute significantly to both planetary albedo and global average temperature. Typically, sources of condensation nuclei in marine air are sulfate aerosols formed from the biogenic production of dimethyl sulfide (DMS) by marine algae. Given that DMS production increases with sea surface temperatures, a negative feedback may result. The central idea in this feedback hypothesis is that warmer waters result in the increased production of condensation nuclei by phytoplankton and thus produce more clouds. Increased cloudiness shades the ocean surface and results in lower temperatures that limit condensation nuclei production. It is estimated that a 30 percent increase in marine condensation nuclei would increase planetary albedo by 0.005 (0.5 percent) or produce a 0.7 percent reduction in solar radiation and a planetary average temperature decrease of 1.3 °C (2.3 °F). The sensitivity of this negative feedback on planetary temperatures remains in active debate.

Biogenic Ice Nuclei

As water vapour condenses onto condensation nuclei, the droplets grow in size. Growth proceeds at relative humidity as low as 70 percent, but the rate of growth is very slow. Growth by condensation is most rapid where the air is slightly supersaturated with water vapour. At this point, cloud droplets typical of the size of fog droplets arise. Should temperatures fall to the level where freezing begins, the temperature difference between the droplet and the surrounding air (the vapour pressure deficit) strongly favours rapid condensation into the crystalline lattice of an ice particle. Ice particles that grow rapidly soon reach sizes where they begin to fall. As they fall, they collide and merge with smaller droplets and thereby grow larger.

The formation of ice is of critical importance. A droplet of pure water, such as distilled water, will automatically freeze in the atmosphere at a temperature of −40 °C (−40 °F). Freezing at warmer temperatures requires a substance upon which ice crystallization can take place. The common clay mineral kaolinite, a contaminant of the droplet, raises this freezing point to around −25 °C (−13 °F). Furthermore, silver iodide, often used in cloud seeding to encourage rainfall, and sea salts also cause ice to form at −25 °C. Freezing at still warmer temperatures is most common with biogenic ice nuclei. Upon ice formation, heat energy on the order of 80 calories per gram of water frozen is released. This energy increases the sensible heat of the air and causes the air to become more buoyant. The process of ice formation encourages convection, cloudiness, and precipitation from clouds.

The decomposition of organic matter is a major source of biogenic ice nuclei. Ice crystal formation has been shown to occur at temperatures as warm as −2 to −3 °C (28.5 to 26.6 °F) when biogenic ice nuclei are involved. The common freezing temperature for biogenic nuclei varies systematically according to biome and latitude. The coldest freezing-temperature nuclei occur above the tropics, whereas the warmest occur above the Arctic. Freezing produces greater buoyancy of the particles and helps them to reach higher vertical velocities within the clouds. The vertical motions and the larger droplet size that occur with biogenic materials favour the charge separation needed to produce lightning. Subsequently, oceanic areas with few biogenic ice nuclei are also areas of low lightning frequency. The production of biogenic nuclei from organic matter decomposition is greatest during the warm months when bacterial decomposition is greatest.

Recycled Rainfall

The water that is transpired into the atmosphere from the biosphere is eventually returned to the surface as precipitation. This vegetation-transpiration component of the hydrologic cycle is referred to as "recycled rainfall." While the oceans are the major source of atmospheric water vapour and rainfall, water from plant transpiration is also significant. For example, in the 1970s and '80s, analyses performed by American meteorologist Michael Garstang on the city of Manaus, Brazil, in the Amazon basin revealed that around 20 percent of the precipitation came from water transpired by vegetation; the remaining 80 percent of this precipitation (an estimate made by German American meteorologist Heinz Lettau in the 1970s) was generated by the Atlantic Ocean. Isotopic studies of rainwater collected at various points in the Amazon basin indicated that nearly half of the total rain came from water originating in the ocean and half transpired through the vegetation. Evidence of the proportion of transpired water in rainfall reaching as high as 88 percent has been reported for the Amazon foothills of the Andes. General climate circulation models indicate that, without transpired water from plants, rainfall in the central regions of the continents would be greatly reduced. As a general rule, the farther the distance from oceanic water sources, the higher the fraction of rainwater originating from transpiration.

Climatology

Climatology is the scientific study of climate. Climatology is regarded as a subdivision of physical geography, atmospheric sciences, and earth sciences in general. Aspects of oceanography and biogeography have also been considered as part of climatology. Climatology focuses on aspects such as atmospheric boundary layer, circulation patterns, heat transfer in the globe, ocean interaction with the atmosphere and land surface, land use and topography.

Climatology has evolved from a simple theoretical bookkeeping activity to the current complex scientific and a practical field. Science is defined as the truths and facts that have been obtained through constant research, systematic methods, evaluation of phenomena, and observation. Climatology is therefore scientific method that involves all the aspects that define science. Other than the above stated aspects, to obtain climatic patterns, several scales and gauges are employed in the climatic research. Climatology is not only concerned with the climate of a place but it also establishes the reason for the fluctuation of climate in the area, how human activities lead to climatic variations, effects of the climate on human activities, and the characteristics of the climate. Climate also depends on the layers of the earth and atmosphere, a further manifestation of its scientific nature.

Sub-fields of Climatology

According to the area of specialization, climatology has been divided into smaller sub-fields: paleoclimatology, paleotempestology, historical climatology, metrology, and bioclimatology. Paleoclimatology focuses on establishing the past climatic patterns of a place by studying ice cores and tree rings. Paleotempestology uses ancient data to determine the frequency and magnitude of past hurricanes. Historical climatology focuses on establishing the climate of a place after studying the activities that the ancient dwellers of the particular place engaged in. Metrology, which is often confused with climatology, deals with weather, which runs for a maximum of probably a week or a month. Bioclimatology deals with the effects the climate has on the living organisms.

Importance of Climatology

Climatology is important in determining the climatic patterns of a particular region. Establishing the climatic pattern is significant in deciding the economic activities that would thrive in that particular region. If the climate of a region is established to be cool and wet, it would be safe to conclude that agriculture might thrive in the region. Having a clear climate pattern makes it easier for people to understand the seasons of engaging in particular tasks. This is especially most important to tourists and farmers. Infrastructure development, especially buildings are dependent on climate. After a climatic pattern has been established, the engineers recommend the use of materials that would not only withstand the conditions but also protect the dwellers from any harsh climatic condition. Furthermore, climatology seeks to establish why climate varies from place to another.

Climate Model

Climate models are based on well-documented physical processes to simulate the transfer of energy and materials through the climate system. Climate models, also known as general circulation models or GCMs, use mathematical equations to characterize how energy and matter interact in different parts of the ocean, atmosphere, land. Building and running a climate model is complex process of identifying and quantifying Earth system processes, representing them with mathematical equations, setting variables to represent initial conditions and subsequent changes in climate forcing, and repeatedly solving the equations using powerful supercomputers.

Essentially, climate models are an extension of weather forecasting. But whereas weather models make predictions over specific areas and short timespans, climate models are broader and analyze long timespans. They predict how average conditions will change in a region over the coming decades.

Climate models include more atmospheric, oceanic and land processes than weather models do—such as ocean circulation and melting glaciers. These models are typically generated from mathematical equations that use thousands of data points to simulate the transfer of energy and water that takes place in climate systems.

Scientists use climate models to understand complex earth systems. These models allow them to test hypotheses and draw conclusions on past and future climate systems. This can help them determine whether abnormal weather events or storms are a result of changes in climate or just part of the routine climate variation. For example, when predicting tropical cyclones during hurricane season, scientists can use climate models to predict the number of tropical storms that may form off the coast and in what regions they are likely to make landfall.

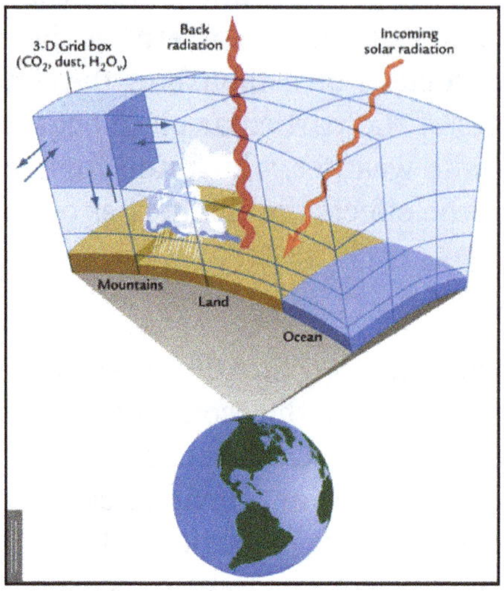

The three-dimensional grid of a climate model.

When creating climate models, scientists use one of three common types of simple climate models: energy balance models, intermediate complexity models, and general circulation models. These models use numbers to simplify the complexities that exist when taking into account all the factors that affect climate, like atmospheric mixing and ocean current.

Energy balance models help to forecast climate changes as a result of Earth's energy budget. This model takes into account surface temperatures from solar energy, albedo or reflectivity, and the natural cooling from the earth emitting heat back out into space. To predict climate, scientists use an equation that represents the amount of energy coming in versus going out, to understand the changes in heat storage—for example, as more heat-absorbing CO_2 fills up the atmosphere. Scientists then take this equation and plug it into box models that represent a square of land within a three-dimensional grid, to express climate in a region or even across a continent.

Intermediate complexity models are similar to energy balance models but they include and combine several of Earth's geographical structures—land, oceans, and ice features, for instance. These geographical features allow intermediate complexity models to simulate large-scale climate scenarios such as glacial fluctuations, ocean current shifts, and atmospheric composition changes over long timescales. Intermediate complexity models describe the climate with less spatial and time-specific detail, so they are best used for large-scale and low-frequency variations in the earth's climate system.

General circulation models are the most complex and precise models for understanding climate systems and predicting climate change. These models include information regarding the atmospheric chemistry, land type, carbon cycle, ocean circulation and glacial makeup of the isolated area. This type of model also uses a three-dimensional grid, with each box representing around 100 square kilometers of land, air, or sea, which is better resolution than the typical 200 to 600 kilometers per box. This model is more sophisticated than the energy balance and intermediate complexity models, but it does require a larger amount of computing time—each simulation could take several weeks to run.

An ice core.

For many decades, scientists have been collecting data on climate using cores from ice, trees, and coral, as well as carbon dating. From this research they have discovered details about past human activity, temperature changes in our oceans, periods of extreme drought, and much more.

As more data points are collected, they increase the accuracy of existing climate models. This enhances climate forecasting, because past climate data helps to establish a baseline for typical climate systems. From there, researchers establish climate variables that they want to keep the same, like cloud cover, and variables they want to test, like increased carbon dioxide, to evaluate hypotheses about future changes. These could estimate anything from sea level rise to increased temperatures and risk of drought and forest fires.

Accuracy of Climate Models

Since the world can't afford to wait decades to measure the accuracy of climate model predictions, scientists test a model's accuracy using past events. If the model accurately predicts past events that we know happened, then it should be pretty good at predicting the future, too. And the more we learn about past and present conditions, the more accurate these models become.

Climate models are complex because of the all the elements that are in flux within Earth's systems. If our atmosphere was like the moon's climate modeling would be fairly easy because the moon barely has an atmosphere. On Earth, climate scientists must account for temperature fluctuations, wind patterns, ocean currents, land surface characteristics and much more. Because of this, the models always consider some level of uncertainty – but models measuring smaller areas with higher resolutions produce more accurate models. Despite a small amount of uncertainty, scientists find climate models of the 21st century to be pretty accurate because they are based on well-founded physical principles of earth system processes. This basis solidifies the confidence of the scientific community that human emissions are changing the climate, which will impact the entire planet.

Importance of Climate Models

Understanding past, present and future climate helps us to understand how Earth's systems naturally function. This information, combined with climate models, allows us to determine how both natural and manmade influences have and will impact changes in our climate. These predictions and results can also suggest how to mitigate the worst effects of climate change, and they help decision-makers to prioritize environmental issues based on scientific evidence.

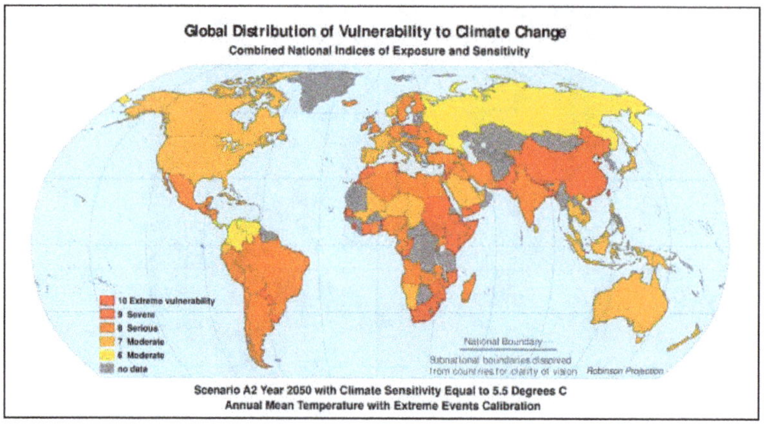

Climate change vulnerability.

Numerous models have shown that the climate is changing. Increased greenhouse gas emissions from human activities are resulting in positive feedbacks in our climate systems. These positive feedbacks can result in not so positive changes in earth systems, like melting glacial ice, rising ocean temperatures, increasing odds of severe flooding and drought, and climbing surface temperatures.

It is crucial that we continue to collect data and improve models, increasing their accuracy to refining our knowledge of climate and weather. It is also imperative that we recognize the importance of data-driven results and science-backed facts as they influence how communities and policy-makers plan for the future. Climate and weather models both have the ability to advance the way we plan our cities, influence business opportunities, and even how we plan out our day. These models are our best chance at finding ways to mitigate the dangerous effects of climate change.

Types of Models

Simplifications are unavoidable when designing a climate model as the processes that should be taken into account range from the scale of centimetres (for instance for atmospheric turbulence) to that of the Earth itself. The involved time scales also vary widely from the order of seconds for some waves, to billions of years when analysing the evolution of the climate since the formation of Earth. It is thus an important skill for a modeller to be able to select the processes that must be explicitly included compared to those that can be neglected or represented in a simplified way. This choice is of course based on the scientific goal of the study. However, it also depends on technical issues since the most sophisticated models require a lot of computational power: even on the largest computer presently available, the models cannot be routinely used for periods longer than a few centuries to millennia. On longer time scales, or when quite a large number of experiments are needed, it is thus necessary to user simpler and faster models. Furthermore, it is often very illuminating to deliberately design a model that includes only the most important properties, so as to understand in depth the nature of a feedback or the complex interaction between the various components of the system. This is also the reason why simple models are often used to analyse the results of more complex models in which the fundamental characteristics of the system could be hidden by the number of processes represented and the details provided.

Modellers have first to decide the variables or processes to be taken into account and those that will be taken as constants. This provides a method of classifying the models as a function of the components that are represented interactively. In the majority of climate studies, at least the physical behaviour of the atmosphere, ocean and sea ice must be represented. In addition, the terrestrial and marine carbon cycles, the dynamic vegetation and the ice sheet components are more and more regularly included, leading to what are called Earth-system models.

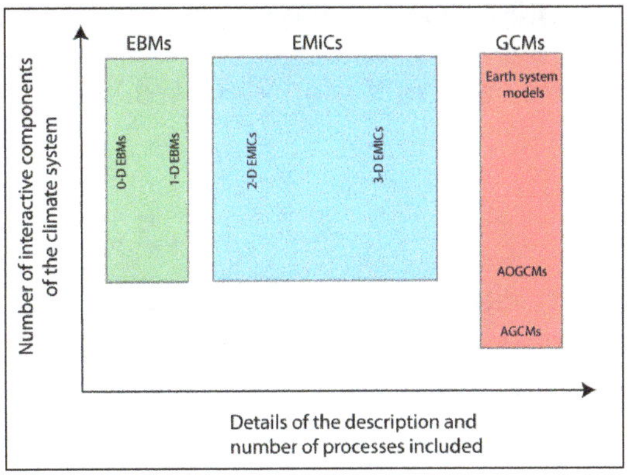

Types of climate model.

A second way of differentiating between models is related to the complexity of the processes that are included. At one end of the spectrum, General Circulation Models (GCMs) try to account for all the important properties of the system at the highest affordable resolution. The term GCM was introduced because one of the first goals of these models is to simulate the three dimensional structure of winds and currents realistically. They have classically been divided into Atmospheric General Circulation Models (AGCMs) and Ocean General Circulation Models (OGCMs). For climate studies using interactive atmospheric and oceanic components, the acronyms AOGCM (Atmosphere Ocean General Circulation Model) and the broader CGCM (Coupled General Circulation Model) are generally used.

At the other end of the spectrum, simple climate models (such as the Energy Balance Models, or EBMs, see section) propose a highly simplified version of the dynamic of the climate system. The variables are averaged over large regions, sometimes over the whole Earth, and many processes are not represented or accounted for by the parameterisations. EBMs thus include a relatively small number of degree of freedom.

EMICs (Earth Models of Intermediate Complexity) are located between those two extremes. They are based on a more complex representation of the system than EBMs but include simplifications and parameterisations for some processes that are explicitly accounted for in GCMs. The EMICs form the broadest category of models. Some of them are relatively close to simple models, while others are slightly degraded GCMs.

When employed correctly, all the model types can produce useful information on the behaviour of the climate system. There is no perfect model, suitable for all purposes. This is why a wide range of climate models exists, forming what is called the spectrum or the hierarchy of models. Depending on the objective or the question, one type of models could be selected. The best type of model to use depends on the objective or the question. On the other hand, combining the results from various types of models is often the best way to gain a deep understanding of the dominant processes in action.

A Hierarchy of Models

Energy Balance Models

As indicated by their name, energy balance models estimate the changes in the climate system from an analysis of the energy budget of the Earth. In their simplest form, they do not include any explicit spatial dimension, providing only globally averaged values for the computed variables. They are thus referred to as zero-dimensional EBMs.

Changes in heat storage = absorbed solar radiation - emitted terrestrial radiation,

$$C_E \frac{\partial T_s}{\partial t} = \left((1 - \alpha_p) \frac{S_0}{4} - A \uparrow \right)$$

Where, C_E is the effective heat capacity of the media (measured in J m-2 K^{-1}), T_s the surface temperature, t the time, α_p the planetary albedo, S_0 the Total Solar Irradiace (TSI) and A↑ the total amount of energy that is emitted by a 1 m^2 surface of the Earth. A↑ could be represented on the basis of the Stefan-Boltzmann law, using a factor τ_a to represent the infrared transmissivity of the atmosphere (including the greenhouse gas effect), as,

$$A \uparrow = \varepsilon \sigma T_s^4 \tau_a$$

Where ε is the emissivity of the surface. Using an albedo of 0.3, an emissivity of 0.97, and a value of τ_a of 0.64 leads to an equilibrium temperature T_s = 287K, which is close to the observed one. In some EBMs, Eq. is linearised to give an even simpler formulation of the model. On the other hand, τ_a and αp are often parameterised as a function of the temperature, in particular to take into account the fact that cooling increases the surface area covered by ice and snow, and thus increases the planetary albedo.

In order to take the geographical distribution of temperature at the Earth's surface into account, zero-dimensional EBMs can be extended to include one (generally the latitude) or two horizontal dimensions. An additional term Δ tran sp is then included in Equation representing the net effect of heat input and output associated with horizontal transport:

$$C_E \frac{\partial T_{s,i}}{\partial t} = \left((1 - \alpha_p) \frac{S_0}{4} - A \uparrow \right) + \Delta transp$$

An index i has been added to the surface temperature to indicate that the variable corresponds to the region i. The simplest form for the transport is to treat it as a linear function of temperature, but more sophisticated parameterisations are also used, including, for instance, a diffusion term.

Box models have clear similarities to EBMs as they represent large areas or an entire component of the system by an average which describes the mean over one "box". The exchanges between the compartments are then parameterised as a function of the characteristics of the different boxes. The exact definition of the boxes depends on the purpose of the model. For instance, some box models have a compartment for the atmosphere, the land surface, the ocean surface layers and the deep ocean, possibly making a distinction between the two hemispheres. Others include additional components allowing a description of the carbon cycle and thus have boxes corresponding to the various reservoirs described in section.

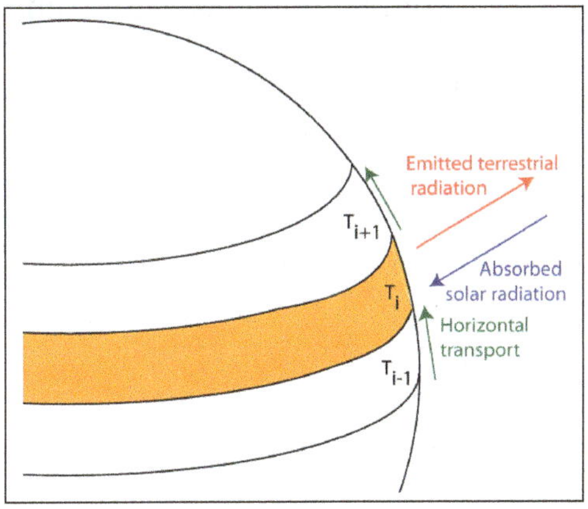

Representation of a one-dimensional EBM for which the temperature T_i is averaged over a band of longitude.

Intermediate Complexity Models

Like EBMs, EMICs involve some simplifications, but they always include a representation of the Earth's geography, i.e. they provide more than averages over the whole Earth or large boxes. Secondly, they include many more degrees of freedom than EBMs. As a consequence, the parameters of EMICs cannot easily be adjusted to reproduce the observed characteristics of the climate system, as can be done with some simpler models.

The level of approximation involved in the development of the model varies widely between different EMICs. Some models use a very simple representation of the geography, with a zonally averaged representation of the atmosphere and ocean. A distinction is always made between the Atlantic, Pacific and Indian basins because of the strong differences between them in the circulation. As the atmospheric and oceanic circulations are fundamentally three-dimensional, some parameterisations of the meridional transport are required. Those developed for EMICs are generally more complex and physically based than the ones employed in one-dimensional EBMs.

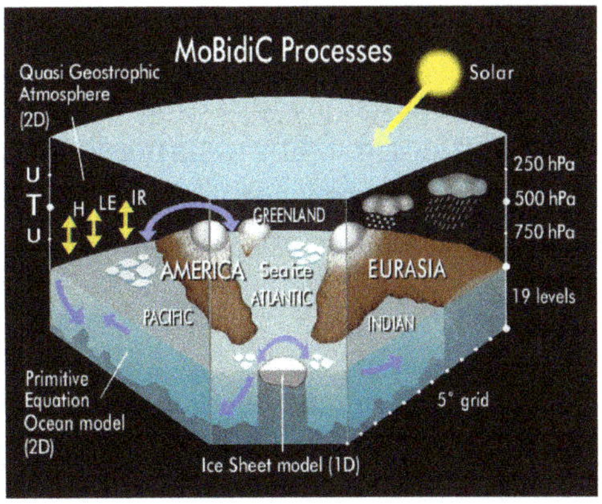

Figure: Schematic illustration of the structure of the climate model of intermediate complexity MOBIDIC that includes a zonally averaged atmosphere, a 3-basin zonal oceanic model (corresponding to the Atlantic, the Pacific and the Indian Oceans) and simplified ice sheets.

On the other hand, some EMICs include components that are very similar to those developed for GCMs, although a coarser numerical grid is used so that the computations proceed fast enough to allow a large number of relatively long simulations to be run. Some other components are simplified, usually including the atmosphere because this is the component that is most depending on computer time in coupled climate models.

General Circulation Models

General circulation models provide the most precise and complex description of the climate system. Currently, their grid resolution is typically of the order of 100 to 200 km. As a consequence, compared to EMICs (which have a grid resolution between 300 km and thousands of kilometres), they provide much more detailed information on a regional scale. A few years ago, GCMs only included a representation of the atmosphere, the land surface, sometimes the ocean circulation, and a very simplified version of the sea ice. Nowadays, GCMs take more and more components into account, and many new models now also include sophisticated models of the sea ice, the carbon cycle, ice sheet dynamics and even atmospheric chemistry.

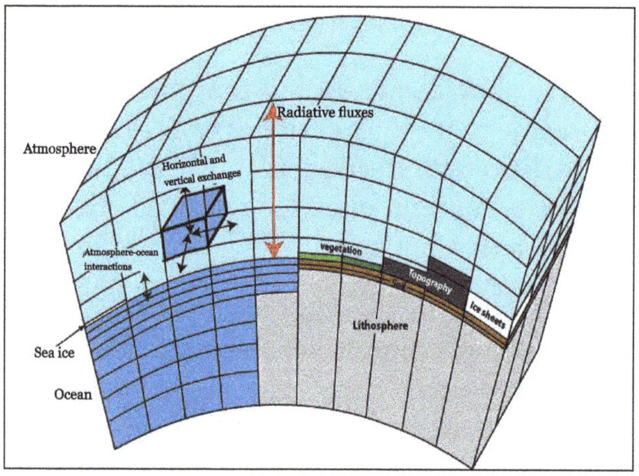

A simplified representation of part of the domain of a general circulation model. The curvature of the Earth has been amplified, the horizontal and vertical coordinates are not to scale and the number of grid points has been reduced compared to state-of-the-art models.

Because of the large number of processes included and their relatively high resolution, GCM simulations require a large amount of computer time. For instance, an experiment covering one century typically takes several weeks to run on the fastest computers. As computing power increases, longer simulations with a higher resolution become affordable, providing more regional details than the previous generation of models.

Climate Change

Climate change refers to the periodic modification of Earth's climate brought about as a result of changes in the atmosphere as well as interactions between the atmosphere and various other geologic, chemical, biological, and geographic factors within the Earth system.

The atmosphere is a dynamic fluid that is continually in motion. Both its physical properties and its rate and direction of motion are influenced by a variety of factors, including solar radiation, the geographic position of continents, ocean currents, the location and orientation of mountain ranges, atmospheric chemistry, and vegetation growing on the land surface. All these factors change through time. Some factors, such as the distribution of heat within the oceans, atmospheric chemistry, and surface vegetation, change at very short timescales. Others, such as the position of continents and the location and height of mountain ranges, change over very long timescales. Therefore, climate, which results from the physical properties and motion of the atmosphere, varies at every conceivable timescale.

Climate is often defined loosely as the average weather at a particular place, incorporating such features as temperature, precipitation, humidity, and windiness. A more specific definition would state that climate is the mean state and variability of these features over some extended time period. Both definitions acknowledge that the weather is always changing, owing to instabilities in the atmosphere. And as weather varies from day to day, so too does climate vary, from daily day-and-night cycles up to periods of geologic time hundreds of millions of years long. In a very real sense, climate variation is a redundant expression—climate is always varying. No two years are exactly alike, nor are any two decades, any two centuries, or any two millennia.

Evidence for Climate Change

All historical sciences share a problem: As they probe farther back in time, they become more reliant on fragmentary and indirect evidence. Earth system history is no exception. High-quality instrumental records spanning the past century exist for most parts of the world, but the records become sparse in the 19th century, and few records predate the late 18th century. Other historical documents, including ship's logs, diaries, court and church records, and tax rolls, can sometimes be used. Within strict geographic contexts, these sources can provide information on frosts, droughts, floods, sea ice, the dates of monsoons, and other climatic features—in some cases up to several hundred years ago.

Fortunately, climatic change also leaves a variety of signatures in the natural world. Climate influences the growth of trees and corals, the abundance and geographic distribution of plant and animal species, the chemistry of oceans and lakes, the accumulation of ice in cold regions, and the erosion and deposition of materials on Earth's surface. Paleoclimatologists study the traces of these effects, devising clever and subtle ways to obtain information about past climates. Most of the evidence of past climatic change is circumstantial, so paleoclimatology involves a great deal of investigative work. Wherever possible, paleoclimatologists try to use multiple lines of evidence to cross-check their conclusions. They are frequently confronted with conflicting evidence, but this, as in other sciences, usually leads to an enhanced understanding of the Earth system and its complex history. New sources of data, analytical tools, and instruments are becoming available, and the field is moving quickly. Revolutionary changes in the understanding of Earth's climate history have occurred since the 1990s, and coming decades will bring many new insights and interpretations.

Ongoing climatic changes are being monitored by networks of sensors in space, on the land surface, and both on and below the surface of the world's oceans. Climatic changes of the past 200–300 years, especially since the early 1900s, are documented by instrumental records and other archives. These written documents and records provide information about climate change in some locations for the past few hundred years. Some very rare records date back over 1,000 years. Researchers studying climatic changes predating the instrumental record rely increasingly on natural archives, which are biological or geologic processes that record some aspect of past climate. These natural archives, often referred to as proxy evidence, are extraordinarily diverse; they include, but are not limited to, fossil records of past plant and animal distributions, sedimentary and geochemical indicators of former conditions of oceans and continents, and land surface features characteristic of past climates. Paleoclimatologists study these natural archives by collecting cores, or cylindrical samples, of sediments from lakes, bogs, and oceans; by studying surface features and geological strata; by examining tree ring patterns from cores or sections of living and dead trees; by drilling into marine corals and cave stalagmites; by drilling into the ice sheets of Antarctica and Greenland and the high-elevation glaciers of the Plateau of Tibet, the Andes, and other montane regions; and by a wide variety of other means. Techniques for extracting paleoclimatic information are continually being developed and refined, and new kinds of natural archives are being recognized and exploited.

Causes of Climate Change

It is much easier to document the evidence of climate variability and past climate change than it is to determine their underlying mechanisms. Climate is influenced by a multitude of factors that operate at timescales ranging from hours to hundreds of millions of years. Many of the causes of climate change are external to the Earth system. Others are part of the Earth system but external to the atmosphere. Still others involve interactions between the atmosphere and other components of the Earth system and are collectively described as feedbacks within the Earth system. Feedbacks are among the most recently discovered and challenging causal factors to study. Nevertheless, these factors are increasingly recognized as playing fundamental roles in climate variation.

Solar Variability

The luminosity, or brightness, of the Sun has been increasing steadily since its formation. This phenomenon is important to Earth's climate, because the Sun provides the energy to drive atmospheric circulation and constitutes the input for Earth's heat budget. Low solar luminosity during Precambrian time underlies the faint young Sun paradox.

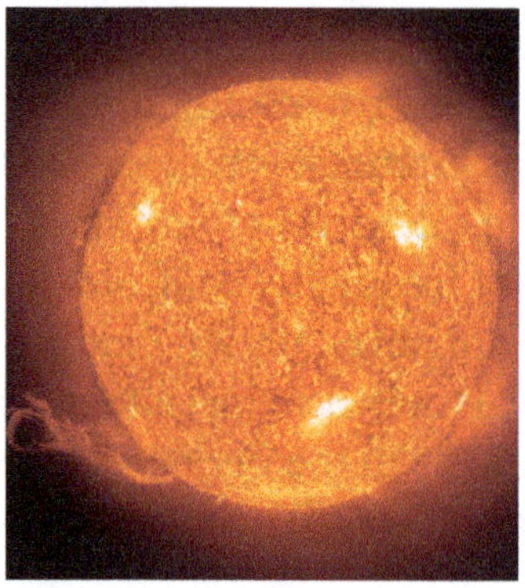

The Sun as imaged in extreme ultraviolet light by the Earth-orbiting Solar and Hemispheric Observatory (SOHO) satellite. A massive loop-shaped eruptive prominence is visible at the lower left. Nearly white areas are the hottest; deeper reds indicate cooler temperatures.

Radiative energy from the Sun is variable at very small timescales, owing to solar storms and other disturbances, but variations in solar activity, particularly the frequency of sunspots, are also documented at decadal to millennial timescales and probably occur at longer timescales as well. The "Maunder minimum," a period of drastically reduced sunspot activity between AD 1645 and 1715, has been suggested as a contributing factor to the Little Ice Age.

Volcanic Activity

Volcanic activity can influence climate in a number of ways at different timescales. Individual volcanic eruptions can release large quantities of sulfur dioxide and other aerosols into the stratosphere, reducing atmospheric transparency and thus the amount of solar radiation reaching Earth's surface and troposphere. A recent example is the 1991 eruption in the Philippines of Mount Pinatubo, which had measurable influences on atmospheric circulation and heat budgets. The 1815 eruption of Mount Tambora on the island of Sumbawa had more dramatic consequences, as the spring and summer of the following year (1816, known as "the year without a summer") were unusually cold over much of the world. New England and Europe experienced snowfalls and frosts throughout the summer of 1816.

A column of gas and ash rising from Mount Pinatubo in the Philippines on June 12, 1991, just days before the volcano's climactic explosion on June 15.

Volcanoes and related phenomena, such as ocean rifting and subduction, release carbon dioxide into both the oceans and the atmosphere. Emissions are low; even a massive volcanic eruption such as Mount Pinatubo releases only a fraction of the carbon dioxide emitted by fossil-fuel combustion in a year. At geologic timescales, however, release of this greenhouse gas can have important effects. Variations in carbon dioxide release by volcanoes and ocean rifts over millions of years can alter the chemistry of the atmosphere. Such changeability in carbon dioxide concentrations probably accounts for much of the climatic variation that has taken place during the Phanerozoic Eon.

Tectonic Activity

Tectonic movements of Earth's crust have had profound effects on climate at timescales of millions to tens of millions of years. These movements have changed the shape, size, position, and elevation of the continental masses as well as the bathymetry of the oceans. Topographic and bathymetric changes in turn have had strong effects on the circulation of both the atmosphere and the oceans. For example, the uplift of the Tibetan Plateau during the Cenozoic Era affected atmospheric circulation patterns, creating the South Asian monsoon and influencing climate over much of the rest of Asia and neighbouring regions.

Tectonic activity also influences atmospheric chemistry, particularly carbon dioxide concentrations. Carbon dioxide is emitted from volcanoes and vents in rift zones and subduction zones. Variations in the rate of spreading in rift zones and the degree of volcanic activity near plate margins have influenced atmospheric carbon dioxide concentrations throughout Earth's history. Even the chemical weathering of rock constitutes an important sink for carbon dioxide. (A carbon sink is any process that removes carbon dioxide from the atmosphere by the chemical conversion of CO2 to organic or inorganic carbon compounds.) Carbonic acid, formed from carbon dioxide and water, is a reactant in dissolution of silicates and other minerals. Weathering rates are related to the mass, elevation, and exposure of bedrock. Tectonic uplift can increase all these factors and thus lead to increased weathering and carbon dioxide absorption. For example, the chemical weathering of the rising Tibetan Plateau may have played an important role in depleting the atmosphere of carbon dioxide during a global cooling period in the late Cenozoic Era.

Orbital (Milankovich) variations

The orbital geometry of Earth is affected in predictable ways by the gravitational influences of other planets in the solar system. Three primary features of Earth's orbit are affected, each in a cyclic, or regularly recurring, manner. First, the shape of Earth's orbit around the Sun, varies from nearly circular to elliptical (eccentric), with periodicities of 100,000 and 413,000 years. Second, the tilt of Earth's axis with respect to the Sun, which is primarily responsible for Earth's seasonal climates, varies between 22.1° and 24.5° from the plane of Earth's rotation around the Sun. This variation occurs on a cycle of 41,000 years. In general, the greater the tilt, the greater the solar radiation received by hemispheres in summer and the less received in winter. The third cyclic change to Earth's orbital geometry results from two combined phenomena: (1) Earth's axis of rotation wobbles, changing the direction of the axis with respect to the Sun, and (2) the orientation of Earth's orbital ellipse rotates slowly. These two processes create a 26,000-year cycle, called precession of the equinoxes, in which the position of Earth at the equinoxes and solstices changes. Today Earth is closest to the Sun (perihelion) near the December solstice, whereas 9,000 years ago perihelion occurred near the June solstice.

These orbital variations cause changes in the latitudinal and seasonal distribution of solar radiation, which in turn drive a number of climate variations. Orbital variations play major roles in pacing glacial-interglacial and monsoonal patterns. Their influences have been identified in climatic changes over much of the Phanerozoic. For example, cyclothems—which are interbedded marine, fluvial, and coal beds characteristic of the Pennsylvanian Subperiod (318.1 million to 299 million years ago)—appear to represent Milankovitch-driven changes in mean sea level.

Greenhouse Gases

Greenhouse gases are gas molecules that have the property of absorbing infrared radiation (net heat energy) emitted from Earth's surface and reradiating it back to Earth's surface, thus contributing to the phenomenon known as the greenhouse effect. Carbon dioxide, methane, and water vapour are the most important greenhouse gases, and they have a profound effect on the energy budget of the Earth system despite making up only a fraction of all atmospheric gases. Concentrations of greenhouse gases have varied substantially during Earth's history, and these variations have driven substantial climate changes at a wide range of timescales. In general, greenhouse gas concentrations have been particularly high during warm periods and low during cold phases. A number of processes influence greenhouse gas concentrations. Some, such as tectonic activities, operate at timescales of millions of years, whereas others, such as vegetation, soil, wetland, and ocean sources and sinks, operate at timescales of hundreds to thousands of years. Human activities—especially fossil-fuel combustion since the Industrial Revolution—are responsible for steady increases in atmospheric concentrations of various greenhouse gases, especially carbon dioxide, methane, ozone, and chlorofluorocarbons (CFCs).

Perhaps the most intensively discussed and researched topic in climate variability is the role of interactions and feedbacks among the various components of the Earth system. The feedbacks involve different components that operate at different rates and timescales. Ice sheets, sea ice, terrestrial vegetation, ocean temperatures, weathering rates, ocean circulation, and greenhouse gas concentrations are all influenced either directly or indirectly by the atmosphere; however, they also all feed back into the atmosphere, thereby influencing it in important ways. For example, different forms and densities of vegetation on the land surface influence the albedo, or reflectivity, of Earth's surface, thus affecting the overall radiation budget at local to regional scales. At the same time, the transfer of water molecules from soil to the atmosphere is mediated by vegetation, both directly (from transpiration through plant stomata) and indirectly (from shading and temperature influences on direct evaporation from soil). This regulation of latent heat flux by vegetation can influence climate at local to global scales. As a result, changes in vegetation, which are partially controlled by climate, can in turn influence the climate system. Vegetation also influences greenhouse gas concentrations; living plants constitute an important sink for atmospheric carbon dioxide, whereas they act as sources of carbon dioxide when they are burned by wildfires or undergo decomposition. These and other feedbacks among the various components of the Earth system are critical for both understanding past climate changes and predicting future ones.

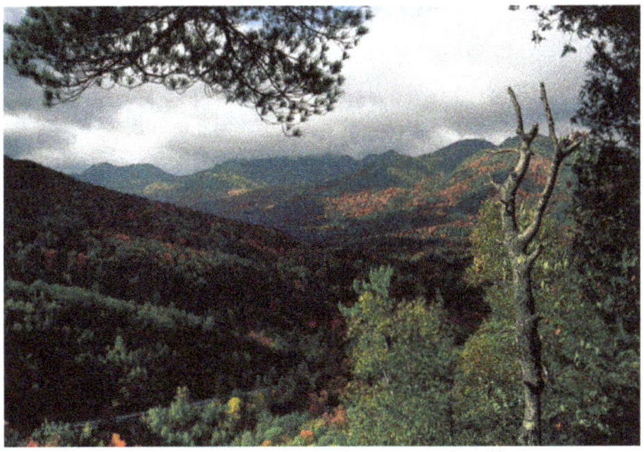

Mixed evergreen and hardwood forest on the slopes of the Adirondack Mountains near Keene Valley, New York.

Human Activities

Recognition of global climate change as an environmental issue has drawn attention to the climatic impact of human activities. Most of this attention has focused on carbon dioxide emission via fossil-fuel combustion and deforestation. Human activities also yield releases of other greenhouse gases, such as methane (from rice cultivation, livestock, landfills, and other sources) and chlorofluorocarbons (from industrial sources). There is little doubt among climatologists that these greenhouse gases affect the radiation budget of Earth; the nature and magnitude of the climatic response are a subject of intense research activity. Paleoclimate records from tree rings, coral, and ice cores indicate a clear warming trend spanning the entire 20th century and the first decade of the 21st century. In fact, the 20th century was the warmest of the past 10 centuries, and the decade 2001–10 was the warmest decade since the beginning of modern instrumental record keeping. Many climatologists have pointed to this warming pattern as clear evidence of human-induced climate change resulting from the production of greenhouse gases.

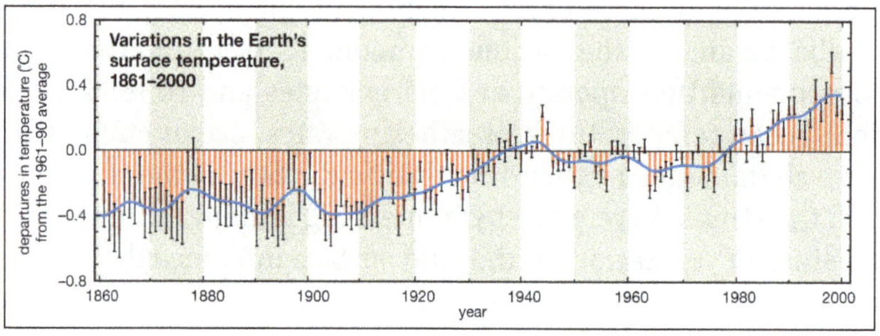

The global average surface temperature range for each year from 1861 to 2000 is shown by solid red bars, with the confidence range in the data for each year shown by thin whisker bars. The average change over time is shown by the solid curve.

A second type of human impact, the conversion of vegetation by deforestation, afforestation, and agriculture, is receiving mounting attention as a further source of climate change. It is becoming increasingly clear that human impacts on vegetation cover can have local, regional, and even global effects on climate, due to changes in the sensible and latent heat flux to the atmosphere and the distribution of energy within the climate system. The extent to which these factors contribute to recent and ongoing climate change is an important, emerging area of study.

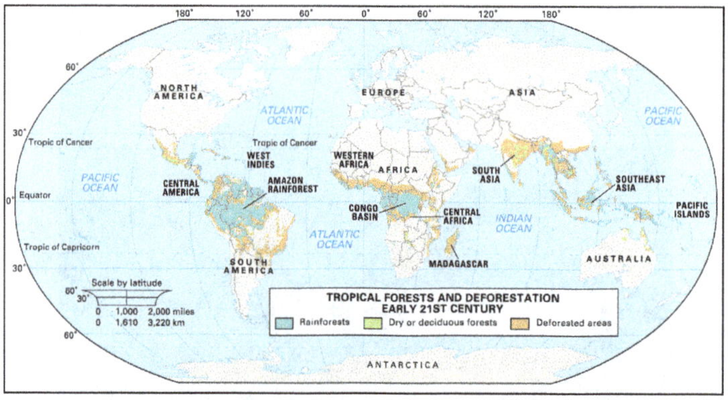

Tropical forests and deforestation: Tropical forests and deforestation in the early 21st century.

Climate Change within a Human Life Span

Regardless of their locations on the planet, all humans experience climate variability and change within their lifetimes. The most familiar and predictable phenomena are the seasonal cycles, to which people adjust their clothing, outdoor activities, thermostats, and agricultural practices. However, no two summers or winters are exactly alike in the same place; some are warmer, wetter, or stormier than others. This interannual variation in climate is partly responsible for year-to-year variations in fuel prices, crop yields, road maintenance budgets, and wildfire hazards. Single-year, precipitation-driven floods can cause severe economic damage, such as those of the upper Mississippi River drainage basin during the summer of 1993, and loss of life, such as those that devastated much of Bangladesh in the summer of 1998. Similar damage and loss of life can also occur as the result of wildfires, severe storms, hurricanes, heat waves, and other climate-related events.

Climate variation and change may also occur over longer periods, such as decades. Some locations experience multiple years of drought, floods, or other harsh conditions. Such decadal variation of climate poses challenges to human activities and planning. For example, multiyear droughts can disrupt water supplies, induce crop failures, and cause economic and social dislocation, as in the case of the Dust Bowl droughts in the midcontinent of North America during the 1930s. Multiyear droughts may even cause widespread starvation, as in the Sahel drought that occurred in northern Africa during the 1970s and '80s.

Abandoned farmstead showing the effects of wind erosion in the Dust Bowl

Seasonal variation

Every place on Earth experiences seasonal variation in climate (though the shift can be slight in some tropical regions). This cyclic variation is driven by seasonal changes in the supply of solar radiation to Earth's atmosphere and surface. Earth's orbit around the Sun is elliptical; it is closer to the Sun (147 million km [about 91 million miles]) near the winter solstice and farther from the Sun (152 million km [about 94 million miles]) near the summer solstice in the Northern Hemisphere. Furthermore, Earth's axis of rotation occurs at an oblique angle (23.5°) with respect to its orbit. Thus, each hemisphere is tilted away from the Sun during its winter period and toward the Sun in its summer period. When a hemisphere is tilted away from the Sun, it receives less solar radiation than the opposite hemisphere, which at that time is pointed toward the Sun. Thus, despite the closer proximity of the Sun at the winter solstice, the Northern Hemisphere receives less solar radiation during the winter than it does during the summer. Also as a consequence of the tilt, when the Northern Hemisphere experiences winter, the Southern Hemisphere experiences summer.

Earth's climate system is driven by solar radiation; seasonal differences in climate ultimately result from the seasonal changes in Earth's orbit. The circulation of air in the atmosphere and water in the oceans responds to seasonal variations of available energy from the Sun. Specific seasonal changes in climate occurring at any given location on Earth's surface largely result from the transfer of energy from atmospheric and oceanic circulation. Differences in surface heating taking place between summer and winter cause storm tracks and pressure centres to shift position and strength. These heating differences also drive seasonal changes in cloudiness, precipitation, and wind.

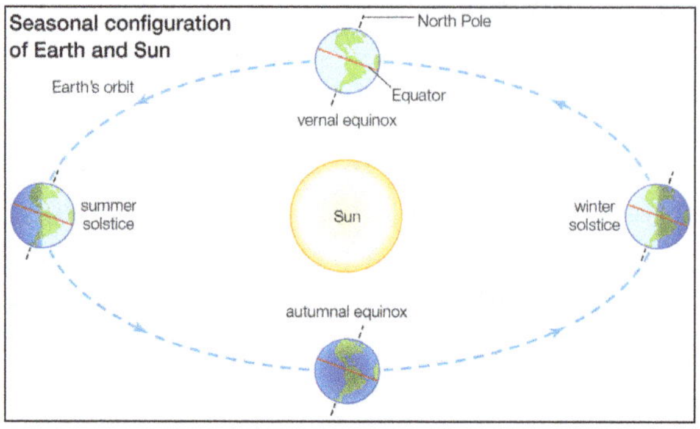

A diagram shows the position of Earth at the beginning of each season in the Northern Hemisphere.

Seasonal responses of the biosphere (especially vegetation) and cryosphere (glaciers, sea ice, snowfields) also feed into atmospheric circulation and climate. Leaf fall by deciduous trees as they go into winter dormancy increases the albedo (reflectivity) of Earth's surface and may lead to greater local and regional cooling. Similarly, snow accumulation also increases the albedo of land surfaces and often amplifies winter's effects.

Interannual Variation

Interannual climate variations, including droughts, floods, and other events, are caused by a complex array of factors and Earth system interactions. One important feature that plays a role in these variations is the periodic change of atmospheric and oceanic circulation patterns in the tropical Pacific region, collectively known as El Niño–Southern Oscillation (ENSO) variation. Although its primary climatic effects are concentrated in the tropical Pacific, ENSO has cascading effects that often extend to the Atlantic Ocean region, the interior of Europe and Asia, and the polar regions. These effects, called teleconnections, occur because alterations in low-latitude atmospheric circulation patterns in the Pacific region influence atmospheric circulation in adjacent and downstream systems. As a result, storm tracks are diverted and atmospheric pressure ridges (areas of high pressure) and troughs (areas of low pressure) are displaced from their usual patterns.

As an example, El Niño events occur when the easterly trade winds in the tropical Pacific weaken or reverse direction. This shuts down the upwelling of deep, cold waters off the west coast of South America, warms the eastern Pacific, and reverses the atmospheric pressure gradient in the western Pacific. As a result, air at the surface moves eastward from Australia and Indonesia toward the central Pacific and the Americas. These changes produce high rainfall and flash floods along the normally arid coast of Peru and severe drought in the normally wet regions of northern Australia and Indonesia. Particularly severe El Niño events lead to monsoon failure in the Indian Ocean region, resulting in intense drought in India and East Africa. At the same time, the westerlies and storm tracks are displaced toward the Equator, providing California and the desert Southwest of the United States with wet, stormy winter weather and causing winter conditions in the Pacific Northwest, which are typically wet, to become warmer and drier. Displacement of the westerlies also results in drought in northern China and from northeastern Brazil through sections of Venezuela. Long-term records of ENSO variation from historical documents, tree rings, and reef corals indicate that El Niño events occur, on average, every two to seven years. However, the frequency and intensity of these events vary through time.

The North Atlantic Oscillation (NAO) is another example of an interannual oscillation that produces important climatic effects within the Earth system and can influence climate throughout the Northern Hemisphere. This phenomenon results from variation in the pressure gradient, or the difference in atmospheric pressure between the subtropical high, usually situated between the Azores and Gibraltar, and the Icelandic low, centred between Iceland and Greenland. When the pressure gradient is steep due to a strong subtropical high and a deep Icelandic low (positive phase), northern Europe and northern Asia experience warm, wet winters with frequent strong winter storms. At the same time, southern Europe is dry. The eastern United States also experiences warmer, less snowy winters during positive NAO phases, although the effect is not as great as in Europe. The pressure gradient is dampened when NAO is in a negative mode—that is, when a weaker pressure gradient exists from the presence of a weak subtropical high and Icelandic low. When this happens, the Mediterranean region receives abundant winter rainfall, while northern Europe is cold and dry. The eastern United States is typically colder and snowier during a negative NAO phase.

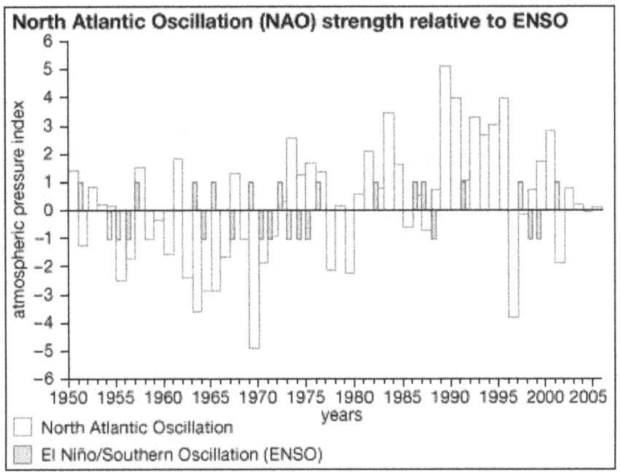

During years when the North Atlantic Oscillation (NAO) is in its positive phase, the eastern United States, southeastern Canada, and northwestern Europe experience warmer winter temperatures, whereas colder temperatures are found in these locations during its negative phase. When the El Niño/Southern Oscillation (ENSO) and NAO are both in their positive phase, European winters tend to be wetter and less severe; however, beyond this general tendency, the influence of the ENSO upon the NAO is not well understood.

The ENSO and NAO cycles are driven by feedbacks and interactions between the oceans and atmosphere. Interannual climate variation is driven by these and other cycles, interactions among cycles, and perturbations in the Earth system, such as those resulting from large injections of aerosols from volcanic eruptions. One example of a perturbation due to volcanism is the 1991 eruption of Mount Pinatubo in the Philippines, which led to a decrease in the average global temperature of approximately 0.5 °C (0.9 °F) the following summer.

Decadal variation

Climate varies on decadal timescales, with multiyear clusters of wet, dry, cool, or warm conditions. These multiyear clusters can have dramatic effects on human activities and welfare. For instance, a severe three-year drought in the late 16th century probably contributed to the destruction of Sir Walter Raleigh's "Lost Colony" at Roanoke Island in what is now North Carolina, and a subsequent seven-year drought (1606–12) led to high mortality at the Jamestown Colony in Virginia. Also, some scholars have implicated persistent and severe droughts as the main reason for the collapse of the Maya civilization in Mesoamerica between AD 750 and 950; however, discoveries in the early 21st century suggest that war-related trade disruptions played a role, possibly interacting with famines and other drought-related stresses.

Although decadal-scale climate variation is well documented, the causes are not entirely clear. Much decadal variation in climate is related to interannual variations. For example, the frequency and magnitude of ENSO change through time. The early 1990s were characterized by repeated El Niño events, and several such clusters have been identified as having taken place during the 20th century. The steepness of the NAO gradient also changes at decadal timescales; it has been particularly steep since the 1970s.

Recent research has revealed that decadal-scale variations in climate result from interactions between the ocean and the atmosphere. One such variation is the Pacific Decadal Oscillation (PDO), also referred to as the Pacific Decadal Variability (PDV), which involves changing sea surface temperatures (SSTs) in the North Pacific Ocean. The SSTs influence the strength and position of the Aleutian Low, which in turn strongly affects precipitation patterns along the Pacific Coast of North America. PDO variation consists of an alternation between "cool-phase" periods, when coastal Alaska is relatively dry and the Pacific Northwest relatively wet, and "warm-phase" periods, characterized by relatively high precipitation in coastal Alaska and low precipitation in the Pacific Northwest. Tree ring and coral records, which span at least the last four centuries, document PDO variation.

A similar oscillation, the Atlantic Multidecadal Oscillation (AMO), occurs in the North Atlantic and strongly influences precipitation patterns in eastern and central North America. A warm-phase AMO (relatively warm North Atlantic SSTs) is associated with relatively high rainfall in Florida and low rainfall in much of the Ohio Valley. However, the AMO interacts with the PDO, and both interact with interannual variations, such as ENSO and NAO, in complex ways. Such interactions may lead to the amplification of droughts, floods, or other climatic anomalies. For example, severe droughts over much of the conterminous United States in the first few years of the 21st century were associated with warm-phase AMO combined with cool-phase PDO. The mechanisms underlying decadal variations, such as PDO and AMO, are poorly understood, but they are probably related to ocean-atmosphere interactions with larger time constants than interannual variations. Decadal climatic variations are the subject of intense study by climatologists and paleoclimatologists.

Climate Change since the Emergence of Civilization

Human societies have experienced climate change since the development of agriculture some 10,000 years ago. These climate changes have often had profound effects on human cultures and societies. They include annual and decadal climate fluctuations such as those described above, as well as large-magnitude changes that occur over centennial to multimillennial timescales. Such changes are believed to have influenced and even stimulated the initial cultivation and domestication of crop plants, as well as the domestication and pastoralization of animals. Human societies have changed adaptively in response to climate variations, although evidence abounds that certain societies and civilizations have collapsed in the face of rapid and severe climatic changes.

Centennial-scale variation

Historical records as well as proxy records (particularly tree rings, corals, and ice cores) indicate that climate has changed during the past 1,000 years at centennial timescales; that is, no two centuries have been exactly alike. During the past 150 years, the Earth system has emerged from a period called the Little Ice Age, which was characterized in the North Atlantic region and elsewhere by relatively cool temperatures. The 20th century in particular saw a substantial pattern of warming in many regions. Some of this warming may be attributable to the transition from the Little Ice Age or other natural causes. However, many climate scientists believe that much of the 20th-century warming, especially in the later decades, resulted from atmospheric accumulation of greenhouse gases (especially carbon dioxide, CO_2).

The Little Ice Age is best known in Europe and the North Atlantic region, which experienced relatively cool conditions between the early 14th and mid-19th centuries. This was not a period of uniformly cool climate, since interannual and decadal variability brought many warm years. Furthermore, the coldest periods did not always coincide among regions; some regions experienced relatively warm conditions at the same time others were subjected to severely cold conditions. Alpine glaciers advanced far below their previous (and present) limits, obliterating farms, churches, and villages in Switzerland, France, and elsewhere. Frequent cold winters and cool, wet summers ruined wine harvests and led to crop failures and famines over much of northern and central Europe. The North Atlantic cod fisheries declined as ocean temperatures fell in the 17th century. The Norse colonies on the coast of Greenland were cut off from the rest of Norse civilization during the early 15th century as pack ice and storminess increased in the North Atlantic. The western colony of Greenland collapsed through starvation, and the eastern colony was abandoned. In addition, Iceland became increasingly isolated from Scandinavia.

The Little Ice Age was preceded by a period of relatively mild conditions in northern and central Europe. This interval, known as the Medieval Warm Period, occurred from approximately AD 1000 to the first half of the 13th century. Mild summers and winters led to good harvests in much of Europe. Wheat cultivation and vineyards flourished at far higher latitudes and elevations than today. Norse colonies in Iceland and Greenland prospered, and Norse parties fished, hunted, and explored the coast of Labrador and Newfoundland. The Medieval Warm Period is well documented in much of the North Atlantic region, including ice cores from Greenland. Like the Little Ice Age, this time was neither a climatically uniform period nor a period of uniformly warm temperatures everywhere in the world. Other regions of the globe lack evidence for high temperatures during this period.

Much scientific attention continues to be devoted to a series of severe droughts that occurred between the 11th and 14th centuries. These droughts, each spanning several decades, are well documented in tree-ring records across western North America and in the peatland records of the Great Lakes region. The records appear to be related to ocean temperature anomalies in the Pacific and Atlantic basins, but they are still inadequately understood. The information suggests that much of the United States is susceptible to persistent droughts that would be devastating for water resources and agriculture.

Millennial and Multimillennial variation

The climatic changes of the past thousand years are superimposed upon variations and trends at both millennial timescales and greater. Numerous indicators from eastern North America and Europe show trends of increased cooling and increased effective moisture during the past 3,000 years. For example, in the Great Lakes–St. Lawrence regions along the U.S.-Canadian border, water levels of the lakes rose, peatlands developed and expanded, moisture-loving trees such as beech and hemlock expanded their ranges westward, and populations of boreal trees, such as spruce and tamarack, increased and expanded southward. These patterns all indicate a trend of increased effective moisture, which may indicate increased precipitation, decreased evapotranspiration due to cooling, or both. The patterns do not necessarily indicate a monolithic cooling event; more complex climatic changes probably occurred. For example, beech expanded northward and spruce southward during the past 3,000 years in both eastern North America and western Europe. The beech expansions may indicate milder winters or longer growing seasons, whereas the spruce expansions appear related to cooler, moister summers. Paleoclimatologists are applying a variety of approaches and proxies to help identify such changes in seasonal temperature and moisture during the Holocene Epoch.

Just as the Little Ice Age was not associated with cool conditions everywhere, so the cooling and moistening trend of the past 3,000 years was not universal. Some regions became warmer and drier during the same time period. For example, northern Mexico and the Yucatan experienced decreasing moisture in the past 3,000 years. Heterogeneity of this type is characteristic of climatic change, which involves changing patterns of atmospheric circulation. As circulation patterns change, the transport of heat and moisture in the atmosphere also changes. This fact explains the apparent paradox of opposing temperature and moisture trends in different regions.

The trends of the past 3,000 years are just the latest in a series of climatic changes that occurred over the past 11,700 years or so—the interglacial period referred to as the Holocene Epoch. At the start of the Holocene, remnants of continental glaciers from the last glaciation still covered much of eastern and central Canada and parts of Scandinavia. These ice sheets largely disappeared by 6,000 years ago. Their absence— along with increasing sea surface temperatures, rising sea levels (as glacial meltwater flowed into the world's oceans), and especially changes in the radiation budget of Earth's surface owing to Milankovitch variations (changes in the seasons resulting from periodic adjustments of Earth's orbit around the Sun)—affected atmospheric circulation. The diverse changes of the past 10,000 years across the globe are difficult to summarize in capsule, but some general highlights and large-scale patterns are worthy of note. These include the presence of early to mid-Holocene thermal maxima in various locations, variation in ENSO patterns, and an early to mid-Holocene amplification of the Indian Ocean monsoon.

Thermal Maxima

Many parts of the globe experienced higher temperatures than today some time during the early to mid-Holocene. In some cases the increased temperatures were accompanied by decreased moisture availability. Although the thermal maximum has been referred to in North America and elsewhere as a single widespread event (variously referred to as the "Altithermal," "Xerothermic Interval," "Climatic Optimum," or "Thermal Optimum"), it is now recognized that the periods of maximum temperatures varied among regions. For example, northwestern Canada experienced its highest temperatures several thousand years earlier than central or eastern North America. Similar heterogeneity is seen in moisture records. For instance, the record of the prairie-forest boundary in the Midwestern region of the United States shows eastward expansion of prairie in Iowa and Illinois 6,000 years ago (indicating increasingly dry conditions), whereas Minnesota's forests expanded westward into prairie regions at the same time (indicating increasing moisture). The Atacama Desert, located primarily in present-day Chile and Bolivia, on the western side of South America, is one of the driest places on Earth today, but it was much wetter during the early Holocene when many other regions were at their driest.

The primary driver of changes in temperature and moisture during the Holocene was orbital variation, which slowly changed the latitudinal and seasonal distribution of solar radiation on Earth's surface and atmosphere. However, the heterogeneity of these changes was caused by changing patterns of atmospheric circulation and ocean currents.

ENSO variation in the Holocene

Because of the global importance of ENSO variation today, Holocene variation in ENSO patterns and intensity is under serious study by paleoclimatologists. The record is still fragmentary, but evidence from fossil corals, tree rings, lake records, climate modeling, and other approaches is accumulating that suggests that (1) ENSO variation was relatively weak in the early Holocene, (2) ENSO has undergone centennial to millennial variations in strength during the past 11,700 years, and (3) ENSO patterns and strength similar to those currently in place developed within the past 5,000 years. This evidence is particularly clear when comparing ENSO variation over the past 3,000 years to today's patterns. The causes of long-term ENSO variation are still being explored, but changes in solar radiation owing to Milankovitch variations are strongly implicated by modeling studies.

Amplification of the Indian Ocean Monsoon

Much of Africa, the Middle East, and the Indian subcontinent are under the strong influence of an annual climatic cycle known as the Indian Ocean monsoon. The climate of this region is highly seasonal, alternating between clear skies with dry air (winter) and cloudy skies with abundant rainfall (summer). Monsoon intensity, like other aspects of climate, is subject to interannual, decadal, and centennial variations, at least some of which are related to ENSO and other cycles. Abundant evidence exists for large variations in monsoon intensity during the Holocene Epoch. Paleontological and paleoecological studies show that large portions of the region experienced much greater precipitation during the early Holocene (11,700–6,000 years ago) than today. Lake and wetland sediments dating to this period have been found under the sands of parts of the Sahara Desert. These sediments contain fossils of elephants, crocodiles, hippopotamuses, and giraffes, together with pollen evidence of forest and woodland vegetation. In arid and semiarid parts of Africa, Arabia, and India, large and deep freshwater lakes occurred in basins that are now dry or are occupied by shallow, saline lakes. Civilizations based on plant cultivation and grazing animals, such as the Harappan civilization of northwestern India and adjacent Pakistan, flourished in these regions, which have since become arid.

These and similar lines of evidence, together with paleontological and geochemical data from marine sediments and climate-modeling studies, indicate that the Indian Ocean monsoon was greatly amplified during the early Holocene, supplying abundant moisture far inland into the African and Asian continents. This amplification was driven by high solar radiation in summer, which was approximately 7 percent higher 11,700 years ago than today and resulted from orbital forcing (changes in Earth's eccentricity, precession, and axial tilt). High summer insolation resulted in warmer summer air temperatures and lower surface pressure over continental regions and, hence, increased inflow of moisture-laden air from the Indian Ocean to the continental interiors. Modeling studies indicate that the monsoonal flow was further amplified by feedbacks involving the atmosphere, vegetation, and soils. Increased moisture led to wetter soils and lusher vegetation, which in turn led to increased precipitation and greater penetration of moist air into continental interiors. Decreasing summer insolation during the past 4,000–6,000 years led to the weakening of the Indian Ocean monsoon.

Climate Change since the Advent of Humans

The history of humanity—from the initial appearance of genus Homo over 2,000,000 years ago to the advent and expansion of the modern human species (Homo sapiens) beginning some 150,000 years ago—is integrally linked to climate variation and change. Homo sapiens has experienced nearly two full glacial-interglacial cycles, but its global geographical expansion, massive population increase, cultural diversification, and worldwide ecological domination began only during the last glacial period and accelerated during the last glacial-interglacial transition. The first bipedal apes appeared in a time of climatic transition and variation, and Homo erectus, an extinct species possibly ancestral to modern humans, originated during the colder Pleistocene Epoch and survived both the transition period and multiple glacial-interglacial cycles. Thus, it can be said that climate variation has been the midwife of humanity and its various cultures and civilizations.

Recent Glacial and Interglacial Periods

The Most Recent Glacial Phase

With glacial ice restricted to high latitudes and altitudes, Earth 125,000 years ago was in an interglacial period similar to the one occurring today. During the past 125,000 years, however, the Earth system went through an entire glacial-interglacial cycle, only the most recent of many taking place over the last million years. The most recent period of cooling and glaciation began approximately 120,000 years ago. Significant ice sheets developed and persisted over much of Canada and northern Eurasia.

After the initial development of glacial conditions, the Earth system alternated between two modes, one of cold temperatures and growing glaciers and the other of relatively warm temperatures (although much cooler than today) and retreating glaciers. These Dansgaard-Oeschger (DO) cycles, recorded in both ice cores and marine sediments, occurred approximately every 1,500 years. A lower-frequency cycle, called the Bond cycle, is superimposed on the pattern of DO cycles; Bond cycles occurred every 3,000–8,000 years. Each Bond cycle is characterized by unusually cold conditions that take place during the cold phase of a DO cycle, the subsequent Heinrich event (which is a brief dry and cold phase), and the rapid warming phase that follows each Heinrich event. During each Heinrich event, massive fleets of icebergs were released into the North Atlantic, carrying rocks picked up by the glaciers far out to sea. Heinrich events are marked in marine sediments by conspicuous layers of iceberg-transported rock fragments.

Many of the transitions in the DO and Bond cycles were rapid and abrupt, and they are being studied intensely by paleoclimatologists and Earth system scientists to understand the driving mechanisms of such dramatic climatic variations. These cycles now appear to result from interactions between the atmosphere, oceans, ice sheets, and continental rivers that influence thermohaline circulation (the pattern of ocean currents driven by differences in water density, salinity, and temperature, rather than wind). Thermohaline circulation, in turn, controls ocean heat transport, such as the Gulf Stream.

The Last Glacial Maximum

During the past 25,000 years, the Earth system has undergone a series of dramatic transitions. The most recent glacial period peaked 21,500 years ago during the Last Glacial Maximum, or LGM. At that time, the northern third of North America was covered by the Laurentide Ice Sheet, which extended as far south as Des Moines, Iowa; Cincinnati, Ohio; and New York City. The Cordilleran Ice Sheet covered much of western Canada as well as northern Washington, Idaho, and Montana in the United States. In Europe the Scandinavian Ice Sheet sat atop the British Isles, Scandinavia, northeastern Europe, and north-central Siberia. Montane glaciers were extensive in other regions, even at low latitudes in Africa and South America. Global sea level was 125 metres (410 feet) below modern levels, because of the long-term net transfer of water from the oceans to the ice sheets. Temperatures near Earth's surface in unglaciated regions were about 5 °C (9 °F) cooler than today. Many Northern Hemisphere plant and animal species inhabited areas far south of their present ranges. For example, jack pine and white spruce trees grew in northwestern Georgia, 1,000 km (600 miles) south of their modern range limits in the Great Lakes region of North America.

The Last Deglaciation

The continental ice sheets began to melt back about 20,000 years ago. Drilling and dating of submerged fossil coral reefs provide a clear record of increasing sea levels as the ice melted. The most rapid melting began 15,000 years ago. For example, the southern boundary of the Laurentide Ice Sheet in North America was north of the Great Lakes and St. Lawrence regions by 10,000 years ago, and it had completely disappeared by 6,000 years ago.

The warming trend was punctuated by transient cooling events, most notably the Younger Dryas climate interval of 12,800–11,600 years ago. The climatic regimes that developed during the deglaciation period in many areas, including much of North America, have no modern analog (i.e., no regions exist with comparable seasonal regimes of temperature and moisture). For example, in the interior of North America, climates were much more continental (that is, characterized by warm summers and cold winters) than they are today. Also, paleontological studies indicate assemblages of plant, insect, and vertebrate species that do not occur anywhere today. Spruce trees grew with temperate hardwoods (ash, hornbeam, oak, and elm) in the upper Mississippi River and Ohio River regions. In Alaska, birch and poplar grew in woodlands, and there were very few of the spruce trees that dominate the present-day Alaskan landscape. Boreal and temperate mammals, whose geographic ranges are widely separated today, coexisted in central North America and Russia during this period of deglaciation. These unparalleled climatic conditions probably resulted from the combination of a unique orbital pattern that increased summer insolation and reduced winter insolation in the Northern Hemisphere and the continued presence of Northern Hemisphere ice sheets, which themselves altered atmospheric circulation patterns.

Climate Change and the Emergence of Agriculture

The first known examples of animal domestication occurred in western Asia between 11,000 and 9,500 years ago when goats and sheep were first herded, whereas examples of plant domestication date to 9,000 years ago when wheat, lentils, rye, and barley were first cultivated. This phase of technological increase occurred during a time of climatic transition that followed the last glacial period. A number of scientists have suggested that, although climate change imposed stresses on hunter-gatherer-forager societies by causing rapid shifts in resources, it also provided opportunities as new plant and animal resources appeared.

Glacial and Interglacial Cycles of the Pleistocene

The glacial period that peaked 21,500 years ago was only the most recent of five glacial periods in the last 450,000 years. In fact, the Earth system has alternated between glacial and interglacial regimes for more than two million years, a period of time known as the Pleistocene. The duration and severity of the glacial periods increased during this period, with a particularly sharp change occurring between 900,000 and 600,000 years ago. Earth is currently within the most recent interglacial period, which started 11,700 years ago and is commonly known as the Holocene Epoch.

The continental glaciations of the Pleistocene left signatures on the landscape in the form of glacial deposits and landforms; however, the best knowledge of the magnitude and timing of the various glacial and interglacial periods comes from oxygen isotope records in ocean sediments. These records provide both a direct measure of sea level and an indirect measure of global ice volume. Water molecules composed of a lighter isotope of oxygen, 16O, are evaporated more readily than molecules bearing a heavier isotope, 18O. Glacial periods are characterized by high 18O concentrations and represent a net transfer of water, especially with 16O, from the oceans to the ice sheets. Oxygen isotope records indicate that interglacial periods have typically lasted 10,000–15,000 years, and maximum glacial periods were of similar length. Most of the past 500,000 years—approximately 80 percent—have been spent within various intermediate glacial states that were warmer than glacial maxima but cooler than interglacials. During these intermediate times, substantial glaciers occurred over much of Canada and probably covered Scandinavia as well. These intermediate states were not constant; they were characterized by continual, millennial-scale climate variation. There has been no average or typical state for global climate during Pleistocene and Holocene times; the Earth system has been in continual flux between interglacial and glacial patterns.

The cycling of the Earth system between glacial and interglacial modes has been ultimately driven by orbital variations. However, orbital forcing is by it insufficient to explain all of this variation, and Earth system scientists are focusing their attention on the interactions and feedbacks between the myriad components of the Earth system. For example, the initial development of a continental ice sheet increases albedo over a portion of Earth, reducing surface absorption of sunlight and leading to further cooling. Similarly, changes in terrestrial vegetation, such as the replacement of forests by tundra, feed back into the atmosphere via changes in both albedo and latent heat flux from evapotranspiration. Forests—particularly those of tropical and temperate areas, with their large leaf area—release great amounts of water vapour and latent heat through transpiration. Tundra plants, which are much smaller, possess tiny leaves designed to slow water loss; they release only a small fraction of the water vapour that forests do.

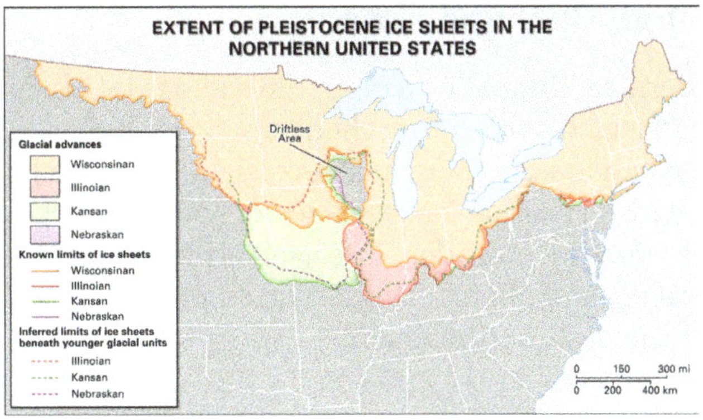

The blue areas are those that were covered by ice sheets in the past.

The Kansan and Nebraskan sheets overlapped almost the same areas, and the Wisconsin and Illinoisan sheets covered approximately the same territory. In the high altitudes of the West are the Cordilleran ice sheets. An area at the junction of Wisconsin, Minnesota, Iowa, and Illinois was never entirely covered with ice.

The discovery in ice core records that atmospheric concentrations of two potent greenhouse gases, carbon dioxide and methane, have decreased during past glacial periods and peaked during interglacials indicates important feedback processes in the Earth system. Reduction of greenhouse gas concentrations during the transition to a glacial phase would reinforce and amplify cooling already under way. The reverse is true for transition to interglacial periods. The glacial carbon sink remains a topic of considerable research activity. A full understanding of glacial-interglacial carbon dynamics requires knowledge of the complex interplay among ocean chemistry and circulation, ecology of marine and terrestrial organisms, ice sheet dynamics, and atmospheric chemistry and circulation.

The Last Great Cooling

The Earth system has undergone a general cooling trend for the past 50 million years, culminating in the development of permanent ice sheets in the Northern Hemisphere about 2.75 million years ago. These ice sheets expanded and contracted in a regular rhythm, with each glacial maximum separated from adjacent ones by 41,000 years (based on the cycle of axial tilt). As the ice sheets waxed and waned, global climate drifted steadily toward cooler conditions characterized by increasingly severe glaciations and increasingly cool interglacial phases. Beginning around 900,000 years ago, the glacial-interglacial cycles shifted frequency. Ever since, the glacial peaks have been 100,000 years apart, and the Earth system has spent more time in cool phases than before. The 41,000-year periodicity has continued, with smaller fluctuations superimposed on the 100,000-year cycle. In addition, a smaller, 23,000-year cycle has occurred through both the 41,000-year and 100,000-year cycles.

The 23,000-year and 41,000-year cycles are driven ultimately by two components of Earth's orbital geometry: the equinoctial precession cycle (23,000 years) and the axial-tilt cycle (41,000 years). Although the third parameter of Earth's orbit, eccentricity, varies on a 100,000-year cycle, its magnitude is insufficient to explain the 100,000-year cycles of glacial and interglacial periods of the past 900,000 years. The origin of the periodicity present in Earth's eccentricity is an important question in current paleoclimate research.

Climate Change through Geologic Time

The Earth system has undergone dramatic changes throughout its 4.5-billion-year history. These have included climatic changes diverse in mechanisms, magnitudes, rates, and consequences. Many of these past changes are obscure and controversial, and some have been discovered only recently. Nevertheless, the history of life has been strongly influenced by these changes, some of which radically altered the course of evolution. Life itself is implicated as a causative agent of some of these changes, as the processes of photosynthesis and respiration have largely shaped the chemistry of Earth's atmosphere, oceans, and sediments.

Cenozoic Climates

The Cenozoic Era—encompassing the past 65.5 million years, the time that has elapsed since the mass extinction event marking the end of the Cretaceous Period—has a broad range of climatic variation characterized by alternating intervals of global warming and cooling. Earth has experienced both extreme warmth and extreme cold during this period. These changes have been driven by tectonic forces, which have altered the positions and elevations of the continents as well as ocean passages and bathymetry. Feedbacks between different components of the Earth system (atmosphere, biosphere, lithosphere, cryosphere, and oceans in the hydrosphere) are being increasingly recognized as influences of global and regional climate. In particular, atmospheric concentrations of carbon dioxide have varied substantially during the Cenozoic for reasons that are poorly understood, though its fluctuation must have involved feedbacks between Earth's spheres.

Orbital forcing is also evident in the Cenozoic, although, when compared on such a vast era-level timescale, orbital variations can be seen as oscillations against a slowly changing backdrop of lower-frequency climatic trends. Descriptions of the orbital variations have evolved according to the growing understanding of tectonic and biogeochemical changes. A pattern emerging from recent paleoclimatologic studies suggests that the climatic effects of eccentricity, precession, and axial tilt have been amplified during cool phases of the Cenozoic, whereas they have been dampened during warm phases.

The meteor impact that occurred at or very close to the end of the Cretaceous came at a time of global warming, which continued into the early Cenozoic. Tropical and subtropical flora and fauna occurred at high latitudes until at least 40 million years ago, and geochemical records of marine sediments have indicated the presence of warm oceans. The interval of maximum temperature occurred during the late Paleocene and early Eocene epochs (58.7 million to 40.4 million years ago). The highest global temperatures of the Cenozoic occurred during the Paleocene-Eocene Thermal Maximum (PETM), a short interval lasting approximately 100,000 years. Although the underlying causes are unclear, the onset of the PETM about 56 million years ago was rapid, occurring within a few thousand years, and ecological consequences were large, with widespread extinctions in both marine and terrestrial ecosystems. Sea surface and continental air temperatures increased by more than 5 °C (9 °F) during the transition into the PETM. Sea surface temperatures in the high-latitude Arctic may have been as warm as 23 °C (73 °F), comparable to modern subtropical and warm-temperate seas. Following the PETM, global temperatures declined to pre-PETM levels, but they gradually increased to near-PETM levels over the next few million years during a period known as the Eocene Optimum. This temperature maximum was followed by a steady decline in global temperatures toward the Eocene-Oligocene boundary, which occurred about 33.9 million years ago. These changes are well-represented in marine sediments and in paleontological records from the continents, where vegetation zones moved Equator-ward. Mechanisms underlying the cooling trend are under study, but it is most likely that tectonic movements played an important role. This period saw the gradual opening of the sea passage between Tasmania and Antarctica, followed by the opening of the Drake Passage between South America and Antarctica. The latter, which isolated Antarctica within a cold polar sea, produced global effects on atmospheric and oceanic circulation. Recent evidence suggests that decreasing atmospheric concentrations of carbon dioxide during this period may have initiated a steady and irreversible cooling trend over the next few million years.

A continental ice sheet developed in Antarctica during the Oligocene Epoch, persisting until a rapid warming event took place 27 million years ago. The late Oligocene and early to mid-Miocene epochs (28.4 million to 13.8 million years ago) were relatively warm, though not nearly as warm as the Eocene. Cooling resumed 15 million years ago, and the Antarctic Ice Sheet expanded again to cover much of the continent. The cooling trend continued through the late Miocene and accelerated into the early Pliocene Epoch, 5.3 million years ago. During this period the Northern Hemisphere remained ice-free, and paleobotanical studies show cool-temperate Pliocene floras at high latitudes on Greenland and the Arctic Archipelago. The Northern Hemisphere glaciation, which began 3.2 million years ago, was driven by tectonic events, such as the closing of the Panama seaway and the uplift of the Andes, the Tibetan Plateau, and western parts of North America. These tectonic events led to changes in the circulation of the oceans and the atmosphere, which in turn fostered the development of persistent ice at high northern latitudes. Small-magnitude variations in carbon dioxide concentrations, which had been relatively low since at least the mid-Oligocene (28.4 million years ago), are also thought to have contributed to this glaciation.

Phanerozoic Climates

The Phanerozoic Eon (542 million years ago to the present), which includes the entire span of complex, multicellular life on Earth, has witnessed an extraordinary array of climatic states and transitions. The sheer antiquity of many of these regimes and events renders them difficult to understand in detail. However, a number of periods and transitions are well known, owing to good geological records and intense study by scientists. Furthermore, a coherent pattern of low-frequency climatic variation is emerging, in which the Earth system alternates between warm ("greenhouse") phases and cool ("icehouse") phases. The warm phases are characterized by high temperatures, high sea levels, and an absence of continental glaciers. Cool phases in turn are marked by low temperatures, low sea levels, and the presence of continental ice sheets, at least at high latitudes. Superimposed on these alternations are higher-frequency variations, where cool periods are embedded within greenhouse phases and warm periods are embedded within icehouse phases. For example, glaciers developed for a brief period (between 1 million and 10 million years) during the late Ordovician and early Silurian, in the middle of the early Paleozoic greenhouse phase (542 million to 350 million years ago). Similarly, warm periods with glacial retreat occurred within the late Cenozoic cool period during the late Oligocene and early Miocene epochs.

The Earth system has been in an icehouse phase for the past 30 million to 35 million years, ever since the development of ice sheets on Antarctica. The previous major icehouse phase occurred between about 350 million and 250 million years ago, during the Carboniferous and Permian periods of the late Paleozoic Era. Glacial sediments dating to this period have been identified in much of Africa as well as in the Arabian Peninsula, South America, Australia, India, and Antarctica. At the time, all these regions were part of Gondwana, a high-latitude supercontinent in the Southern Hemisphere. The glaciers atop Gondwana extended to at least 45° S latitude, similar to the latitude reached by Northern Hemisphere ice sheets during the Pleistocene. Some late Paleozoic glaciers extended even further Equator-ward—to 35° S. One of the most striking features of this time period are cyclothems, repeating sedimentary beds of alternating sandstone, shale, coal, and limestone. The great coal deposits of North America's Appalachian region, the American Midwest, and northern Europe are interbedded in these cyclothems, which may represent repeated transgressions (producing limestone) and retreats (producing shales and coals) of ocean shorelines in response to orbital variations.

The two most prominent warm phases in Earth history occurred during the Mesozoic and early Cenozoic eras (approximately 250 million to 35 million years ago) and the early and mid-Paleozoic (approximately 500 million to 350 million years ago). Climates of each of these greenhouse periods were distinct; continental positions and ocean bathymetry were very different, and terrestrial vegetation was absent from the continents until relatively late in the Paleozoic warm period. Both of these periods experienced substantial long-term climate variation and change; increasing evidence indicates brief glacial episodes during the mid-Mesozoic.

Understanding the mechanisms underlying icehouse-greenhouse dynamics is an important area of research, involving an interchange between geologic records and the modeling of the Earth system and its components. Two processes have been implicated as drivers of Phanerozoic climate change. First, tectonic forces caused changes in the positions and elevations of continents and the bathymetry of oceans and seas. Second, variations in greenhouse gases were also important drivers of climate, though at these long timescales they were largely controlled by tectonic processes, in which sinks and sources of greenhouse gases varied.

Climates of early Earth

The pre-Phanerozoic interval, also known as Precambrian time, comprises some 88 percent of the time elapsed since the origin of Earth. The pre-Phanerozoic is a poorly understood phase of Earth system history. Much of the sedimentary record of the atmosphere, oceans, biota, and crust of the early Earth has been obliterated by erosion, metamorphosis, and subduction. However, a number of pre-Phanerozoic records have been found in various parts of the world, mainly from the later portions of the period. Pre-Phanerozoic Earth system history is an extremely active area of research, in part because of its importance in understanding the origin and early evolution of life on Earth. Furthermore, the chemical composition of Earth's atmosphere and oceans largely developed during this period, with living organisms playing an active role. Geologists, paleontologists, microbiologists, planetary geologists, atmospheric scientists, and geochemists are focusing intense efforts on understanding this period. Three areas of particular interest and debate are the "faint young Sun paradox," the role of organisms in shaping Earth's atmosphere, and the possibility that Earth went through one or more "snowball" phases of global glaciation.

Faint Young Sun Paradox

Astrophysical studies indicate that the luminosity of the Sun was much lower during Earth's early history than it has been in the Phanerozoic. In fact, radiative output was low enough to suggest that all surface water on Earth should have been frozen solid during its early history, but evidence shows that it was not. The solution to this "faint young Sun paradox" appears to lie in the presence of unusually high concentrations of greenhouse gases at the time, particularly methane and carbon dioxide. As solar luminosity gradually increased through time, concentrations of greenhouse gases would have to have been much higher than today. This circumstance would have caused Earth to heat up beyond life-sustaining levels. Therefore, greenhouse gas concentrations must have decreased proportionally with increasing solar radiation, implying a feedback mechanism to regulate greenhouse gases. One of these mechanisms might have been rock weathering, which is temperature-dependent and serves as an important sink for, rather than source of, carbon dioxide by removing sizable amounts of this gas from the atmosphere. Scientists are also looking to biological processes (many of which also serve as carbon dioxide sinks) as complementary or alternative regulating mechanisms of greenhouse gases on the young Earth.

Photosynthesis and Atmospheric Chemistry

The evolution by photosynthetic bacteria of a new photosynthetic pathway, substituting water (H_2O) for hydrogen sulfide (H_2S) as a reducing agent for carbon dioxide, had dramatic consequences for Earth system geochemistry. Molecular oxygen (O_2) is given off as a by-product of photosynthesis using the H_2O pathway, which is energetically more efficient than the more primitive H_2S pathway. Using H_2O as a reducing agent in this process led to the large-scale deposition of banded-iron formations, or BIFs, a source of 90 percent of present-day iron ores. Oxygen present in ancient oceans oxidized dissolved iron, which precipitated out of solution onto the ocean floors. This deposition process, in which oxygen was used up as fast as it was produced, continued for millions of years until most of the iron dissolved in the oceans was precipitated. By approximately 2 billion years ago, oxygen was able to accumulate in dissolved form in seawater and to outgas to the atmosphere. Although oxygen does not have greenhouse gas properties, it plays important indirect roles in Earth's climate, particularly in phases of the carbon cycle. Scientists are studying the role of oxygen and other contributions of early life to the development of the Earth system.

Snowball Earth Hypothesis

Geochemical and sedimentary evidence indicates that Earth experienced as many as four extreme cooling events between 750 million and 580 million years ago. Geologists have proposed that Earth's oceans and land surfaces were covered by ice from the poles to the Equator during these events. This "Snowball Earth" hypothesis is a subject of intense study and discussion. Two important questions arise from this hypothesis. First, how, once frozen, could Earth thaw? Second, how could life survive periods of global freezing? A proposed solution to the first question involves the outgassing of massive amounts of carbon dioxide by volcanoes, which could have warmed the planetary surface rapidly, especially given that major carbon dioxide sinks (rock weathering and photosynthesis) would have been dampened by a frozen Earth. A possible answer to the second question may lie in the existence of present-day life-forms within hot springs and deep-sea vents, which would have persisted long ago despite the frozen state of Earth's surface.

A counter-premise known as the "Slushball Earth" hypothesis contends that Earth was not completely frozen over. Rather, in addition to massive ice sheets covering the continents, parts of the planet (especially ocean areas near the Equator) could have been draped only by a thin, watery layer of ice amid areas of open sea. Under this scenario, photosynthetic organisms in low-ice or ice-free regions could continue to capture sunlight efficiently and survive these periods of extreme cold.

Abrupt Climate Changes in Earth

An important new area of research, abrupt climate change, has developed since the 1980s. This research has been inspired by the discovery, in the ice core records of Greenland and Antarctica, of evidence for abrupt shifts in regional and global climates of the past. These events, which have also been documented in ocean and continental records, involve sudden shifts of Earth's climate system from one equilibrium state to another. Such shifts are of considerable scientific concern because they can reveal something about the controls and sensitivity of the climate system. In particular, they point out nonlinearities, the so-called "tipping points," where small, gradual changes in one component of the system can lead to a large change in the entire system. Such nonlinearities arise from the complex feedbacks between components of the Earth system. For example, during the Younger Dryas event a gradual increase in the release of fresh water to the North Atlantic Ocean led to an abrupt shutdown of the thermohaline circulation in the Atlantic basin. Abrupt climate shifts are of great societal concern, for any such shifts in the future might be so rapid and radical as to outstrip the capacity of agricultural, ecological, industrial, and economic systems to respond and adapt. Climate scientists are working with social scientists, ecologists, and economists to assess society's vulnerability to such "climate surprises."

The Younger Dryas event (12,800 to 11,600 years ago) is the most intensely studied and best-understood example of abrupt climate change. The event took place during the last deglaciation, a period of global warming when the Earth system was in transition from a glacial mode to an interglacial one. The Younger Dryas was marked by a sharp drop in temperatures in the North Atlantic region; cooling in northern Europe and eastern North America is estimated at 4 to 8 °C (7.2 to 14.4 °F). Terrestrial and marine records indicate that the Younger Dryas had detectable effects of lesser magnitude over most other regions of Earth. The termination of the Younger Dryas was very rapid, occurring within a decade. The Younger Dryas resulted from an abrupt shutdown of the thermohaline circulation in the North Atlantic, which is critical for the transport of heat from equatorial regions northward (today the Gulf Stream is a part of that circulation). The cause of the shutdown of the thermohaline circulation is under study; an influx of large volumes of freshwater from melting glaciers into the North Atlantic has been implicated, although other factors probably played a role.

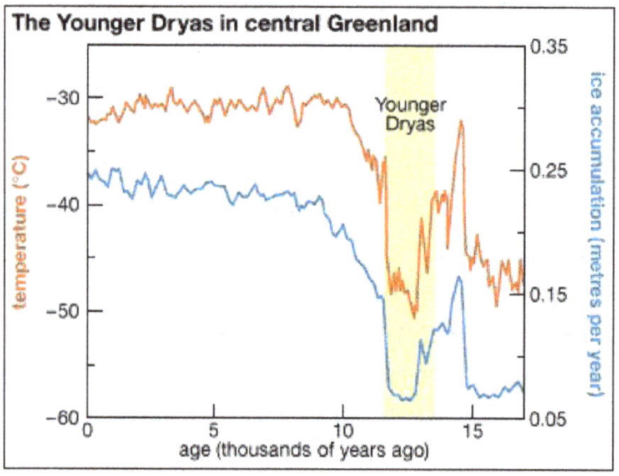

The Younger Dryas event was characterized by a substantial and relatively sudden drop in temperature between 12,800 and 11,600 years ago. In addition to cold regions, the evidence of this temperature change has been discovered in tropical and subtropical regions.

Paleoclimatologists are devoting increasing attention to identifying and studying other abrupt changes. The Dansgaard-Oeschger cycles of the last glacial period are now recognized as representing alternation between two climate states, with rapid transitions from one state to the other. A 200-year-long cooling event in the Northern Hemisphere approximately 8,200 years ago resulted from the rapid draining of glacial Lake Agassiz into the North Atlantic via the Great Lakes and St. Lawrence drainage. This event, characterized as a miniature version of the Younger Dryas, had ecological impacts in Europe and North America that included a rapid decline of hemlock populations in New England forests. In addition, evidence of another such transition, marked by a rapid drop in the water levels of lakes and bogs in eastern North America, occurred 5,200 years ago. It is recorded in ice cores from glaciers at high altitudes in tropical regions as well as tree-ring, lake-level, and peatland samples from temperate regions.

Abrupt climatic changes occurring before the Pleistocene have also been documented. A transient thermal maximum has been documented near the Paleocene-Eocene boundary (55.8 million years ago), and evidence of rapid cooling events are observed near the boundaries between both the Eocene and Oligocene epochs (33.9 million years ago) and the Oligocene and Miocene epochs (23 million years ago). All three of these events had global ecological, climatic, and biogeochemical consequences. Geochemical evidence indicates that the warm event occurring at the Paleocene-Eocene boundary was associated with a rapid increase in atmospheric carbon dioxide concentrations, possibly resulting from the massive outgassing and oxidation of methane hydrates (a compound whose chemical structure traps methane within a lattice of ice) from the ocean floor. The two cooling events appear to have resulted from a transient series of positive feedbacks among the atmosphere, oceans, ice sheets, and biosphere, similar to those observed in the Pleistocene. Other abrupt changes, such as the Paleocene-Eocene Thermal Maximum, are recorded at various points in the Phanerozoic.

Abrupt climate changes can evidently be caused by a variety of processes. Rapid changes in an external factor can push the climate system into a new mode. Outgassing of methane hydrates and the sudden influx of glacial meltwater into the ocean are examples of such external forcing. Alternatively, gradual changes in external factors can lead to the crossing of a threshold; the climate system is unable to return to the former equilibrium and passes rapidly to a new one. Such nonlinear system behaviour is a potential concern as human activities, such as fossil-fuel combustion and land-use change, alter important components of Earth's climate system.

Humans and other species have survived countless climatic changes in the past, and humans are a notably adaptable species. Adjustment to climatic changes, whether it is biological (as in the case of other species) or cultural (for humans), is easiest and least catastrophic when the changes are gradual and can be anticipated to large extent. Rapid changes are more difficult to adapt to and incur more disruption and risk. Abrupt changes, especially unanticipated climate surprises, put human cultures and societies, as well as both the populations of other species and the ecosystems they inhabit, at considerable risk of severe disruption. Such changes may well be within humanity's capacity to adapt, but not without paying severe penalties in the form of economic, ecological, agricultural, human health, and other disruptions. Knowledge of past climate variability provides guidelines on the natural variability and sensitivity of the Earth system. This knowledge also helps identify the risks associated with altering the Earth system with greenhouse gas emissions and regional to global-scale changes in land cover.

Global Warming

Global warming is the phenomenon of increasing average air temperatures near the surface of Earth over the past one to two centuries. Climate scientists have since the mid-20th century gathered detailed observations of various weather phenomena (such as temperatures, precipitation, and storms) and of related influences on climate (such as ocean currents and the atmosphere's chemical composition). These data indicate that Earth's climate has changed over almost every conceivable timescale since the beginning of geologic time and that the influence of human activities since at least the beginning of the Industrial Revolution has been deeply woven into the very fabric of climate change the phenomenon of increasing average air temperatures near the surface of Earth over the past one to two centuries. Climate scientists have since the mid-20th century gathered detailed observations of various weather phenomena (such as temperatures, precipitation, and storms) and of related influences on climate (such as ocean currents and the atmosphere's chemical composition). These data indicate that Earth's climate has changed over almost every conceivable timescale since the beginning of geologic time and that the influence of human activities since at least the beginning of the Industrial Revolution has been deeply woven into the very fabric of climate change.

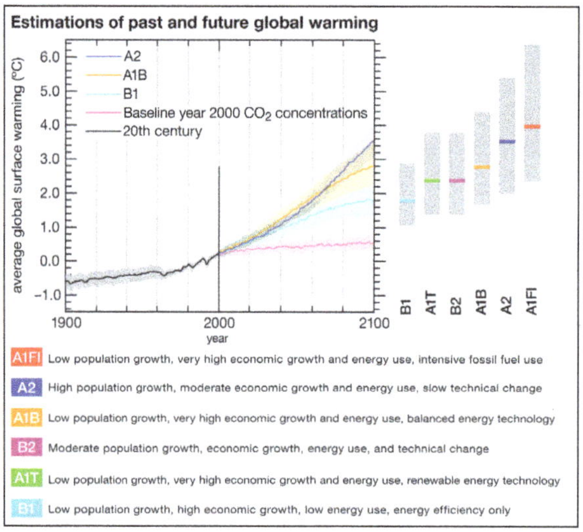

Global warming scenarios: Graph of the predicted increase in Earth's average surface temperature according to a series of climate change scenarios that assume different levels of economic development, population growth, and fossil fuel use. The assumptions made by each scenario are given at the bottom of the graph.

A special report produced by the IPCC in 2018 honed this estimate further, noting that human beings and human activities have been responsible for a worldwide average temperature increase of between 0.8 and 1.2 °C (1.4 and 2.2 °F) of global warming since preindustrial times, and most of the warming observed over the second half of the 20th century could be attributed to human activities. It predicted that the global mean surface temperature would increase between 3 and 4 °C (5.4 and 7.2 °F) by 2100 relative to the 1986–2005 average should carbon emissions continue at their current rate. The predicted rise in temperature was based on a range of possible scenarios that accounted for future greenhouse gas emissions and mitigation (severity reduction) measures and on uncertainties in the model projections. Some of the main uncertainties include the precise role of feedback processes and the impacts of industrial pollutants known as aerosols, which may offset some warming.

Many climate scientists agree that significant societal, economic, and ecological damage would result if global average temperatures rose by more than 2 °C (3.6 °F) in such a short time. Such damage would include increased extinction of many plant and animal species, shifts in patterns of agriculture, and rising sea levels. By 2015 all but a few national governments had begun the process of instituting carbon reduction plans as part of the Paris Agreement, a treaty designed to help countries keep global warming to 1.5 °C (2.7 °F) above preindustrial levels in order to avoid the worst of the predicted effects. Authors of a special report published by the IPCC in 2018 noted that should carbon emissions continue at their present rate, the increase in average near-surface air temperatures would reach 1.5 °C sometime between 2030 and 2052. Past IPCC assessments reported that the global average sea level rose by some 19–21 cm (7.5–8.3 inches) between 1901 and 2010 and that sea levels rose faster in the second half of the 20th century than in the first half. It also predicted, again depending on a wide range of scenarios, that the global average sea level would rise 26–77 cm (10.2–30.3 inches) relative to the 1986–2005 average by 2100 for global warming of 1.5 °C, an average of 10 cm (3.9 inches) less than what would be expected if warming rose to 2 °C (3.6 °F) above preindustrial levels.

The scenarios referred to above depend mainly on future concentrations of certain trace gases, called greenhouse gases that have been injected into the lower atmosphere in increasing amounts through the burning of fossil fuels for industry, transportation, and residential uses. Modern global warming is the result of an increase in magnitude of the so-called greenhouse effect, a warming of Earth's surface and lower atmosphere caused by the presence of water vapour, carbon dioxide, methane, nitrous oxides, and other greenhouse gases. In 2014 the IPCC reported that concentrations of carbon dioxide, methane, and nitrous oxides in the atmosphere surpassed those found in ice cores dating back 800,000 years.

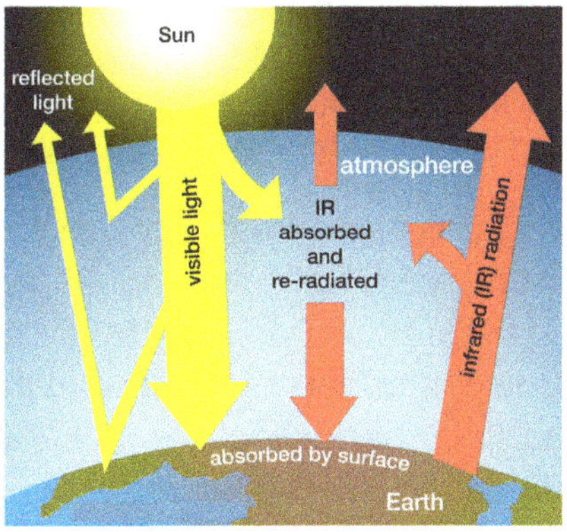

Greenhouse effect on Earth.

The greenhouse effect on Earth - Some incoming sunlight is reflected by Earth's atmosphere and surface, but most is absorbed by the surface, which is warmed. Infrared (IR) radiation is then emitted from the surface. Some IR radiation escapes to space, but some is absorbed by the atmosphere's greenhouse gases (especially water vapour, carbon dioxide, and methane) and reradiated in all directions, some to space and some back toward the surface, where it further warms the surface and the lower atmosphere.

Of all these gases, carbon dioxide is the most important, both for its role in the greenhouse effect and for its role in the human economy. It has been estimated that, at the beginning of the industrial age in the mid-18th century, carbon dioxide concentrations in the atmosphere were roughly 280 parts per million (ppm). By the middle of 2018 they had risen to 406 ppm, and, if fossil fuels continue to be burned at current rates, they are projected to reach 550 ppm by the mid-21st century—essentially, a doubling of carbon dioxide concentrations in 300 years.

A vigorous debate is in progress over the extent and seriousness of rising surface temperatures, the effects of past and future warming on human life, and the need for action to reduce future warming and deal with its consequences.

Climatic Variation since the Last Glaciation

Global warming is related to the more general phenomenon of climate change, which refers to changes in the totality of attributes that define climate. In addition to changes in air temperature, climate change involves changes to precipitation patterns, winds, ocean currents, and other measures of Earth's climate. Normally, climate change can be viewed as the combination of various natural forces occurring over diverse timescales. Since the advent of human civilization, climate change has involved an "anthropogenic," or exclusively human-caused, element, and this anthropogenic element has become more important in the industrial period of the past two centuries. The term global warming is used specifically to refer to any warming of near-surface air during the past two centuries that can be traced to anthropogenic causes.

Grinnell Glacier National Park, Montana (from left to right): the Grinnell Glacier filled the entire area at the bottom of the image. By 2006 it had largely disappeared from this view.

To define the concepts of global warming and climate change properly, it is first necessary to recognize that the climate of Earth has varied across many timescales, ranging from an individual human life span to billions of years. This variable climate history is typically classified in terms of "regimes" or "epochs." For instance, the Pleistocene glacial epoch (about 2,600,000 to 11,700 years ago) was marked by substantial variations in the global extent of glaciers and ice sheets. These variations took place on timescales of tens to hundreds of millennia and were driven by changes in the distribution of solar radiation across Earth's surface. The distribution of solar radiation is known as the insolation pattern, and it is strongly affected by the geometry of Earth's orbit around the Sun and by the orientation, or tilt, of Earth's axis relative to the direct rays of the Sun.

Worldwide, the most recent glacial period, or ice age, culminated about 21,000 years ago in what is often called the Last Glacial Maximum. During this time, continental ice sheets extended well into the middle latitude regions of Europe and North America, reaching as far south as present-day London and New York City. Global annual mean temperature appears to have been about 4–5 °C

(7–9 °F) colder than in the mid-20th century. It is important to remember that these figures are a global average. In fact, during the height of this last ice age, Earth's climate was characterized by greater cooling at higher latitudes (that is, toward the poles) and relatively little cooling over large parts of the tropical oceans (near the Equator). This glacial interval terminated abruptly about 11,700 years ago and was followed by the subsequent relatively ice-free period known as the Holocene Epoch. The modern period of Earth's history is conventionally defined as residing within the Holocene. However, some scientists have argued that the Holocene Epoch terminated in the relatively recent past and that Earth currently resides in a climatic interval that could justly be called the Anthropocene Epoch—that is, a period during which humans have exerted a dominant influence over climate.

Though less dramatic than the climate changes that occurred during the Pleistocene Epoch, significant variations in global climate have nonetheless taken place over the course of the Holocene. During the early Holocene, roughly 9,000 years ago, atmospheric circulation and precipitation patterns appear to have been substantially different from those of today. For example, there is evidence for relatively wet conditions in what is now the Sahara Desert. The change from one climatic regime to another was caused by only modest changes in the pattern of insolation within the Holocene interval as well as the interaction of these patterns with large-scale climate phenomena such as monsoons and El Niño/Southern Oscillation (ENSO).

During the middle Holocene, some 5,000–7,000 years ago, conditions appear to have been relatively warm—indeed, perhaps warmer than today in some parts of the world and during certain seasons. For this reason, this interval is sometimes referred to as the Mid-Holocene Climatic Optimum. The relative warmth of average near-surface air temperatures at this time, however, is somewhat unclear. Changes in the pattern of insolation favoured warmer summers at higher latitudes in the Northern Hemisphere, but these changes also produced cooler winters in the Northern Hemisphere and relatively cool conditions year-round in the tropics. Any overall hemispheric or global mean temperature changes thus reflected a balance between competing seasonal and regional changes. In fact, recent theoretical climate model studies suggest that global mean temperatures during the middle Holocene were probably 0.2–0.3 °C (0.4–0.5 °F) colder than average late 20th-century conditions.

Over subsequent millennia, conditions appear to have cooled relative to middle Holocene levels. This period has sometimes been referred to as the "Neoglacial." In the middle latitudes this cooling trend was associated with intermittent periods of advancing and retreating mountain glaciers reminiscent of (though far more modest than) the more substantial advance and retreat of the major continental ice sheets of the Pleistocene climate epoch.

Causes of Global Warming

The Greenhouse Effect

The average surface temperature of Earth is maintained by a balance of various forms of solar and terrestrial radiation. Solar radiation is often called "shortwave" radiation because the frequencies of the radiation are relatively high and the wavelengths relatively short—close to the visible portion of the electromagnetic spectrum. Terrestrial radiation, on the other hand, is often called "longwave" radiation because the frequencies are relatively low and the wavelengths relatively long—somewhere in the infrared part of the spectrum. Downward-moving solar energy is typically measured in watts per square metre. The energy of the total incoming solar radiation at the top of Earth's atmosphere (the so-called "solar constant") amounts roughly to 1,366 watts per square metre annually. Adjusting for the fact that only one-half of the planet's surface receives solar radiation at any given time, the average surface insolation is 342 watts per square metre annually.

The amount of solar radiation absorbed by Earth's surface is only a small fraction of the total solar radiation entering the atmosphere. For every 100 units of incoming solar radiation, roughly 30 units are reflected back to space by either clouds, the atmosphere, or reflective regions of Earth's surface. This reflective capacity is referred to as Earth's planetary albedo, and it need not remain fixed over time, since the spatial extent and distribution of reflective formations, such as clouds and ice cover, can change. The 70 units of solar radiation that are not reflected may be absorbed by the atmosphere, clouds, or the surface. In the absence of further complications, in order to maintain thermodynamic equilibrium, Earth's surface and atmosphere must radiate these same 70 units back to space. Earth's surface temperature (and that of the lower layer of the atmosphere essentially in contact with the surface) is tied to the magnitude of this emission of outgoing radiation according to the Stefan-Boltzmann law.

Earth's energy budget is further complicated by the greenhouse effect. Trace gases with certain chemical properties—the so-called greenhouse gases, mainly carbon dioxide (CO_2), methane (CH_4), and nitrous oxide (N_2O)—absorb some of the infrared radiation produced by Earth's surface. Because of this absorption, some fraction of the original 70 units does not directly escape to space. Because greenhouse gases emit the same amount of radiation they absorb and because this radiation is emitted equally in all directions (that is, as much downward as upward), the net effect of absorption by greenhouse gases is to increase the total amount of radiation emitted downward toward Earth's surface and lower atmosphere. To maintain equilibrium, Earth's surface and lower atmosphere must emit more radiation than the original 70 units. Consequently, the surface temperature must be higher. This process is not quite the same as that which governs a true greenhouse, but the end effect is similar. The presence of greenhouse gases in the atmosphere leads to a warming of the surface and lower part of the atmosphere (and a cooling higher up in the atmosphere) relative to what would be expected in the absence of greenhouse gases.

It is essential to distinguish the "natural," or background, greenhouse effect from the "enhanced" greenhouse effect associated with human activity. The natural greenhouse effect is associated with surface warming properties of natural constituents of Earth's atmosphere, especially water vapour, carbon dioxide, and methane. The existence of this effect is accepted by all scientists. Indeed, in its absence, Earth's average temperature would be approximately 33 °C (59 °F) colder than today, and Earth would be a frozen and likely uninhabitable planet. What has been subject to controversy is the so-called enhanced greenhouse effect, which is associated with increased concentrations of greenhouse gases caused by human activity. In particular, the burning of fossil fuels raises the concentrations of the major greenhouse gases in the atmosphere, and these higher concentrations have the potential to warm the atmosphere by several degrees.

Radiative Forcing

In light of the greenhouse effect, it is apparent that the temperature of Earth's surface and lower atmosphere may be modified in three ways: (1) through a net increase in the solar radiation entering at the top of Earth's atmosphere, (2) through a change in the fraction of the radiation reaching the surface, and (3) through a change in the concentration of greenhouse gases in the atmosphere. In each case the changes can be thought of in terms of "radiative forcing." As defined by the IPCC, radiative forcing is a measure of the influence a given climatic factor has on the amount of downward-directed radiant energy impinging upon Earth's surface. Climatic factors are divided between those caused primarily by human activity (such as greenhouse gas emissions and aerosol emissions) and those caused by natural forces (such as solar irradiance); then, for each factor, so-called forcing values are calculated for the time period between 1750 and the present day. "Positive forcing" is exerted by climatic factors that contribute to the warming of Earth's surface, whereas "negative forcing" is exerted by factors that cool Earth's surface.

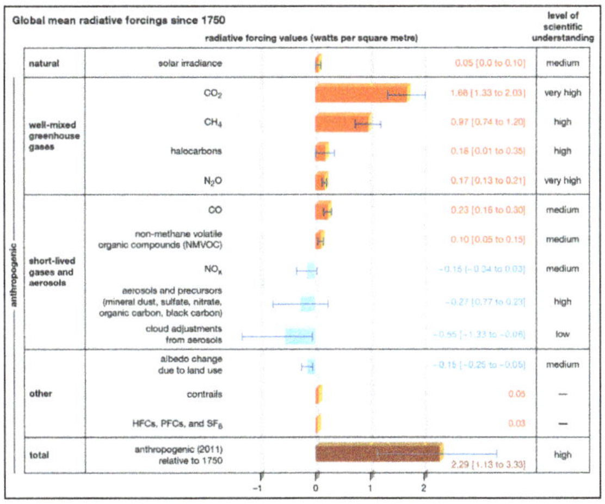
Global mean radiative forcings since 1750
radiative forcing values (watts per square metre)
level of scientific understanding
natural
solar irradiance
0.05 [0.0 to 0.10]
medium
anthropogenic
well-mixed greenhouse gases
CO_2
1.68 [1.33 to 2.03]
very high
CH_4
0.97 [0.74 to 1.20]
high
halocarbons
0.18 [0.01 to 0.35]
high
N_2O
0.17 [0.13 to 0.21]
very high
short-lived gases and aerosols
CO
0.23 [0.16 to 0.30]
medium
non-methane volatile organic compounds (NMVOC)
0.10 [0.05 to 0.15]
medium
NO_x
medium
aerosols and precursors (mineral dust, sulfate, nitrate, organic carbon, black carbon)
high
cloud adjustments from aerosols
low
albedo change due to land use
medium
other
contrails
0.05
—
HFCs, PFCs, and SF_6
0.03
—
total
anthropogenic (2011) relative to 1750
2.29 [1.13 to 3.33]
high
-1
0
1
2

Since 1750 the concentration of carbon dioxide and other greenhouse gases has increased in Earth's atmosphere. As a result of these and other factors, Earth's atmosphere retains more heat than in the past.

On average, about 342 watts of solar radiation strike each square metre of Earth's surface per year, and this quantity can in turn be related to a rise or fall in Earth's surface temperature. Temperatures at the surface may also rise or fall through a change in the distribution of terrestrial radiation (that is, radiation emitted by Earth) within the atmosphere. In some cases, radiative forcing has a natural origin, such as during explosive eruptions from volcanoes where vented gases and ash block some portion of solar radiation from the surface. In other cases, radiative forcing has an anthropogenic, or exclusively human, origin. For example, anthropogenic increases in carbon dioxide, methane, and nitrous oxide are estimated to account for 2.3 watts per square metre of positive radiative forcing. When all values of positive and negative radiative forcing are taken together and all interactions between climatic factors are accounted for, the total net increase in surface radiation due to human activities since the beginning of the Industrial Revolution is 1.6 watts per square metre.

The Influences of Human activity on Climate

Human activity has influenced global surface temperatures by changing the radiative balance governing the Earth on various timescales and at varying spatial scales. The most profound and well-known anthropogenic influence is the elevation of concentrations of greenhouse gases in the atmosphere. Humans also influence climate by changing the concentrations of aerosols and ozone and by modifying the land cover of Earth's surface.

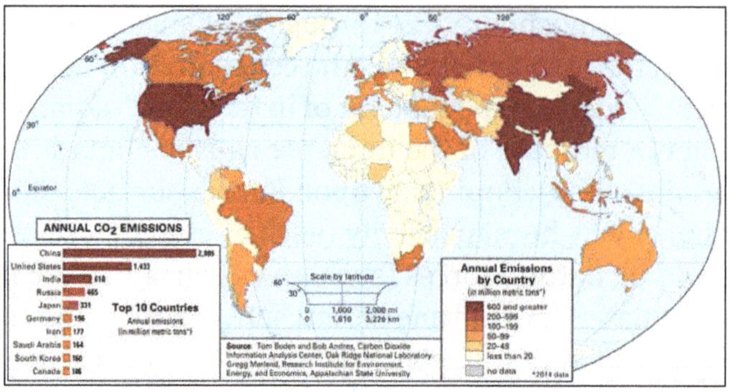

Carbon dioxide emissions: Map of annual carbon dioxide emissions by country in 2014.

Greenhouse Gases

Greenhouse gases warm Earth's surface by increasing the net downward longwave radiation reaching the surface. The relationship between atmospheric concentration of greenhouse gases and the associated positive radiative forcing of the surface is different for each gas. A complicated relationship exists between the chemical properties of each greenhouse gas and the relative amount of longwave radiation that each can absorb. What follows is a discussion of the radiative behaviour of each major greenhouse gas.

Factories that burn fossil fuels help to cause global warming.

Water Vapour

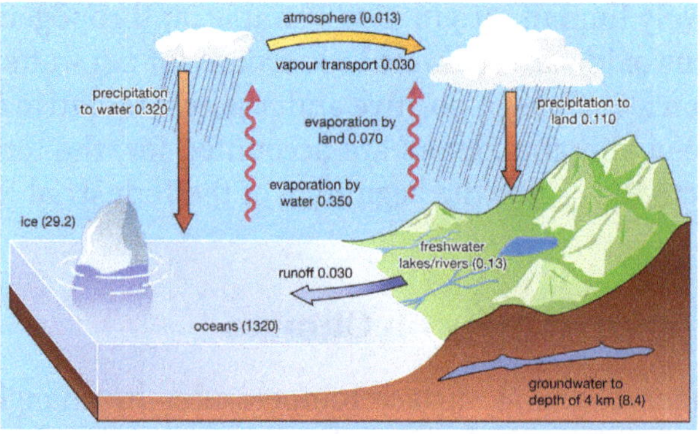

The present-day surface hydrologic cycle.

Water vapour is the most potent of the greenhouse gases in Earth's atmosphere, but its behaviour is fundamentally different from that of the other greenhouse gases. The primary role of water vapour is not as a direct agent of radiative forcing but rather as a climate feedback—that is, as a response within the climate system that influences the system's continued activity. This distinction arises from the fact that the amount of water vapour in the atmosphere cannot, in general, be directly modified by human behaviour but is instead set by air temperatures. The warmer the surface, the greater the evaporation rate of water from the surface. As a result, increased evaporation leads to a greater concentration of water vapour in the lower atmosphere capable of absorbing longwave radiation and emitting it downward.

Water is transferred from the oceans through the atmosphere to the continents and back to the oceans over and beneath the land surface. The values in parentheses following the various forms of water (e.g., ice) refer to volumes in millions of cubic kilometres; those following the processes (e.g., precipitation) refer to their fluxes in millions of cubic kilometres of water per year.

Carbon Dioxide

Of the greenhouse gases, carbon dioxide (CO_2) is the most significant. Natural sources of atmospheric CO_2 include outgassing from volcanoes, the combustion and natural decay of organic matter, and respiration by aerobic (oxygen-using) organisms. These sources are balanced, on average, by a set of physical, chemical, or biological processes, called "sinks," that tend to remove CO_2 from the atmosphere. Significant natural sinks include terrestrial vegetation, which takes up CO_2 during the process of photosynthesis.

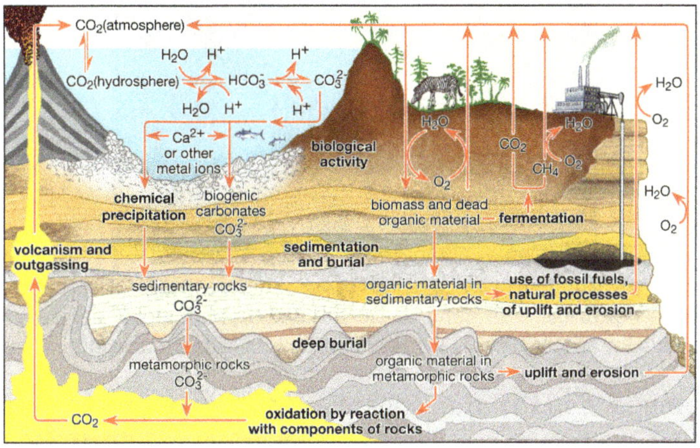

The carbon cycle.

Carbon is transported in various forms through the atmosphere, the hydrosphere, and geologic formations. One of the primary pathways for the exchange of carbon dioxide (CO_2) takes place between the atmosphere and the oceans; there a fraction of the CO_2 combines with water, forming carbonic acid (H_2CO_3) that subsequently loses hydrogen ions (H^+) to form bicarbonate (HCO_3^-) and carbonate (CO_3^{2-}) ions. Mollusk shells or mineral precipitates that form by the reaction of calcium or other metal ions with carbonate may become buried in geologic strata and eventually release CO_2 through volcanic outgassing. Carbon dioxide also exchanges through photosynthesis in plants and through respiration in animals. Dead and decaying organic matter may ferment and release CO_2 or methane (CH_4) or may be incorporated into sedimentary rock, where it is converted to fossil fuels. Burning of hydrocarbon fuels returns CO_2 and water (H_2O) to the atmosphere. The biological and anthropogenic pathways are much faster than the geochemical pathways and, consequently, have a greater impact on the composition and temperature of the atmosphere.

A number of oceanic processes also act as carbon sinks. One such process, called the "solubility pump," involves the descent of surface seawater containing dissolved CO_2. Another process, the "biological pump," involves the uptake of dissolved CO_2 by marine vegetation and phytoplankton (small free-floating photosynthetic organisms) living in the upper ocean or by other marine organisms that use CO_2 to build skeletons and other structures made of calcium carbonate ($CaCO_3$). As these organisms expire and fall to the ocean floor, the carbon they contain is transported downward and eventually buried at depth. A long-term balance between these natural sources and sinks leads to the background, or natural, level of CO_2 in the atmosphere.

In contrast, human activities increase atmospheric CO_2 levels primarily through the burning of fossil fuels—principally oil and coal and secondarily natural gas, for use in transportation, heating, and the generation of electrical power—and through the production of cement. Other anthropogenic sources include the burning of forests and the clearing of land. Anthropogenic emissions currently account for the annual release of about 7 gigatons (7 billion tons) of carbon into the atmosphere. Anthropogenic emissions are equal to approximately 3 percent of the total emissions of CO_2 by natural sources, and this amplified carbon load from human activities far exceeds the offsetting capacity of natural sinks (by perhaps as much as 2–3 gigatons per year).

Deforestation Smoldering remains of a plot of deforested land in the Amazon Rainforest of Brazil. Annually, it is estimated that net global deforestation accounts for about two gigatons of carbon emissions to the atmosphere.

CO_2 consequently accumulated in the atmosphere at an average rate of 1.4 ppm per year between 1959 and 2006 and roughly 2.0 ppm per year between 2006 and 2018. Overall, this rate of accumulation has been linear (that is, uniform over time). However, certain current sinks, such as the oceans, could become sources in the future. This may lead to a situation in which the concentration of atmospheric CO_2 builds at an exponential rate (that is, its rate of increase is also increasing).

The natural background level of carbon dioxide varies on timescales of millions of years because of slow changes in outgassing through volcanic activity. For example, roughly 100 million years ago, during the Cretaceous Period (145 million to 66 million years ago), CO_2 concentrations appear to have been several times higher than they are today (perhaps close to 2,000 ppm). Over the past 700,000 years, CO_2 concentrations have varied over a far smaller range (between roughly 180 and 300 ppm) in association with the same Earth orbital effects linked to the coming and going of the Pleistocene ice ages. By the early 21st century, CO_2 levels had reached 384 ppm, which is approximately 37 percent above the natural background level of roughly 280 ppm that existed at the beginning of the Industrial Revolution. Atmospheric CO_2 levels continued to increase, and by 2018 they had reached 410 ppm. Such levels are believed to be the highest in at least 800,000 years according to ice core measurements and may be the highest in at least 5 million years according to other lines of evidence.

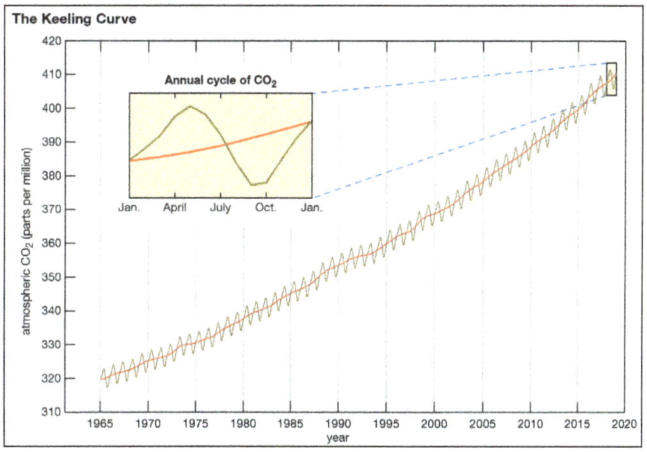

Figure: The Keeling Curve, named after American climate scientist Charles David Keeling, tracks changes in the concentration of carbon dioxide (CO_2) in Earth's atmosphere at a research station on Mauna Loa in Hawaii. Although these concentrations experience small seasonal fluctuations, the overall trend shows that CO_2 is increasing in the atmosphere.

Radiative forcing caused by carbon dioxide varies in an approximately logarithmic fashion with the concentration of that gas in the atmosphere. The logarithmic relationship occurs as the result of a saturation effect wherein it becomes increasingly difficult, as CO_2 concentrations increase, for additional CO_2 molecules to further influence the "infrared window" (a certain narrow band of wavelengths in the infrared region that is not absorbed by atmospheric gases). The logarithmic relationship predicts that the surface warming potential will rise by roughly the same amount for each doubling of CO_2 concentration. At current rates of fossil fuel use, a doubling of CO_2 concentrations over preindustrial levels is expected to take place by the middle of the 21st century (when CO_2 concentrations are projected to reach 560 ppm). A doubling of CO_2 concentrations would represent an increase of roughly 4 watts per square metre of radiative forcing. Given typical estimates of "climate sensitivity" in the absence of any offsetting factors, this energy increase would lead to a warming of 2 to 5 °C (3.6 to 9 °F) over preindustrial times. The total radiative forcing by anthropogenic CO_2 emissions since the beginning of the industrial age is approximately 1.66 watts per square metre.

Methane

Methane (CH_4) is the second most important greenhouse gas. CH_4 is more potent than CO_2 because the radiative forcing produced per molecule is greater. In addition, the infrared window is less saturated in the range of wavelengths of radiation absorbed by CH_4, so more molecules may fill in the region. However, CH_4 exists in far lower concentrations than CO_2 in the atmosphere, and its concentrations by volume in the atmosphere are generally measured in parts per billion (ppb) rather than ppm. CH_4 also has a considerably shorter residence time in the atmosphere than CO_2 (the residence time for CH_4 is roughly 10 years, compared with hundreds of years for CO_2).

Natural sources of methane include tropical and northern wetlands, methane-oxidizing bacteria that feed on organic material consumed by termites, volcanoes, seepage vents of the seafloor in regions rich with organic sediment, and methane hydrates trapped along the continental shelves of the oceans and in polar permafrost. The primary natural sink for methane is the atmosphere itself, as methane reacts readily with the hydroxyl radical (OH) within the troposphere to form CO_2 and water vapour (H_2O). When CH_4 reaches the stratosphere, it is destroyed. Another natural sink is soil, where methane is oxidized by bacteria.

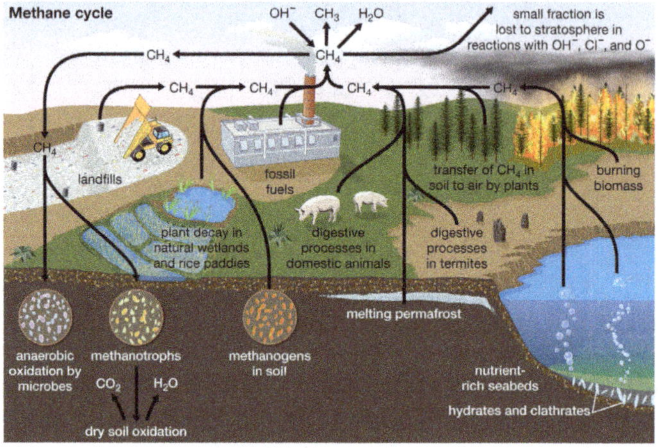

Methane cycle.

As with CO_2, human activity is increasing the CH_4 concentration faster than it can be offset by natural sinks. Anthropogenic sources currently account for approximately 70 percent of total annual emissions, leading to substantial increases in concentration over time. The major anthropogenic sources of atmospheric CH_4 are rice cultivation, livestock farming, the burning of coal and natural gas, the combustion of biomass, and the decomposition of organic matter in landfills. Future trends are particularly difficult to anticipate. This is in part due to an incomplete understanding of the climate feedbacks associated with CH_4 emissions. In addition it is difficult to predict how, as human populations grow, possible changes in livestock raising, rice cultivation, and energy utilization will influence CH_4 emissions.

It is believed that a sudden increase in the concentration of methane in the atmosphere was responsible for a warming event that raised average global temperatures by 4–8 °C (7.2–14.4 °F) over a few thousand years during the so-called Paleocene-Eocene Thermal Maximum, or PETM. This episode took place roughly 55 million years ago, and the rise in CH_4 appears to have been related to a massive volcanic eruption that interacted with methane-containing flood deposits. As a result, large amounts of gaseous CH_4 were injected into the atmosphere. It is difficult to know precisely how high these concentrations were or how long they persisted. At very high concentrations, residence times of CH_4 in the atmosphere can become much greater than the nominal 10-year residence time that applies today. Nevertheless, it is likely that these concentrations reached several ppm during the PETM.

Methane concentrations have also varied over a smaller range (between roughly 350 and 800 ppb) in association with the Pleistocene ice age cycles. Preindustrial levels of CH_4 in the atmosphere were approximately 700 ppb, whereas levels exceeded 1,867 ppb in late 2018. (These concentrations are well above the natural levels observed for at least the past 650,000 years.) The net radiative forcing by anthropogenic CH_4 emissions is approximately 0.5 watt per square metre—or roughly one-third the radiative forcing of CO_2.

Surface-level Ozone and other Compounds

The next most significant greenhouse gas is surface, or low-level, ozone (O_3). Surface O_3 is a result of air pollution; it must be distinguished from naturally occurring stratospheric O_3, which has a very different role in the planetary radiation balance. The primary natural source of surface O_3 is the subsidence of stratospheric O_3 from the upper atmosphere . In contrast, the primary anthropogenic source of surface O_3 is photochemical reactions involving the atmospheric pollutant carbon monoxide (CO). The best estimates of the natural concentration of surface O_3 are 10 ppb, and the net radiative forcing due to anthropogenic emissions of surface O_3 is approximately 0.35 watt per square metre. Ozone concentrations can rise above unhealthy levels (that is, conditions where concentrations meet or exceed 70 ppb for eight hours or longer) in cities prone to photochemical smog.

Nitrous Oxides and Fluorinated Gases

Additional trace gases produced by industrial activity that have greenhouse properties include nitrous oxide (N_2O) and fluorinated gases (halocarbons), the latter including sulfur hexafluoride, hydrofluorocarbons (HFCs), and perfluorocarbons (PFCs). Nitrous oxide is responsible for 0.16 watt per square metre radiative forcing, while fluorinated gases are collectively responsible for 0.34 watt per square metre. Nitrous oxides have small background concentrations due to natural biological reactions in soil and water, whereas the fluorinated gases owe their existence almost entirely to industrial sources.

Aerosols

The production of aerosols represents an important anthropogenic radiative forcing of climate. Collectively, aerosols block—that is, reflect and absorb—a portion of incoming solar radiation, and this creates a negative radiative forcing. Aerosols are second only to greenhouse gases in relative importance in their impact on near-surface air temperatures. Unlike the decade-long residence times of the "well-mixed" greenhouse gases, such as CO_2 and CH_4, aerosols are readily flushed out of the atmosphere within days, either by rain or snow (wet deposition) or by settling out of the air (dry deposition). They must therefore be continually generated in order to produce a steady effect on radiative forcing. Aerosols have the ability to influence climate directly by absorbing or reflecting incoming solar radiation, but they can also produce indirect effects on climate by modifying cloud formation or cloud properties. Most aerosols serve as condensation nuclei (surfaces upon which water vapour can condense to form clouds); however, darker-coloured aerosols may hinder cloud formation by absorbing sunlight and heating up the surrounding air. Aerosols can be transported thousands of kilometres from their sources of origin by winds and upper-level circulation in the atmosphere.

Perhaps the most important type of anthropogenic aerosol in radiative forcing is sulfate aerosol. It is produced from sulfur dioxide (SO_2) emissions associated with the burning of coal and oil. Since the late 1980s, global emissions of SO_2 have decreased from about 151.5 million tonnes (167.0 million tons) to less than 100 million tonnes (110.2 million tons) of sulfur per year.

Nitrate aerosol is not as important as sulfate aerosol, but it has the potential to become a significant source of negative forcing. One major source of nitrate aerosol is smog (the combination of ozone with oxides of nitrogen in the lower atmosphere) released from the incomplete burning of fuel in internal-combustion engines. Another source is ammonia (NH_3), which is often used in fertilizers or released by the burning of plants and other organic materials. If greater amounts of atmospheric nitrogen are converted to ammonia and agricultural ammonia emissions continue to increase as projected, the influence of nitrate aerosols on radiative forcing is expected to grow.

Both sulfate and nitrate aerosols act primarily by reflecting incoming solar radiation, thereby reducing the amount of sunlight reaching the surface. Most aerosols, unlike greenhouse gases, impart a cooling rather than warming influence on Earth's surface. One prominent exception is carbonaceous aerosols such as carbon black or soot, which are produced by the burning of fossil fuels and biomass. Carbon black tends to absorb rather than reflect incident solar radiation, and so it has a warming impact on the lower atmosphere, where it resides. Because of its absorptive properties, carbon black is also capable of having an additional indirect effect on climate. Through its deposition in snowfall, it can decrease the albedo of snow cover. This reduction in the amount of solar radiation reflected back to space by snow surfaces creates a minor positive radiative forcing.

Natural forms of aerosol include windblown mineral dust generated in arid and semiarid regions and sea salt produced by the action of waves breaking in the ocean. Changes to wind patterns as a result of climate modification could alter the emissions of these aerosols. The influence of climate change on regional patterns of aridity could shift both the sources and the destinations of dust clouds. In addition, since the concentration of sea salt aerosol, or sea aerosol, increases with the strength of the winds near the ocean surface, changes in wind speed due to global warming and climate change could influence the concentration of sea salt aerosol. For example, some studies suggest that climate change might lead to stronger winds over parts of the North Atlantic Ocean. Areas with stronger winds may experience an increase in the concentration of sea salt aerosol.

Other natural sources of aerosols include volcanic eruptions, which produce sulfate aerosol, and biogenic sources (e.g., phytoplankton), which produce dimethyl sulfide (DMS). Other important biogenic aerosols, such as terpenes, are produced naturally by certain kinds of trees or other plants. For example, the dense forests of the Blue Ridge Mountains of Virginia in the United States emit terpenes during the summer months, which in turn interact with the high humidity and warm temperatures to produce a natural photochemical smog. Anthropogenic pollutants such as nitrate and ozone, both of which serve as precursor molecules for the generation of biogenic aerosol, appear to have increased the rate of production of these aerosols severalfold. This process appears to be responsible for some of the increased aerosol pollution in regions undergoing rapid urbanization.

Human activity has greatly increased the amount of aerosol in the atmosphere compared with the background levels of preindustrial times. In contrast to the global effects of greenhouse gases, the impact of anthropogenic aerosols is confined primarily to the Northern Hemisphere, where most of the world's industrial activity occurs. The pattern of increases in anthropogenic aerosol over time is also somewhat different from that of greenhouse gases. During the middle of the 20th century, there was a substantial increase in aerosol emissions. This appears to have been at least partially responsible for a cessation of surface warming that took place in the Northern Hemisphere from the 1940s through the 1970s. Since that time, aerosol emissions have leveled off due to antipollution measures undertaken in the industrialized countries since the 1960s. Aerosol emissions may rise in the future, however, as a result of the rapid emergence of coal-fired electric power generation in China and India.

The total radiative forcing of all anthropogenic aerosols is approximately −1.2 watts per square metre. Of this total, −0.5 watt per square metre comes from direct effects (such as the reflection of solar energy back into space), and −0.7 watt per square metre comes from indirect effects (such as the influence of aerosols on cloud formation). This negative radiative forcing represents an offset of roughly 40 percent from the positive radiative forcing caused by human activity. However, the relative uncertainty in aerosol radiative forcing (approximately 90 percent) is much greater than that of greenhouse gases. In addition, future emissions of aerosols from human activities, and the influence of these emissions on future climate change, are not known with any certainty. Nevertheless, it can be said that, if concentrations of anthropogenic aerosols continue to decrease as they have since the 1970s, a significant offset to the effects of greenhouse gases will be reduced, opening future climate to further warming.

Land-use Change

There are a number of ways in which changes in land use can influence climate. The most direct influence is through the alteration of Earth's albedo, or surface reflectance. For example, the replacement of forest by cropland and pasture in the middle latitudes over the past several centuries has led to an increase in albedo, which in turn has led to greater reflection of incoming solar radiation in those regions. This replacement of forest by agriculture has been associated with a change in global average radiative forcing of approximately −0.2 watt per square metre since 1750. In Europe and other major agricultural regions, such land-use conversion began more than 1,000 years ago and has proceeded nearly to completion. For Europe, the negative radiative forcing due to land-use change has probably been substantial, perhaps approaching −5 watts per square metre. The influence of early land use on radiative forcing may help to explain a long period of cooling in Europe that followed a period of relatively mild conditions roughly 1,000 years ago. It is generally believed that the mild temperatures of this "medieval warm period," which was followed by a long period of cooling, rivaled those of 20th-century Europe.

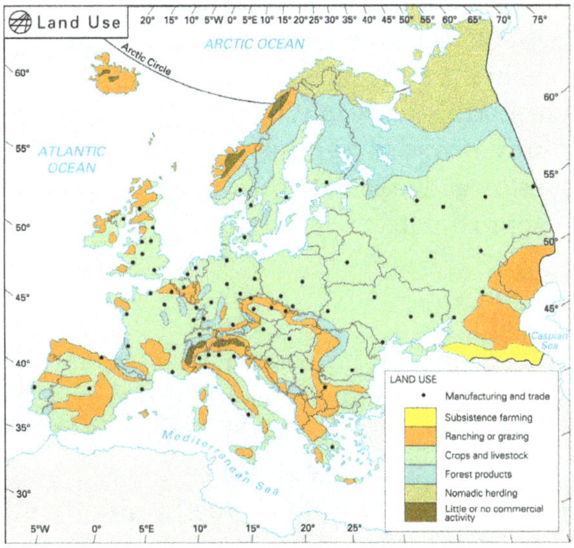

Europe: Land use in Europe.

Land-use changes can also influence climate through their influence on the exchange of heat between Earth's surface and the atmosphere. For example, vegetation helps to facilitate the evaporation of water into the atmosphere through evapotranspiration. In this process, plants take up liquid water from the soil through their root systems. Eventually this water is released through transpiration into the atmosphere, as water vapour through the stomata in leaves. While deforestation generally leads to surface cooling due to the albedo factor, the land surface may also be warmed as a result of the release of latent heat by the evapotranspiration process. The relative importance of these two factors, one exerting a cooling effect and the other a warming effect, varies by both season and region. While the albedo effect is likely to dominate in middle latitudes, especially during the period from autumn through spring, the evapotranspiration effect may dominate during the summer in the midlatitudes and year-round in the tropics. The latter case is particularly important in assessing the potential impacts of continued tropical deforestation.

The rate at which tropical regions are deforested is also relevant to the process of carbon sequestration, the long-term storage of carbon in underground cavities and biomass rather than in the atmosphere. By removing carbon from the atmosphere, carbon sequestration acts to mitigate global warming. Deforestation contributes to global warming, as fewer plants are available to take up carbon dioxide from the atmosphere. In addition, as fallen trees, shrubs, and other plants are burned or allowed to slowly decompose, they release as carbon dioxide the carbon they stored during their lifetimes. Furthermore, any land-use change that influences the amount, distribution, or type of vegetation in a region can affect the concentrations of biogenic aerosols, though the impact of such changes on climate is indirect and relatively minor.

Stratospheric Ozone Depletion

Since the 1970s the loss of ozone (O_3) from the stratosphere has led to a small amount of negative radiative forcing of the surface. This negative forcing represents a competition between two distinct effects caused by the fact that ozone absorbs solar radiation. In the first case, as ozone levels in the stratosphere are depleted, more solar radiation reaches Earth's surface. In the absence of any other influence, this rise in insolation would represent a positive radiative forcing of the surface. However, there is a second effect of ozone depletion that is related to its greenhouse properties. As the amount of ozone in the stratosphere is decreased, there is also less ozone to absorb longwave radiation emitted by Earth's surface. With less absorption of radiation by ozone, there is a corresponding decrease in the downward reemission of radiation. This second effect overwhelms the first and results in a modest negative radiative forcing of Earth's surface and a modest cooling of the lower stratosphere by approximately 0.5 °C (0.9 °F) per decade since the 1970s.

Natural Influences on Climate

There are a number of natural factors that influence Earth's climate. These factors include external influences such as explosive volcanic eruptions, natural variations in the output of the Sun, and slow changes in the configuration of Earth's orbit relative to the Sun. In addition, there are natural oscillations in Earth's climate that alter global patterns of wind circulation, precipitation, and surface temperatures. One such phenomenon is the El Niño/Southern Oscillation (ENSO), a coupled atmospheric and oceanic event that occurs in the Pacific Ocean every three to seven years. In addition, the Atlantic Multidecadal Oscillation (AMO) is a similar phenomenon that occurs over decades in the North Atlantic Ocean. Other types of oscillatory behaviour that produce dramatic shifts in climate may occur across timescales of centuries and millennia.

Volcanic Aerosols

Explosive volcanic eruptions have the potential to inject substantial amounts of sulfate aerosols into the lower stratosphere. In contrast to aerosol emissions in the lower troposphere, aerosols that enter the stratosphere may remain for several years before settling out, because of the relative absence of turbulent motions there. Consequently, aerosols from explosive volcanic eruptions have the potential to affect Earth's climate. Less-explosive eruptions, or eruptions that are less vertical in orientation, have a lower potential for substantial climate impact. Furthermore, because of large-scale circulation patterns within the stratosphere, aerosols injected within tropical regions tend to spread out over the globe, whereas aerosols injected within midlatitude and polar regions tend to remain confined to the middle and high latitudes of that hemisphere. Tropical eruptions, therefore, tend to have a greater climatic impact than eruptions occurring toward the poles. In 1991 the moderate eruption of Mount Pinatubo in the Philippines provided a peak forcing of approximately −4 watts per square metre and cooled the climate by about 0.5 °C (0.9 °F) over the following few years. By comparison, the 1815 Mount Tambora eruption in present-day Indonesia, typically implicated for the 1816 "year without a summer" in Europe and North America, is believed to have been associated with a radiative forcing of approximately −6 watts per square metre.

A column of gas and ash rising from Mount Pinatubo in the Philippines on June 12, 1991, just days before the volcano's climactic explosion on June 15.

While in the stratosphere, volcanic sulfate aerosol actually absorbs longwave radiation emitted by Earth's surface, and absorption in the stratosphere tends to result in a cooling of the troposphere below. This vertical pattern of temperature change in the atmosphere influences the behaviour of winds in the lower atmosphere, primarily in winter. Thus, while there is essentially a global cooling effect for the first few years following an explosive volcanic eruption, changes in the winter patterns of surface winds may actually lead to warmer winters in some areas, such as Europe. Some modern examples of major eruptions include Krakatoa (Indonesia) in 1883, El Chichón (Mexico) in 1982, and Mount Pinatubo in 1991. There is also evidence that volcanic eruptions may influence other climate phenomena such as ENSO.

Variations in Solar Output

Direct measurements of solar irradiance, or solar output, have been available from satellites only since the late 1970s. These measurements show a very small peak-to-peak variation in solar irradiance (roughly 0.1 percent of the 1,366 watts per square metre received at the top of the atmosphere, for approximately 1.4 watts per square metre). However, indirect measures of solar activity are available from historical sunspot measurements dating back through the early 17th century. Attempts have been made to reconstruct graphs of solar irradiance variations from historical sunspot data by calibrating them against the measurements from modern satellites. However, since the modern measurements span only a few of the most recent 11-year solar cycles, estimates of solar output variability on 100-year and longer timescales are poorly correlated. Different assumptions regarding the relationship between the amplitudes of 11-year solar cycles and long-period solar output changes can lead to considerable differences in the resulting solar reconstructions. These differences in turn lead to fairly large uncertainty in estimating positive forcing by changes in solar irradiance since 1750. (Estimates range from 0.06 to 0.3 watt per square metre.) Even more challenging, given the lack of any modern analog, is the estimation of solar irradiance during the so-called Maunder Minimum, a period lasting from the mid-17th century to the early 18th century when very few sunspots were observed. While it is likely that solar irradiance was reduced at this time, it is difficult to calculate by how much. However, additional proxies of solar output exist that match reasonably well with the sunspot-derived records following the Maunder Minimum; these may be used as crude estimates of the solar irradiance variations.

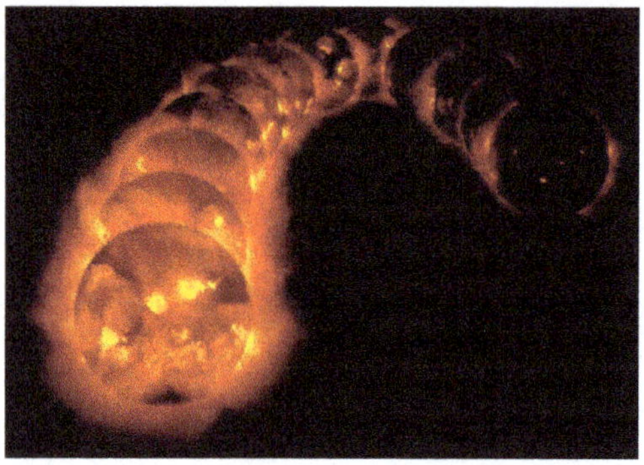

Twelve solar X-ray images. The solar coronal brightness decreases by a factor of about 100 during a solar cycle as the Sun goes from an "active" state (left) to a less active state (right).

In theory it is possible to estimate solar irradiance even farther back in time, over at least the past millennium, by measuring levels of cosmogenic isotopes such as carbon-14 and beryllium-10. Cosmogenic isotopes are isotopes that are formed by interactions of cosmic rays with atomic nuclei in the atmosphere and that subsequently fall to Earth, where they can be measured in the annual layers found in ice cores. Since their production rate in the upper atmosphere is modulated by changes in solar activity, cosmogenic isotopes may be used as indirect indicators of solar irradiance. However, as with the sunspot data, there is still considerable uncertainty in the amplitude of past solar variability implied by these data.

Solar forcing also affects the photochemical reactions that manufacture ozone in the stratosphere. Through this modulation of stratospheric ozone concentrations, changes in solar irradiance (particularly in the ultraviolet portion of the electromagnetic spectrum) can modify how both shortwave and longwave radiation in the lower stratosphere are absorbed. As a result, the vertical temperature profile of the atmosphere can change, and this change can in turn influence phenomena such as the strength of the winter jet streams.

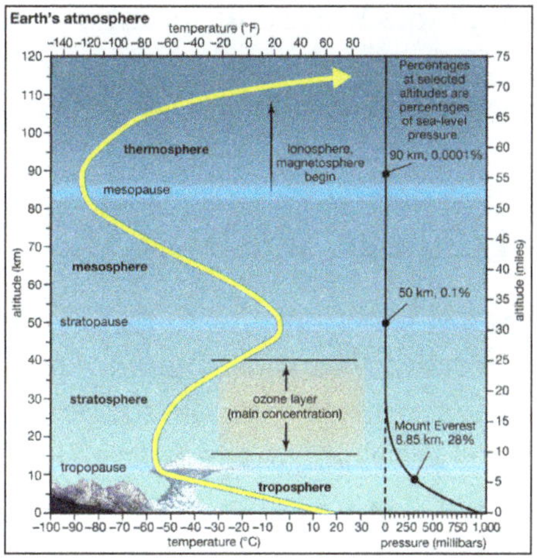

The layers of Earth's atmosphere: The yellow line shows the response of air temperature to increasing height.

Variations in Earth's Orbit

On timescales of tens of millennia, the dominant radiative forcing of Earth's climate is associated with slow variations in the geometry of Earth's orbit about the Sun. These variations include the precession of the equinoxes (that is, changes in the timing of summer and winter), occurring on a roughly 26,000-year timescale; changes in the tilt angle of Earth's rotational axis relative to the plane of Earth's orbit around the Sun, occurring on a roughly 41,000-year timescale; and changes in the eccentricity (the departure from a perfect circle) of Earth's orbit around the Sun, occurring on a roughly 100,000-year timescale. Changes in eccentricity slightly influence the mean annual solar radiation at the top of Earth's atmosphere, but the primary influence of all the orbital variations listed above is on the seasonal and latitudinal distribution of incoming solar radiation over Earth's surface. The major ice ages of the Pleistocene Epoch were closely related to the influence of these variations on summer insolation at high northern latitudes. Orbital variations thus exerted a primary control on the extent of continental ice sheets. However, Earth's orbital changes are generally believed to have had little impact on climate over the past few millennia, and so they are not considered to be significant factors in present-day climate variability.

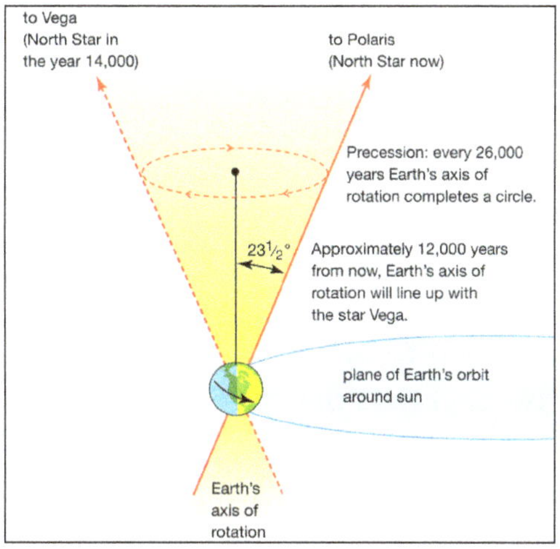

Earth's axis of rotation itself rotates, or precesses, completing one circle every 26,000 years.

Consequently, Earth's North Pole points toward different stars (and sometimes toward empty space) as it travels in this circle. This precession is so slow that it is not noticeable in a person's lifetime, though astronomers must consider its effect when studying ancient sites such as Stonehenge.

Feedback Mechanisms and Climate Sensitivity

There are a number of feedback processes important to Earth's climate system and, in particular, its response to external radiative forcing. The most fundamental of these feedback mechanisms involves the loss of longwave radiation to space from the surface. Since this radiative loss increases with increasing surface temperatures according to the Stefan-Boltzmann law, it represents a stabilizing factor (that is, a negative feedback) with respect to near-surface air temperature.

Climate sensitivity can be defined as the amount of surface warming resulting from each additional watt per square metre of radiative forcing. Alternatively, it is sometimes defined as the warming that would result from a doubling of CO_2 concentrations and the associated addition of 4 watts per square metre of radiative forcing. In the absence of any additional feedbacks, climate sensitivity would be approximately 0.25 °C (0.45 °F) for each additional watt per square metre of radiative forcing. Stated alternatively, if the CO2 concentration of the atmosphere present at the start of the industrial age (280 ppm) were doubled (to 560 ppm), the resulting additional 4 watts per square metre of radiative forcing would translate into a 1 °C (1.8 °F) increase in air temperature. However, there are additional feedbacks that exert a destabilizing, rather than stabilizing, influence, and these feedbacks tend to increase the sensitivity of climate to somewhere between 0.5 and 1.0 °C (0.9 and 1.8 °F) for each additional watt per square metre of radiative forcing.

Water Vapour Feedback

Unlike concentrations of other greenhouse gases, the concentration of water vapour in the atmosphere cannot freely vary. Instead, it is determined by the temperature of the lower atmosphere and surface through a physical relationship known as the Clausius-Clapeyron equation, named for 19th-century German physicist Rudolf Clausius and 19th-century French engineer Émile Clapeyron. Under the assumption that there is a liquid water surface in equilibrium with the atmosphere, this relationship indicates that an increase in the capacity of air to hold water vapour is a function of increasing temperature of that volume of air. This assumption is relatively good over the oceans, where water is plentiful, but not over the continents. For this reason the relative humidity (the percent of water vapour the air contains relative to its capacity) is approximately 100 percent over ocean regions and much lower over continental regions (approaching 0 percent in arid regions). Not surprisingly, the average relative humidity of Earth's lower atmosphere is similar to the fraction of Earth's surface covered by the oceans (that is, roughly 70 percent). This quantity is expected to remain approximately constant as Earth warms or cools. Slight changes to global relative humidity may result from human land-use modification, such as tropical deforestation and irrigation, which can affect the relative humidity over land areas up to regional scales.

The amount of water vapour in the atmosphere will rise as the temperature of the atmosphere rises. Since water vapour is a very potent greenhouse gas, even more potent than CO_2, the net greenhouse effect actually becomes stronger as the surface warms, which leads to even greater warming. This positive feedback is known as the "water vapour feedback." It is the primary reason that climate sensitivity is substantially greater than the previously stated theoretical value of 0.25 °C (0.45 °F) for each increase of 1 watt per square metre of radiative forcing.

Cloud Feedbacks

It is generally believed that as Earth's surface warms and the atmosphere's water vapour content increases, global cloud cover increases. However, the effects on near-surface air temperatures are complicated. In the case of low clouds, such as marine stratus clouds, the dominant radiative feature of the cloud is its albedo. Here any increase in low cloud cover acts in much the same way as an increase in surface ice cover: more incoming solar radiation is reflected and Earth's surface cools. On the other hand, high clouds, such as the towering cumulus clouds that extend up to the boundary between the troposphere and stratosphere, have a quite different impact on the surface radiation balance. The tops of cumulus clouds are considerably higher in the atmosphere and colder than their undersides. Cumulus cloud tops emit less longwave radiation out to space than the warmer cloud bottoms emit downward toward the surface. The end result of the formation of high cumulus clouds is greater warming at the surface.

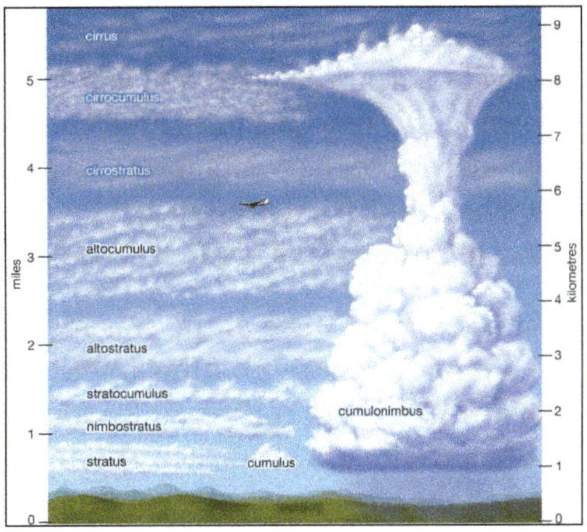

Different types of clouds form at different heights.

The net feedback of clouds on rising surface temperatures is therefore somewhat uncertain. It represents a competition between the impacts of high and low clouds, and the balance is difficult to determine. Nonetheless, most estimates indicate that clouds on the whole represent a positive feedback and thus additional warming.

Ice Albedo Feedback

Another important positive climate feedback is the so-called ice albedo feedback. This feedback arises from the simple fact that ice is more reflective (that is, has a higher albedo) than land or water surfaces. Therefore, as global ice cover decreases, the reflectivity of Earth's surface decreases, more incoming solar radiation is absorbed by the surface, and the surface warms. This feedback is considerably more important when there is relatively extensive global ice cover, such as during the height of the last ice age, roughly 25,000 years ago. On a global scale the importance of ice albedo feedback decreases as Earth's surface warms and there is relatively less ice available to be melted.

Carbon Cycle Feedbacks

Another important set of climate feedbacks involves the global carbon cycle. In particular, the two main reservoirs of carbon in the climate system are the oceans and the terrestrial biosphere. These reservoirs have historically taken up large amounts of anthropogenic CO_2 emissions. Roughly 50–70 percent is removed by the oceans, whereas the remainder is taken up by the terrestrial biosphere. Global warming, however, could decrease the capacity of these reservoirs to sequester atmospheric CO_2. Reductions in the rate of carbon uptake by these reservoirs would increase the pace of CO_2 buildup in the atmosphere and represent yet another possible positive feedback to increased greenhouse gas concentrations.

In the world's oceans, this feedback effect might take several paths. First, as surface waters warm, they would hold less dissolved CO_2. Second, if more CO_2 were added to the atmosphere and taken up by the oceans, bicarbonate ions (HCO_3^-) would multiply and ocean acidity would increase. Since calcium carbonate ($CaCO_3$) is broken down by acidic solutions, rising acidity would threaten ocean-dwelling fauna that incorporate $CaCO_3$ into their skeletons or shells. As it becomes increasingly difficult for these organisms to absorb oceanic carbon, there would be a corresponding decrease in the efficiency of the biological pump that helps to maintain the oceans as a carbon sink. Third, rising surface temperatures might lead to a slowdown in the so-called thermohaline circulation, a global pattern of oceanic flow that partly drives the sinking of surface waters near the poles and is responsible for much of the burial of carbon in the deep ocean. A slowdown in this flow due to an influx of melting fresh water into what are normally saltwater conditions might also cause the solubility pump, which transfers CO_2 from shallow to deeper waters, to become less efficient. Indeed, it is predicted that if global warming continued to a certain point, the oceans would cease to be a net sink of CO_2 and would become a net source.

As large sections of tropical forest are lost because of the warming and drying of regions such as Amazonia, the overall capacity of plants to sequester atmospheric CO_2 would be reduced. As a result, the terrestrial biosphere, though currently a carbon sink, would become a carbon source. Ambient temperature is a significant factor affecting the pace of photosynthesis in plants, and many plant species that are well adapted to their local climatic conditions have maximized their photosynthetic rates. As temperatures increase and conditions begin to exceed the optimal temperature range for both photosynthesis and soil respiration, the rate of photosynthesis would decline. As dead plants decompose, microbial metabolic activity (a CO_2 source) would increase and would eventually outpace photosynthesis.

Under sufficient global warming conditions, methane sinks in the oceans and terrestrial biosphere also might become methane sources. Annual emissions of methane by wetlands might either increase or decrease, depending on temperatures and input of nutrients, and it is possible that wetlands could switch from source to sink. There is also the potential for increased methane release as a result of the warming of Arctic permafrost (on land) and further methane release at the continental margins of the oceans (a few hundred metres below sea level). The current average atmospheric methane concentration of 1,750 ppb is equivalent to 3.5 gigatons (3.5 billion tons) of carbon. There are at least 400 gigatons of carbon equivalent stored in Arctic permafrost and as much as 10,000 gigatons (10 trillion tons) of carbon equivalent trapped on the continental margins of the oceans in a hydrated crystalline form known as clathrate. It is believed that some fraction of this trapped methane could become unstable with additional warming, although the amount and rate of potential emission remain highly uncertain.

Potential Effects of Global Warming

The path of future climate change will depend on what courses of action are taken by society—in particular the emission of greenhouse gases from the burning of fossil fuels. A range of alternative emissions scenarios known as representative concentration pathways (RCPs) were proposed by the IPCC in the Fifth Assessment Report (AR5), which was published in 2014, to examine potential future climate changes. The scenarios depend on various assumptions concerning future rates of human population growth, economic development, energy demand, technological advancement, and other factors. Unlike the scenarios used in previous IPCC assessments, the AR5 RCPs explicitly account for climate change mitigation efforts.

Projected range of sea-level rise by climate change scenario*		
scenario	temperature change (°C) in 2090–99 relative to 1980–99	sea-level rise (m) in 2090–99 relative to 1980–99
B1	1.1–2.9	0.18–0.38
A1T	1.4–3.8	0.20–0.45
B2	1.4–3.8	0.20–0.43
A1B	1.7–4.4	0.21–0.48
A2	2.0–5.4	0.23–0.51
A1Fl	2.4–6.4	0.26–0.59
*Ranges of sea-level rise are based on various models of climate change that exclude the possibility of future rapid changes in ice flow, such as the melting of the Greenland and Antarctic ice caps.		

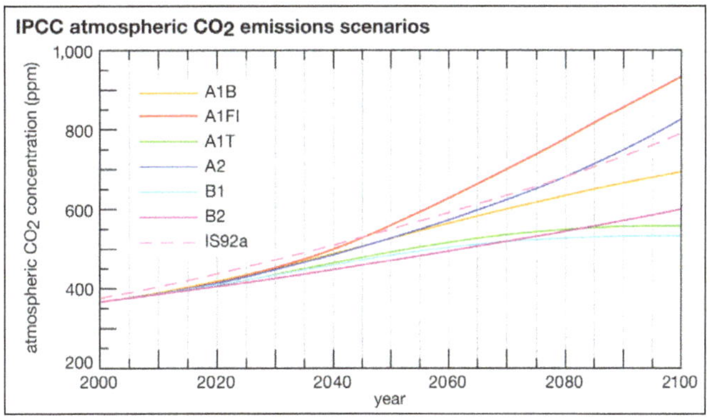

Carbon dioxide: global warming scenarios.

Graph of the predicted increase in the concentration of carbon dioxide (CO_2) in Earth's atmosphere according to a series of climate change scenarios that assume different levels of economic development, population growth, and fossil fuel use.The results of each scenario in the IPCC's Fourth Assessment Report (2007) are depicted in the graph.

The AR5 scenario with the smallest increases in greenhouse gases is RCP 2.6, which denotes the net radiative forcing by 2100 in watts per square metre (a doubling of CO_2 concentrations from preindustrial values of 280 ppm to 560 ppm represents roughly 3.7 watts per square metre). RCP 2.6 assumes substantial improvements in energy efficiency, a rapid transition away from fossil fuel energy, and a global population that peaks at roughly nine billion people in the 21st century. In that scenario CO_2 concentrations remain below 450 ppm and actually fall toward the end of the century (to about 420 ppm) as a result of widespread deployment of carbon-capture technology.

Scenario RCP 8.5, by contrast, might be described as "business as usual." It reflects the assumption of an energy-intensive global economy, high population growth, and a reduced rate of technological development. CO_2 concentrations are more than three times greater than preindustrial levels (roughly 936 ppm) by 2100 and continue to grow thereafter. RCP 4.5 and RCP 6.0 envision intermediate policy choices, resulting in stabilization by 2100 of CO_2 concentrations at 538 and 670 ppm, respectively. In all those scenarios, the cooling effect of industrial pollutants such as sulfate particulates, which have masked some of the past century's warming, is assumed to decline to near zero by 2100 because of policies restricting their industrial production.

Simulations of Future Climate Change

The differences between the various simulations arise from disparities between the various climate models used and from assumptions made by each emission scenario. For example, best estimates of the predicted increases in global surface temperature between the years 2000 and 2100 range from about 0.3 to 4.8 °C (0.5 to 8.6 °F), depending on which emission scenario is assumed and which climate model is used. Relative to preindustrial (i.e., 1750–1800) temperatures, these estimates reflect an overall warming of the globe of 1.4 to 5.0 °C (2.5 to 9.0 °F). These projections are conservative in that they do not take into account potential positive carbon cycle feedbacks. Only the lower-end emissions scenario RCP 2.6 has a reasonable chance (roughly 50 percent) of holding additional global surface warming by 2100 to less than 2.0 °C (3.6 °F)—a level considered by many scientists to be the threshold above which pervasive and extreme climatic effects will occur.

Patterns of Warming

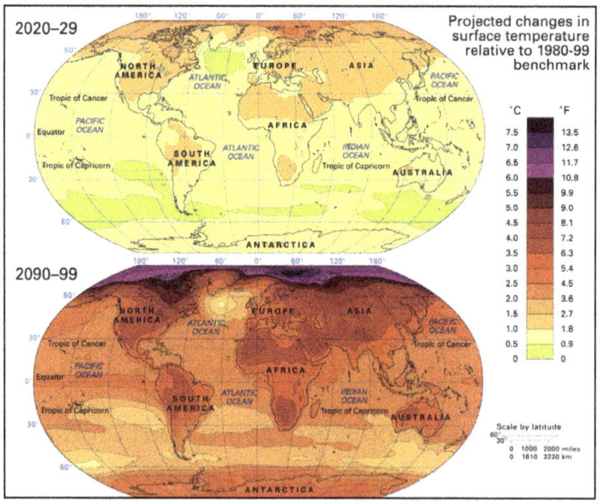

Projected changes in mean surface temperatures by the late 21st century according to the A1B climate change scenario. All values for the period 2090–99 are shown relative to the mean temperature values for the period 1980–99.

The greatest increase in near-surface air temperature is projected to occur over the polar region of the Northern Hemisphere because of the melting of sea ice and the associated reduction in surface albedo. Greater warming is predicted over land areas than over the ocean. Largely due to the delayed warming of the oceans and their greater specific heat, the Northern Hemisphere—with less than 40 percent of its surface area covered by water—is expected to warm faster than the Southern Hemisphere. Some of the regional variation in predicted warming is expected to arise from changes to wind patterns and ocean currents in response to surface warming. For example, the warming of the region of the North Atlantic Ocean just south of Greenland is expected to be slight. This anomaly is projected to arise from a weakening of warm northward ocean currents combined with a shift in the jet stream that will bring colder polar air masses to the region.

Precipitation Patterns

The climate changes associated with global warming are also projected to lead to changes in precipitation patterns across the globe. Increased precipitation is predicted in the polar and subpolar regions, whereas decreased precipitation is projected for the middle latitudes of both hemispheres as a result of the expected poleward shift in the jet streams. Whereas precipitation near the Equator is predicted to increase, it is thought that rainfall in the subtropics will decrease. Both phenomena are associated with a forecasted strengthening of the tropical Hadley cell pattern of atmospheric circulation.

Changes in precipitation patterns are expected to increase the chances of both drought and flood conditions in many areas. Decreased summer precipitation in North America, Europe, and Africa, combined with greater rates of evaporation due to warming surface temperatures, is projected to lead to decreased soil moisture and drought in many regions. Furthermore, since anthropogenic climate change will likely lead to a more vigorous hydrologic cycle with greater rates of both evaporation and precipitation, there will be a greater probability for intense precipitation and flooding in many regions.

Regional Predictions

Regional predictions of future climate change remain limited by uncertainties in how the precise patterns of atmospheric winds and ocean currents will vary with increased surface warming. For example, some uncertainty remains in how the frequency and magnitude of El Niño/Southern Oscillation (ENSO) events will adjust to climate change. Since ENSO is one of the most prominent sources of interannual variations in regional patterns of precipitation and temperature, any uncertainty in how it will change implies a corresponding uncertainty in certain regional patterns of climate change. For example, increased El Niño activity would likely lead to more winter precipitation in some regions, such as the desert southwest of the United States. This might offset the drought predicted for those regions, but at the same time it might lead to less precipitation in other regions. Rising winter precipitation in the desert southwest of the United States might exacerbate drought conditions in locations as far away as South Africa.

Ice Melt and Sea Level Rise

A warming climate holds important implications for other aspects of the global environment. Because of the slow process of heat diffusion in water, the world's oceans are likely to continue to warm for several centuries in response to increases in greenhouse concentrations that have taken place so far. The combination of seawater's thermal expansion associated with this warming and the melting of mountain glaciers is predicted to lead to an increase in global sea level of 0.45–0.82 metre (1.4–2.7 feet) by 2100 under the RCP 8.5 emissions scenario. However, the actual rise in sea level could be considerably greater than this. It is probable that the continued warming of Greenland will cause its ice sheet to melt at accelerated rates. In addition, this level of surface warming may also melt the ice sheet of West Antarctica. Paleoclimatic evidence suggests that an additional 2 °C (3.6 °F) of warming could lead to the ultimate destruction of the Greenland Ice Sheet, an event that would add another 5 to 6 metres (16 to 20 feet) to predicted sea level rise. Such an increase would submerge a substantial number of islands and lowland regions. Coastal lowland regions vulnerable to sea level rise include substantial parts of the U.S. Gulf Coast and Eastern Seaboard (including roughly the lower third of Florida), much of the Netherlands and Belgium (two of the European Low Countries), and heavily populated tropical areas such as Bangladesh. In addition, many of the world's major cities—such as Tokyo, New York, Mumbai, Shanghai, and Dhaka—are located in lowland regions vulnerable to rising sea levels. With the loss of the West Antarctic ice sheet, additional sea level rise would approach 10.5 metres (34 feet).

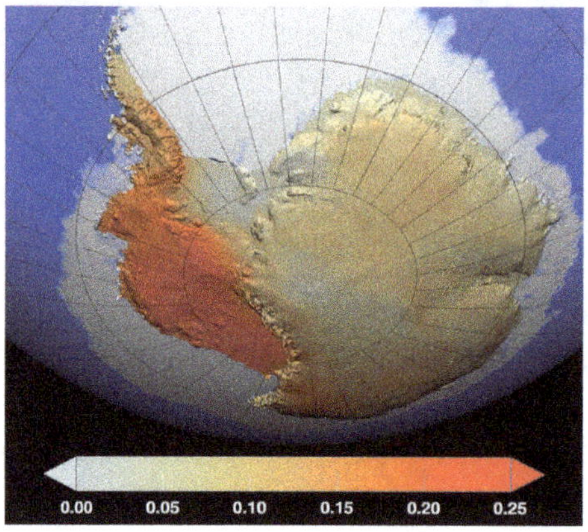

NASA image showing locations on Antarctica where temperatures had increased between 1959 and 2009. Red represents areas where temperatures had increased the most over the period, particularly in West Antarctica, while dark blue represents areas with a lesser degree of warming. Temperature changes are measured in degrees Celsius.

While the current generation of models predicts that such global sea level changes might take several centuries to occur, it is possible that the rate could accelerate as a result of processes that tend to hasten the collapse of ice sheets. One such process is the development of moulins—large vertical shafts in the ice that allow surface meltwater to penetrate to the base of the ice sheet. A second process involves the vast ice shelves off Antarctica that buttress the grounded continental ice sheet of Antarctica's interior. If those ice shelves collapse, the continental ice sheet could become unstable, slide rapidly toward the ocean, and melt, thereby further increasing mean sea level. Thus far, neither process has been incorporated into the theoretical models used to predict sea level rise.

Ocean Circulation Changes

Another possible consequence of global warming is a decrease in the global ocean circulation system known as the "thermohaline circulation" or "great ocean conveyor belt." This system involves the sinking of cold saline waters in the subpolar regions of the oceans, an action that helps to drive warmer surface waters poleward from the subtropics. As a result of this process, a warming influence is carried to Iceland and the coastal regions of Europe that moderates the climate in those regions. Some scientists believe that global warming could shut down this ocean current system by creating an influx of fresh water from melting ice sheets and glaciers into the subpolar North Atlantic Ocean. Since fresh water is less dense than saline water, a significant intrusion of fresh water would lower the density of the surface waters and thus inhibit the sinking motion that drives the large-scale thermohaline circulation. It has also been speculated that, as a consequence of large-scale surface warming, such changes could even trigger colder conditions in regions surrounding the North Atlantic. Experiments with modern climate models suggest that such an event would be unlikely. Instead, a moderate weakening of the thermohaline circulation might occur that would lead to a dampening of surface warming—rather than actual cooling—in the higher latitudes of the North Atlantic Ocean.

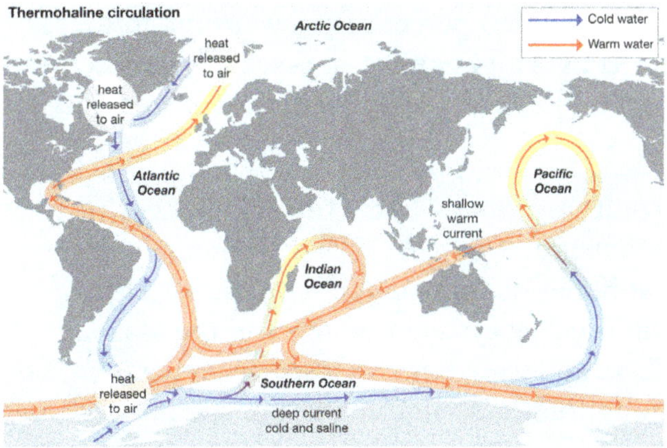

Thermohaline circulation transports and mixes the water of the oceans. In the process it transports heat, which influences regional climate patterns. The density of seawater is determined by the temperature and salinity of a volume of seawater at a particular location. The difference in density between one location and another drives the thermohaline circulation.

Tropical Cyclones

One of the more controversial topics in the science of climate change involves the impact of global warming on tropical cyclone activity. It appears likely that rising tropical ocean temperatures associated with global warming will lead to an increase in the intensity (and the associated destructive potential) of tropical cyclones. In the Atlantic a close relationship has been observed between rising ocean temperatures and a rise in the strength of hurricanes. Trends in the intensities of tropical cyclones in other regions, such as in the tropical Pacific and Indian oceans, are more uncertain due to a paucity of reliable long-term measurements.

While the warming of oceans favours increased tropical cyclone intensities, it is unclear to what extent rising temperatures affects the number of tropical cyclones that occur each year. Other factors, such as wind shear, could play a role. If climate change increases the amount of wind shear—a factor that discourages the formation of tropical cyclones—in regions where such storms tend to form, it might partially mitigate the impact of warmer temperatures. On the other hand, changes in atmospheric winds are themselves uncertain—because of, for example, uncertainties in how climate change will affect ENSO.

Environmental Consequences of Global Warming

Global warming and climate change have the potential to alter biological systems. More specifically, changes to near-surface air temperatures will likely influence ecosystem functioning and thus the biodiversity of plants, animals, and other forms of life. The current geographic ranges of plant and animal species have been established by adaptation to long-term seasonal climate patterns. As global warming alters these patterns on timescales considerably shorter than those that arose in the past from natural climate variability, relatively sudden climatic changes may challenge the natural adaptive capacity of many species.

A large fraction of plant and animal species are likely to be at an increased risk of extinction if global average surface temperatures rise another 1.5 to 2.5 °C (2.7 to 4.5 °F) by the year 2100. Species loss estimates climb to as much as 40 percent for a warming in excess of 4.5 °C (8.1 °F)—a level that could be reached in the IPCC's higher emissions scenarios. A 40 percent extinction rate would likely lead to major changes in the food webs within ecosystems and has a destructive impact on ecosystem function.

Surface warming in temperate regions is likely to lead changes in various seasonal processes—for instance, earlier leaf production by trees, earlier greening of vegetation, altered timing of egg laying and hatching, and shifts in the seasonal migration patterns of birds, fishes, and other migratory animals. In high-latitude ecosystems, changes in the seasonal patterns of sea ice threaten predators such as polar bears and walruses; both species rely on broken sea ice for their hunting activities. Also in the high latitudes, a combination of warming waters, decreased sea ice, and changes in ocean salinity and circulation is likely to lead to reductions or redistributions in populations of algae and plankton. As a result, fish and other organisms that forage upon algae and plankton may be threatened. On land, rising temperatures and changes in precipitation patterns and drought frequencies are likely to alter patterns of disturbance by fires and pests.

Numerous ecologists, conservation biologists, and other scientists studying climate warn that rising surface temperatures will bring about an increased extinction risk. In 2015 one study that examined 130 extinction models developed in previous studies predicted that 5.2 percent of species would be lost with a rise in average temperatures of 2 °C (3.6 °F) above temperature benchmarks from before the onset of the Industrial Revolution. The study also predicted that 16 percent of Earth's species would be lost if surface warming increased to about 4.3 °C (7.7 °F) above preindustrial temperature benchmarks.

Other likely impacts on the environment include the destruction of many coastal wetlands, salt marshes, and mangrove swamps as a result of rising sea levels and the loss of certain rare and fragile habitats that are often home to specialist species that are unable to thrive in other environments. For example, certain amphibians limited to isolated tropical cloud forests either have become extinct already or are under serious threat of extinction. Cloud forests—tropical forests that depend on persistent condensation of moisture in the air—are disappearing as optimal condensation levels move to higher elevations in response to warming temperatures in the lower atmosphere.

In many cases a combination of stresses caused by climate change as well as human activity represents a considerably greater threat than either climatic stresses or nonclimatic stresses alone. A particularly important example is coral reefs, which contain much of the ocean's biodiversity. Rising ocean temperatures increase the tendency for coral bleaching (a condition where zooxanthellae, or yellow-green algae, living in symbiosis with coral either lose their pigments or abandon the coral polyps altogether), and they also raise the likelihood of greater physical damage by progressively more destructive tropical cyclones. In many areas coral is also under stress from increased ocean acidification, marine pollution, runoff from agricultural fertilizer, and physical damage by boat anchors and dredging.

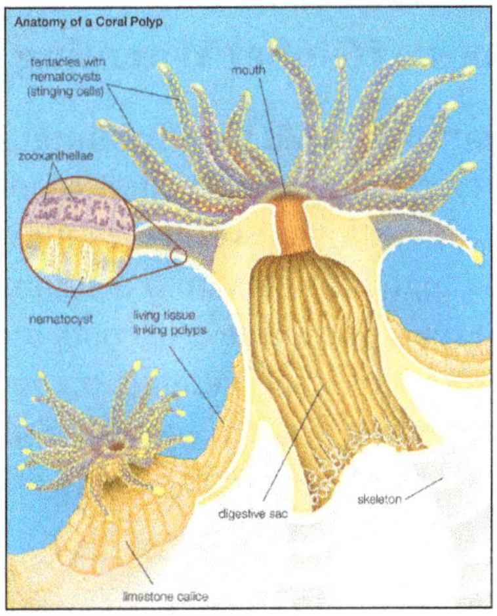

Cross section of a generalized coral polyp.

Another example of how climate and nonclimatic stresses combine is illustrated by the threat to migratory animals. As these animals attempt to relocate to regions with more favourable climate conditions, they are likely to encounter impediments such as highways, walls, artificial waterways, and other man-made structures.

Warmer temperatures are also likely to affect the spread of infectious diseases, since the geographic ranges of carriers, such as insects and rodents, are often limited by climatic conditions. Warmer winter conditions in New York in 1999, for example, appear to have facilitated an outbreak of West Nile virus, whereas the lack of killing frosts in New Orleans during the early 1990s led to an explosion of disease-carrying mosquitoes and cockroaches. Warmer winters in the Korean peninsula and southern Europe have allowed the spread of the Anopheles mosquito, which carries the malaria parasite, whereas warmer conditions in Scandinavia in recent years have allowed for the northward advance of encephalitis.

Anopheles mosquito, carrier of the malarial parasite.

In the southwestern United States, alternations between drought and flooding related in part to the ENSO phenomenon have created conditions favourable for the spread of Hantaviruses by rodents. The spread of mosquito-borne Rift Valley fever in equatorial East Africa has also been related to wet conditions in the region associated with ENSO. Severe weather conditions conducive to rodents or insects have been implicated in infectious disease outbreaks—for instance, the outbreaks of cholera and leptospirosis that occurred after Hurricane Mitch struck Central America in

1998. Global warming could therefore affect the spread of infectious disease through its influence on ENSO or on severe weather conditions.

Socioeconomic Consequences of Global Warming

Socioeconomic impacts of global warming could be substantial, depending on the actual temperature increases over the next century. Models predict that a net global warming of 1 to 3 °C (1.8 to 5.4 °F) beyond the late 20th-century global average would produce economic losses in some regions (particularly the tropics and high latitudes) and economic benefits in others. For warming beyond those levels, benefits would tend to decline and costs increase. For warming in excess of 4 °C (7.2 °F), models predict that costs will exceed benefits on average, with global mean economic losses estimated between 1 and 5 percent of gross domestic product. Substantial disruptions could be expected under those conditions, specifically in the areas of agriculture, food and forest products, water and energy supply, and human health.

Agricultural productivity might increase modestly in temperate regions for some crops in response to a local warming of 1–3 °C (1.8–5.4 °F), but productivity will generally decrease with further warming. For tropical and subtropical regions, models predict decreases in crop productivity for even small increases in local warming. In some cases, adaptations such as altered planting practices are projected to ameliorate losses in productivity for modest amounts of warming. An increased incidence of drought and flood events would likely lead to further decreases in agricultural productivity and to decreases in livestock production, particularly among subsistence farmers in tropical regions. In regions such as the African Sahel, decreases in agricultural productivity have already been observed as a result of shortened growing seasons, which in turn have occurred as a result of warmer and drier climatic conditions. In other regions, changes in agricultural practice, such as planting crops earlier in the growing season, have been undertaken. The warming of oceans is predicted to have an adverse impact on commercial fisheries by changing the distribution and productivity of various fish species, whereas commercial timber productivity may increase globally with modest warming.

Water resources are likely to be affected substantially by global warming. At current rates of warming, a 10–40 percent increase in average surface runoff and water availability has been projected in higher latitudes and in certain wet regions in the tropics by the middle of the 21st century, while decreases of similar magnitude are expected in other parts of the tropics and in the dry regions in the subtropics. This would be particularly severe during the summer season. In many cases water availability is already decreasing or expected to decrease in regions that have been stressed for water resources since the turn of the 21st century. Such regions as the African Sahel, western North America, southern Africa, the Middle East, and western Australia continue to be particularly vulnerable. In these regions drought is projected to increase in both magnitude and extent, which would bring about adverse effects on agriculture and livestock raising. Earlier and increased spring runoff is already being observed in western North America and other temperate regions served by glacial or snow-fed streams and rivers. Fresh water currently stored by mountain glaciers and snow in both the tropics and extratropics is also projected to decline and thus reduce the availability of fresh water for more than 15 percent of the world's population. It is also likely that warming temperatures, through their impact on biological activity in lakes and rivers, may have an adverse impact on water quality, further diminishing access to safe water sources for drinking or farming. For example, warmer waters favour an increased frequency of nuisance algal blooms, which can pose health risks to humans. Risk-management procedures have already been taken by some countries in response to expected changes in water availability.

Energy availability and use could be affected in at least two distinct ways by rising surface temperatures. In general, warmer conditions would favour an increased demand for air-conditioning; however, this would be at least partially offset by decreased demand for winter heating in temperate regions. Energy generation that requires water either directly, as in hydroelectric power, or indirectly, as in steam turbines used in coal-fired power plants or in cooling towers used in nuclear power plants, may become more difficult in regions with reduced water supplies.

As discussed above, it is expected that human health will be further stressed under global warming conditions by potential increases in the spread of infectious diseases. Declines in overall human health might occur with increases in the levels of malnutrition due to disruptions in food production and by increases in the incidence of afflictions. Such afflictions could include diarrhea, cardiorespiratory illness, and allergic reactions in the midlatitudes of the Northern Hemisphere as a result of rising levels of pollen. Rising heat-related mortality, such as that observed in response to the 2003 European heat wave, might occur in many regions, especially in impoverished areas where air-conditioning is not generally available.

The economic infrastructure of most countries is predicted to be severely strained by global warming and climate change. Poor countries and communities with limited adaptive capacities are likely to be disproportionately affected. Projected increases in the incidence of severe weather, heavy flooding, and wildfires associated with reduced summer ground moisture in many regions will threaten homes, dams, transportation networks and other facets of human infrastructure. In high-latitude and mountain regions, melting permafrost is likely to lead to ground instability or rock avalanches, further threatening structures in those regions. Rising sea levels and the increased potential for severe tropical cyclones represent a heightened threat to coastal communities throughout the world. It has been estimated that an additional warming of 1–3 °C (1.8–5.4 °F) beyond the late 20th-century global average would threaten millions more people with the risk of annual flooding. People in the densely populated, poor, low-lying regions of Africa, Asia, and tropical islands would be the most vulnerable, given their limited adaptive capacity. In addition, certain regions in developed countries, such as the Low Countries of Europe and the Eastern Seaboard and Gulf Coast of the United States, would also be vulnerable to the effects of rising sea levels. Adaptive steps are already being taken by some governments to reduce the threat of increased coastal vulnerability through the construction of dams and drainage works.

References

- Climate-meteorology, science: britannica.com, Retrieved 17 January, 2019
- What-is-climatology: worldatlas.com, Retrieved 8 August, 2019
- Climate-models: climate.gov, Retrieved 3 March, 2019
- Climate-change, science: britannica.com, Retrieved 20 June, 2019
- Global-warming, science: britannica.com, Retrieved 24 April, 2019

Permissions

All chapters in this book are published with permission under the Creative Commons Attribution Share Alike License or equivalent. Every chapter published in this book has been scrutinized by our experts. Their significance has been extensively debated. The topics covered herein carry significant information for a comprehensive understanding. They may even be implemented as practical applications or may be referred to as a beginning point for further studies.

We would like to thank the editorial team for lending their expertise to make the book truly unique. They have played a crucial role in the development of this book. Without their invaluable contributions this book wouldn't have been possible. They have made vital efforts to compile up to date information on the varied aspects of this subject to make this book a valuable addition to the collection of many professionals and students.

This book was conceptualized with the vision of imparting up-to-date and integrated information in this field. To ensure the same, a matchless editorial board was set up. Every individual on the board went through rigorous rounds of assessment to prove their worth. After which they invested a large part of their time researching and compiling the most relevant data for our readers.

The editorial board has been involved in producing this book since its inception. They have spent rigorous hours researching and exploring the diverse topics which have resulted in the successful publishing of this book. They have passed on their knowledge of decades through this book. To expedite this challenging task, the publisher supported the team at every step. A small team of assistant editors was also appointed to further simplify the editing procedure and attain best results for the readers.

Apart from the editorial board, the designing team has also invested a significant amount of their time in understanding the subject and creating the most relevant covers. They scrutinized every image to scout for the most suitable representation of the subject and create an appropriate cover for the book.

The publishing team has been an ardent support to the editorial, designing and production team. Their endless efforts to recruit the best for this project, has resulted in the accomplishment of this book. They are a veteran in the field of academics and their pool of knowledge is as vast as their experience in printing. Their expertise and guidance has proved useful at every step. Their uncompromising quality standards have made this book an exceptional effort. Their encouragement from time to time has been an inspiration for everyone.

The publisher and the editorial board hope that this book will prove to be a valuable piece of knowledge for students, practitioners and scholars across the globe.

Index

R

S

T

V

W

www.ingramcontent.com/pod-product-compliance
Lightning Source LLC
Chambersburg PA
CBHW061302190326
41458CB00011B/3742
9781641165648